P9-DEN-536

Praise for Rebecca Newberger Goldstein's

PLATO AT THE GOOGLEPLEX

"The man who gave us philosophy as we know it is back, walking among us . . . engaging with our current struggles. For this we must thank Rebecca Newberger Goldstein's inventiveness and intellectual courage."   —Marcelo Gleiser, NPR

"Philosopher Rebecca Goldstein probes why Plato—and the philosophical enterprise itself—remain a force in science and culture. . . . This thought experiment usefully casts an eye on our turbocharged century. And it shows what survives of this classical titan: an ability to plumb the deep questions we still grapple with, from the nature of knowledge to morality."

—*Nature*

"I have not done justice to the richness and detail of this invigorating book. The combination of historical scholarship, lively presentation, vernacular dialogue, and intellectual passion make it a unique achievement. Plato may have died over two thousand years ago, but he lives on, vibrantly, in these piquant pages."   —Colin McGinn, *The Wall Street Journal*

"*Plato at the Googleplex* merits comparison to two of the best books of its kind in recent years, Kathryn Schulz's *Being Wrong* and Daniel Kahneman's *Thinking, Fast and Slow*, but Goldstein's is, in my opinion, the best of the lot, not because it necessarily has more facts or science, but because it hits more deeply and broadly at the faults of our societal discourse and makes us (well, me at least) feel embarrassed over it."

—David Auerbach, *Slate*

"Books like Rebecca Newberger Goldstein's *Plato at the Googleplex* are of the rare type that contribute to the popularization of knowledge and create appetite for more. After reading this book you will . . . question your views and knowledge about politics, psychology, science, history, and ethics."

—*New York Journal of Books*

"[A] remarkably alive, ruminative new work. . . . *Plato at the Googleplex* delivers loads of colorful, fascinating ancient historical context. . . . [The book is] fun and at times wickedly funny."

—*San Francisco Chronicle*

"*Plato at the Googleplex* makes a compelling case, not just for why undergraduates should take humanities courses, but for the value of a life of genuine introspection over one devoted to the pursuit of material gain. . . . Perhaps the greatest accomplishment is Goldstein's crafting of Plato as a three-dimensional figure. . . . He emerges as a believable public intellectual and kind of a mensch—a man whose genuine curiosity and openness towards all overshadows any ego or attachment to one particular doctrine."

—*The Christian Science Monitor*

"Rebecca Newberger Goldstein manages to be so funny and right."

—Stephen Fry

"*Plato at the Googleplex* is a wonderful book—enjoyably readable, full of stimulating insights and refreshing observations, unintimidatingly erudite, and salted with a gentle wit."

—Harry Frankfurt, author of *On Bullshit*

"This could be one of the best ever demonstrations of the value and utility of philosophy. Richly insightful, beautifully written, it is at once introduction, exploration and application, revealing the fascination and significance of philosophical ideas and their relevance to life. Like the Plato who figures largely here, Goldstein has both literary and philosophical gifts of the highest order: the combination is superb."

—A. C. Grayling, author of *The God Argument*

Rebecca Newberger Goldstein

PLATO AT THE GOOGLEPLEX

Rebecca Newberger Goldstein graduated summa cum laude from Barnard College, Columbia University, and received her doctorate in philosophy from Princeton University. Her award-winning books include the novels *The Mind-Body Problem, Properties of Light, Mazel,* and *36 Arguments for the Existence of God: A Work of Fiction,* as well as studies of Kurt Gödel and Baruch Spinoza. She has received a MacArthur genius award, been designated the Humanist of the Year, and been elected to the American Academy of Arts and Sciences. She lives in Massachusetts.

www.rebeccagoldstein.com

Also by Rebecca Newberger Goldstein

The Mind-Body Problem
The Late-Summer Passion of a Woman of Mind
The Dark Sister
Strange Attractors: Stories
Mazel
Properties of Light: A Novel of Love, Betrayal and Quantum Physics
Incompleteness: The Proof and Paradox of Kurt Gödel
Betraying Spinoza: The Renegade Jew Who Gave Us Modernity
36 Arguments for the Existence of God: A Work of Fiction

PLATO AT THE GOOGLEPLEX

PLATO
AT THE GOOGLEPLEX

Why Philosophy Won't Go Away

184
G624p
2015

Rebecca Newberger Goldstein

Vintage Books
A Division of Random House LLC
New York

FIRST VINTAGE BOOKS EDITION, JANUARY 2015

Copyright © 2014 by Rebecca Goldstein

All rights reserved. Published in the United States by Vintage Books,
a division of Random House LLC, New York, and distributed in Canada
by Random House of Canada Limited, Toronto, Penguin Random House
companies. Originally published in hardcover in the United States by
Pantheon, a division of Random House LLC, New York, in 2014.

Vintage and colophon are registered trademarks of Random House LLC.

Owing to limitations of space, permissions to reprint from previously
published material are listed following the bibliographical note.

The Library of Congress has cataloged the Pantheon edition as follows:
Goldstein, Rebecca.
Plato at the Googleplex : why philosophy won't go away / Rebecca Goldstein.
pages cm
Includes bibliographical references.
1. Plato—Influence.   2. Philosophy—History—21st century.
3. Imaginary conversations.   I. Title.
B395.G4435 2014 184—dc23 2013029660

Vintage Books Trade Paperback ISBN: 978-0-307-45672-4
eBook ISBN: 978-0-307-90887-2

Author photograph © Steven Pinker
Book design by M. Kristen Bearse

www.vintagebooks.com

Printed in the United States of America
10 9 8 7 6 5 4 3 2 1

FOR HARRY AND ROZ PINKER

CONTENTS

Prologue   3

α Man Walks into a Seminar Room   15
β Plato at the Googleplex   59
γ In the Shadow of the Acropolis   121
δ Plato at the 92nd Street Y   163
ε I Don't Know How to Love Him   223
ς xxxPlato   261
ζ Socrates Must Die   281
η Plato on Cable News   335
θ Let the Sunshine In   361
ι Plato in the Magnet   397

Appendix A: Socratic Sources   423
Appendix B: The Two Speeches of Pericles from
    Thucydides' *The History of the Peloponnesian War*   425
Glossary   435
Acknowledgments   439
Bibliographical Note   441
Index   445

PLATO AT THE GOOGLEPLEX

PROLOGUE

A book devoted to a particular thinker often presumes that thinker got everything right. I don't think this is true of Plato. Plato got about as much wrong as we would expect from a philosopher who lived 2,400 years ago. Were this not the case, then philosophy, advancing our knowledge not at all, would be useless. I don't think it's useless, so I'm quite happy to acknowledge how mistaken or confused Plato can often strike us.

Plato is surprisingly relevant to many of our contemporary discussions, but this isn't because he knew as much as we do. Obviously, he didn't know the science that we know. But, less obviously, he didn't know the philosophy that we know, including philosophy that has filtered outward beyond the seminar table. Conclusions that philosophers first establish by way of tortuous reasoning have a way, over time, of leaking into shared knowledge. Such leakage is perhaps more common as regards the questions of morality than other branches of philosophy, since those are questions that constantly test us. We can hardly get through our lives—in fact, it's hard to get through a week—without considering what makes specific actions right and others wrong and debating with ourselves whether that is a difference that must compel the actions we choose. (Okay, it's wrong! I get it! But why should I care?)

Plato's ruminations, as profound as they are, hardly give us the last word on such matters. European thinkers of the Age of Reason and the Enlightenment, coming two millennia after Plato, had much to add to our shared conceptions of morality, particularly as regards individual rights, and we have learned from them and gone on.[*] This is why it is

---

[*] The important point—that the Greek philosophers lacked the idea of individual rights as it was developed by thinkers of the seventeenth and eighteenth centuries—is discussed by Stephen Darwall. See his "Grotius at the Creation of Modern Moral Philosophy," in *Honor,*

impossible for us to read Plato now without occasional disapproval. It's precisely because he initiated a process that has taken us beyond him.

So Plato hardly did all the philosophical work. And yet he did do something so extraordinary as to mark his thinking as one of the pivotal stages in humankind's development. What Plato did was to carve out the field of philosophy itself. It was Plato who first framed the majority of fundamental philosophical questions. He grasped the essence of a peculiar kind of question, the philosophical question, some specimens of which were already afloat in the Athens of his day, and he extended its application. He applied the philosophical question not only to norms of human behavior, as Socrates had done, but to language, to politics, to art, to mathematics, to religion, to love and friendship, to the mind, to personal identity, to the meaning of life and the meaning of death, to the natures of explanation, of rationality, and of knowledge itself. Philosophical questions could be framed in all these far-flung areas of human concern and inquiry, and Plato framed them, often in their definitive form. How did he do it? Why was it he who did it? This is a mystery I've always wanted to unravel. But how do you get close enough to Plato to even attempt to figure him out? Drawing conclusions about which doctrines he meant to assert—or even whether he meant to assert any doctrines at all—is difficult enough, much less hoping to get a glimpse into the soul of the man.

Though Plato is (at least for many of us) an easy philosopher to love, he is also a deucedly difficult philosopher to get close to. Despite his enormous influence, he is one of the most remote figures in the history of thought. His remoteness is not only a matter of his antiquity, but also of the manner in which he gave himself to us by way of his writings. He didn't create treatises, essays, or inquiries that propound positions. Instead, he wrote dialogues, which are not only great works of philosophy but also great works of literature.

His language is that of a consummate artist. Classical scholars affirm that his Greek is the purest and finest of any of the ancient writings that have come down to us. "The lyrical prose of Plato had no peer in the ancient world," writes one scholar in his introduction to Percy Bysshe Shelley's extraordinary translation of Plato's *Symposium,* the

History, and Relationship: Essays in Second-Personal Ethics II (Oxford: Oxford University Press, 2013).

great Romantic pouring his own lyrical gifts into the text.* But, more
to the point, Plato's vivid characters discuss philosophical problems in
so lively and natural a manner that it is difficult to catch the author's
point of view through the engagement of the many voices with one
another. His dialogues allow us to draw a little bit closer to many of his
contemporaries—including Socrates—while Plato holds himself aloof.
Some readers of the dialogues interpret the character of Socrates, who
is often the character who gets the most lines, as a stand-in for Plato,
much as Salviati speaks for Galileo in *Dialogue Concerning the Two Chief
World Systems* and as Philo speaks for David Hume in *Dialogues Concerning Natural Religion;* but this pastes too simple a face over an interpretive chimera.† It is almost as naive to reduce the dialogic Socrates
to a mere sock puppet for the philosopher Plato as it is to reduce Plato
to a mere notetaker for the philosopher Socrates. Plato floats fugitive
between these two reductions.

His elusiveness is comparable to that of another protean writer of
whom it is difficult to catch a glimpse through the genius of the work,
William Shakespeare. In both, it's the capaciousness and vivacity of
points of view animating the text that drives the author into the shadows. In the case of Shakespeare, the remoteness of the author has
provoked some otherwise sober people to contend that the actor born
on Henley Street in Stratford-upon-Avon, who left school at fourteen
and never went to university and married an already pregnant Anne
Hathaway, to whom he willed his "second-best" bed, was merely a front
man for the real author—even a whole *committee* of authors.‡ In the
case of Plato, the remoteness makes itself felt not only in the difficulties of disentangling Plato from Socrates, but, even more dramatically,
in the mutually incompatible characterizations that have been foisted
upon him.

It has been claimed that Plato was an egalitarian; it has been claimed
that he was a totalitarian. It has been claimed that he was a utopian,

---

* *The Symposium of Plato: The Shelley Translation,* edited and introduced by David K. O'Connor
(South Bend, IN: St. Augustine's Press, 2002).
† See *Who Speaks for Plato: Studies in Platonic Anonymity,* edited by Gerald A. Press (Lanham,
MD: Rowman and Littlefield, 2000). The eleven contributors to the volume all argue against
the view that Socrates, or any other character in the dialogues, is a mouthpiece for Plato.
‡ Proposed candidates have included a death-faking Christopher Marlowe; Francis Bacon;
Walter Raleigh; Edmund Spenser; Lord Buckhurst; Edward de Vere, seventeenth Earl of
Oxford; and William Stanley, sixth Earl of Derby.

proposing a universal blueprint for the ideal state; it has been claimed he was an anti-utopian, demonstrating that all political idealism is folly. It has been claimed he was a populist, concerned with the best interests of all citizens; it has been claimed he was an elitist with disturbing eugenicist tendencies. It has been claimed he was other-worldly; it has been claimed he was this-worldly. It has been claimed he was a romantic; it has been claimed he was a prig. It has been claimed that he was a theorizer, with sweeping metaphysical doctrines; it has been claimed he was an anti-theorizing skeptic, always intent on unsettling convictions. It has been claimed he was full of humor and play; it has been claimed he was as solemn as a sermon limning the torments of the damned. It has been claimed he loved his fellow man; it has been claimed he loathed his fellow man. It has been claimed he was a philosopher who used his artistic gifts in the service of philosophy; it has been claimed he was an artist who used philosophy in the service of his art.

Isn't it curious that a figure can exert so much influence throughout the course of Western civilization and escape consensus as to what he was all about? And how in the world can one hope to draw closer to so elusive a figure?

He was an ancient Greek, a citizen of the city-state of Athens during its classical age. How much of Plato's achievement in almost singlehandedly creating philosophy is explained by his having been a Greek? The Greeks have fascinated us for a good long while now. Even the Romans, who vanquished them militarily, were vanquished from within by the fascinating Greeks. After the millennia of obsession, is there anything new to say about them? I think so, and it is this: the preconditions for philosophy were created there in ancient Greece, and most especially in Athens. These preconditions lay not only in a preoccupation with the question of what it is that makes life worth living but in a distinctive approach to this question.

The Greeks were not alone in being preoccupied with the question of human worth and human mattering. Across the Mediterranean was the still-obscure tribe called the *Ivrim*, the Hebrews, from the word for "over," since they were over on the other side of the Jordan. There they worked out their notion of a covenantal relationship with a tribal god whom they eventually elevated to the position of the one and only God, The Master of the Universe who provides the foundation for both the physical world without and the moral world within. To live according to his commandments was to live a life worth living. Our Western culture is still an uneasy mix of the approaches to the question of human worth worked out by these two Mediterranean peoples, the Greeks and the Hebrews. But even they weren't alone in their existential preoccupations. In Persia, Zoroastrianism presented a dualistic version of the forces of good and evil; in China, there was Confucius and Lao Tzu and Chuang Tzu; and in India, there was the Buddha. Each of these approaches adds to the range of choices we have for conceiving the life worth living.

The philosopher Karl Jaspers baptized this normatively* fertile period in human history—which was roughly 800 to 200 B.C.E.—the "Axial Age," because visions forged during that period extend out into our own day, like the axials of a wheel. These ways of normatively framing our lives still resonate with millions of people, including secularists, who are the inheritors of the Greek tradition.

The Greeks themselves can hardly be called secularists. Religious rituals saturated their lives—their gods and goddesses were everywhere and had to be propitiated or something terrible would happen. Their rituals were, by and large, apotropaic, meant to ward off evil. There were public rites associated with the individual city-states and others that were Panhellenic; there were secret rites that belonged to the mystery cults. But what is remarkable about the Greeks—even pre-philosophically—is that, despite the salience of religious rituals in their lives, when it came to the question of what it is that makes an individual human life worth living they didn't look to their immortals but rather approached the question in mortal terms. Their approaching the question of human mattering in human terms is the singularity that creates the conditions for philosophy in ancient Greece, most especially as these conditions were realized in the city-state of Athens.

Their human approach to the question of human mattering meant that the tragic point of view—in fact, several versions of the tragic point

---

*Philosophers use the word "normative" to refer to any propositions that contain the word "ought," as in "You ought to consider the interests of others as well as those of yourself," and "You ought to be rational and consider all the facts, not only those that support your favored hypothesis." Though many normative propositions deal with ethical matters, not all of them do, as my second example demonstrates. In particular, epistemology, which examines the conditions for securing knowledge, raises normative issues. Religion, of course, addresses normative issues, but so, too, does secular philosophy.

of view—were agonizingly distinct possibilities. It is no accident that Athens was the home not only of Socrates, Plato, and Aristotle,* but also of Aeschylus, Sophocles, and Euripides. Their approach to the question of what makes a human life worth living created not only the conditions for the great tragic dramatists but also the audiences for them. Those audiences didn't shrink from confronting the possibility that human life, tragically, is *not* worth living. Perhaps we don't matter and nothing can be done to make us matter. Or, only slightly less tragic, perhaps there is something that must be done in order to achieve a life worth living, something that will redeem that life by singling it out as extraordinary, and only then will it matter. It is only an ordinary life—with nothing to distinguish it from the great masses of other anonymous lives that have come before us and will come after us—that doesn't matter. There is a pronounced pitilessness in this proposition, and there was a pronounced pitilessness in the Greeks. One must *exert* oneself in order to achieve a life that matters. If you don't exert yourself, or if your exertions don't amount to much of anything, then you might as well not have bothered to have shown up for your existence at all.

How many of us harbor something like this attitude, whether vaguely or not, that the ordinary souls among us—by definition, the overwhelming majority of us—don't matter as much as the extraordinary ones do? So, too, did a great many Greeks, at least those among them who had the luxury of worrying over such existential quandaries, Greeks who not only wrote the tragedies but were moved to pity and terror by them. I call their attitude the Ethos of the Extraordinary. It is only by making oneself extraordinary that one can keep from disappearing without a trace, like some poor soul who slips beneath the ocean's waves—an image that called forth an intensity of terror for the seafaring Greeks.† One must live so that one will be spoken about, by as

---

* Aristotle was born in Stagira, near Macedonia, but came to Athens to study in Plato's Academy. He stayed to eventually found his own Athenian school, the Lyceum.

† Telemakhos, whose father, Odysseus, hasn't been heard from since he set sail after the sacking of Troy, mourns a fate he describes as far worse than death. "The gods have made him invisible. If he were dead, I would not grieve for him so much—if he had been killed at Troy, or died in the arms of friends after the war. Then, the Greeks would have made a tomb for him, and he would have won great glory for me, his son, as well as for himself. Instead, the storm fiends have snatched him away and left no word of him. He has perished unseen and unheard of" (*Iliad* I.235ff). That phrase "unseen and unheard of" contains all the terror of a life that, in the end, amounts to nothing.

many speakers as possible and for as long as possible. It is, in the end, the only kind of immortality for which we may hope. And, of course, we *are* still speaking about the ancient Greeks, especially the extraordinary ones among them, of which there were so many.

Plato shared, with radical modifications, in the Ethos of the Extraordinary, and it led him to create philosophy as we know it. The kind of exertion that is required if one is to achieve a life worth living is philosophy as he understood it. It is our exertions in *reason* that make us matter—make us, to the extent that we can be, godlike. And if such exertions don't win the acclaim of the masses, so much the worse for the masses. The kind of extraordinary that matters is likely to go undetected by them—so, in a certain sense, though not in all senses, they really don't matter. This is a harsh statement, but, as already noted, harshness didn't much faze the Greeks, and Plato is no exception here.

Plato opened up his dialogues to many different kinds of people, including those who didn't conventionally count for much in Athenian society. He did the same in the Academy that he established in a grove outside the city center and which became the prototype for the European university. It is reported that even women could study there, which accords with what he has to say about female intellectual potential in the *Republic* and the *Laws*. Nevertheless, his philosophical version of the Ethos of the Extraordinary left many stranded outside of the mattering class, namely all those who aren't able, or inclined, to *do* philosophy, to *do* reason. When, in the *Apology*, his rendition of Socrates' trial in 399 B.C.E., he has Socrates declare that the unexamined life is not worth living, he is both endorsing the Ethos of the Extraordinary shared by many in his culture and, at the same time, modifying it sufficiently to outrage his fellow Athenians. (That trial did not end well for Socrates.) A widely shared Greek presumption slips unexamined into his thinking. It will be pried out when European philosophers, after the centuries of encasing the question of human worth in religious thinking, return once again to consider the question of what makes a human life matter in secular terms, as the Greeks had done.

Here is an irony: the unexamined presumption that led Plato to create philosophy as we know it would eventually be invalidated by philosophy. That's progress. The progress to be made in philosophy is often a matter of discovering presumptions that slip unexamined into reasoning, so why not the unexamined presumption that got the whole self-critical process started? Plato, I would think, could only approve.

But thinking about Plato in these terms only gets us so close to him. Yes, he was an Athenian and, as an Athenian, imported certain preoccupations and preconceptions into his thinking. But that is only part of drawing closer to the remote figure of Plato. The other part is Plato's relationship with Socrates.

We know precious little about the personal life of Plato, but this we do know. The drama of Socrates' life—the true meaning of which was given, for Plato as well as for others, in his death—was personally transformative for Plato. It convinced him to devote his life to philosophy—he tells us this himself in his *Seventh Letter**—which he did with singular effect. His response to the trauma of Socrates' execution by the democratic *polis* of Athens, when Socrates was seventy years old and Plato was in his late twenties, was to create philosophy as we know it, formulating its central questions, questions far beyond any that had, in all probability, occurred to Socrates himself.[†]

But almost until the very end of his life, he kept the figure of Socrates at the center of his work. Plato wrote about philosophy with misgivings. He worried, for one thing, that philosophical writing would take the place of living conversations, for which, in philosophy, there is no substitute. (Philosophy, still, is an unusually gregarious subject.) Having agonized no less about the best way to write (and teach) philosophy than about philosophy itself, Plato created his dialogues, all of which have come down to us. (No commentator ever mentions a work of Plato that we don't have, in contrast to the works of Aristotle.) Twenty-five out of his twenty-six dialogues feature the character of Socrates, who, whether he is carrying the thrust of the argument forward or not—and often he isn't—is central to Plato's conception of philosophy. Socrates

---

* A good many scholars now seem to think the *Seventh Letter* is authentic; but even if it isn't, scholars agree that it was written by someone who was well informed about the private details of Plato's life.

† Plato's dialogues are traditionally divided into the Early, the Middle, and the Late, though there continue to be disagreements on aspects of the chronology, and there are scholars who dispute the entire idea of a set chronology. Plato might well have gone back and edited dialogues until almost the time of his death, somewhat like Henry James rewriting earlier works in his later style. Traditionally, the early dialogues are accepted as most representative of Socrates' practices and preoccupations, and these are confined to moral questions and often end in the impasse of aporia, a conceptual dead-end. It is only in the middle dialogues that Plato raises questions of metaphysics, epistemology, political philosophy, cosmology, philosophy of language, and so on.

is altogether absent only in the *Laws*, written when Plato was an old man, almost a decade older than Socrates had been when he died. But even in his absence, Socrates is significant.

This literary ploy of Plato's makes it difficult to distill out of the dialogues what is historically true of Socrates, the man who wandered barefoot through the Athenian agora in a not terribly clean *chitōn* and persistently asked questions whose points were difficult to grasp, creating a crowd of onlookers around him as he went about thwarting every proffered answer, a busker of dialectics, a philosophical urban guerrilla. Plato is not the only Athenian who wrote Socratic dialogues following Socrates' execution.* But he is the only writer of Socratic dialogues who is a philosophical genius. Plato's attitude toward "his" Socrates doesn't remain static over the course of his long life, any more than his ever self-critical philosophical positions remain static. Tracing the shifts in his attitude toward the philosopher whose death turned him to philosophy is perhaps a way of trying to bring the remote figure of Plato closer to us as a person.

It's hard, not to say presumptuous, to approach Plato as a person. No philosopher more discourages such an approach. Plato seemed to have little sympathy for the merely personal. We become more worthy the more we bend our minds to the impersonal. We become better as we take in the universe, thinking more about the largeness that it is and less about the smallness that is us. Plato often betrays a horror of human nature, seeing it as more beastly than godlike.† Human nature is an ethical and political problem to be solved, and only the universe is adequate to the enormous task.‡ The *Laws*, which features three old men in conversation, twice unflatteringly compares humans

---

* Aristotle writes in his *Poetics* (1447b) of an established genre of Socratic literature, *Sōkratikoi Logoi*, all of which were written after Socrates' death. See Appendix A.

† In the *Phaedo*, he indulges in a riff on the inhuman forms that most people will take after they die—becoming donkeys "and other perverse animals," or predators, "like wolves and hawks and kites," while the "ordinary citizens," the upright and uptight bourgeoisie, will be transformed into busy little bees and ants (81e–82b). It's an amusing passage, as well as telling.

‡ Such a view—setting the universe itself to the task of making us humans better—tends to meld together subjects that we keep resolutely apart. So it is impossible to speak of Plato's ethics or political theory or aesthetics without also speaking of his cosmology, metaphysics, epistemology, and psychology. Our division of domains is foreign to Plato's thought. "Metaphysics, Ethics, and Psychology would have seemed to Plato a meaningless classification and he would certainly have protested against its application to himself. Each of these terms he would have thought to include all the others." G. M. A. Grube, *Plato's Thought: Eight Cardinal Points of Plato's Philosophy as Treated in the Whole of His Works* (Boston: Beacon Press, 1958), p. viii.

to puppets, "though with some touch of reality about them, too," as the Athenian says. The old man from Sparta responds to this, "I must say, sir, you have but a poor estimate of our race," and the Athenian doesn't bother to deny it.

Plato's bleak despair regarding "our race" might have grown more pronounced in his old age, but I suspect Plato took a dim view of humanity even when he was younger. Socrates' fate at the hands of the democracy—his death sentence, like the guilty verdict, was the result of popular vote—might have had as much to do with his dim view of humanity as it did with his turning to philosophy in the first place. Whereas Socrates might laugh out loud at the vulgar jokes of the comic writers, even when he was made the butt of them,[*] Plato's more characteristic reaction toward the riotous and ridiculous aspects of human nature was, I suspect, a shudder. His love for Socrates helped him to repress the shudder. Socrates was, for him, a means of reconciling himself to human life, deformed as it is by ugly contradictions. Socrates, so very human—as Plato takes pains to show us—himself embodied these contradictions. Because there had been such a man as Socrates, Plato could convince himself that human life was worth caring about. But I suspect that for him it did take convincing.

By writing as he did, Plato created a morass of interpretive confusion. But he also created philosophy as a living monument to Socrates. The word "philosophy" has love written into it. It translates as love of wisdom. Love of wisdom is an impersonal sort of love. So it bears mentioning that a very personal love—Plato's love for Socrates—was working itself out in the man who created philosophy as we know it.

All of this adds an element of paradox to the style in which Plato wrote, especially given what Plato will say about philosophical love replacing personal love (the source for our degraded notion of "Platonic love"). But even this tension is put to philosophical use. Plato

---

[*] Socrates was often featured in the dramatic works of his Athenian contemporaries, most notably in those of the comic playwright Aristophanes, though also in other comic writers whose works did not survive. Aristophanes featured Socrates in three of his extant plays: *Clouds, Frogs,* and *Birds. Clouds* came in third at the Athenian literary festival in 423 B.C.E., trailing yet another play that featured a barefoot Socrates. Although his character is mercilessly lampooned—in *Clouds* he hangs from a basket in midair and perorates with impressive absurdity, offering solecisms on how to avoid repaying one's debts and urging the young to beat their know-nothing parents into philosophical submission—Socrates himself is reported to have found his notoriety good fun. In his *Moralia,* Plutarch, the first century C.E. philosopher and historian, quoted Socrates as having said, "When they break a jest upon me in the theater, I feel as if I were at a big party of good friends."

worries about so many dangers tripping us up in our thinking, and one of these dangers is that our thinking might become too reflexive and comfortable with itself. He aims to keep our thinking from becoming thoughtless, and to that end he is never averse to the destabilizing effects of paradox.

The expository chapters of this book alternate with anachronistic dialogues in which Plato himself is a character, taking up our contemporary questions, which are continuous with ones that Plato first raised. The questions in each dialogue are related to ones raised in the expository chapter immediately preceding.

These are, quite literally, dialogues out of time. But there is a way in which the dialogues that Plato wrote are also dialogues out of time. They wrench a person out of time, as Plato believed philosophy must do. In the *Phaedo,* which presents Socrates' death, Plato dramatically puts the detachment of the philosopher from his time this way: *to philosophize is to prepare to die.* (Oddly, philosophy departments have forgone turning this into an enrollment-boosting slogan.)

When I was a child I was addicted to science fiction, and my favorite science fiction required the reader to accept just one preposterous premise, and then everything else made sense. That is what the dialogues of this book ask of the reader. Just accept the one preposterous premise that Plato could turn up in twenty-first-century America, an author on a book tour, and everything else, I hope, makes sense. So here he is at Google headquarters, in Mountain View, California, discussing with his media escort and a software engineer whether crowd-sourcing can answer all ethical questions. And here he is on a panel of child-rearing experts in Manhattan, including a psychoanalyst and a "tiger mom," discussing the question of how to raise a child so that it will shine. And here he is helping out an advice columnist on some of the trickier questions concerning love and sex and revealing the shallowness of our notion of "Platonic love." And here he is on cable news discussing with an aggressive interviewer whether reason has any useful role to play in our moral and political lives. And here he is in a cognitive neuroscience laboratory at a prestigious university, volunteering to have his brain scanned, and discussing with two scientists whether the problems of free will and personal identity can be solved by brain imaging.

As often as I can, I interweave passages from his writings into the conversations he has with our contemporaries, giving the citations. His words sound natural in conversations that will be familiar to the reader, and this is a testament to the surprising relevance he still has—but not because his intuitions always ring true to us. His relevance derives overwhelmingly from the questions he asked and from his insistence that they cannot be easily dispensed with in the ways that people often think. One of the peculiar features of philosophical questions is how eager people are to offer solutions that miss the point of the questions. Sometimes these failed solutions are scientific, and sometimes they are religious, and sometimes they are based on what is called plain common sense. Plato composed some of the most definitive rebuttals of uncomprehending answers to philosophical questions that have ever been made, and one can (and I do) fit these smoothly into conversations he has with neuroscientists and software engineers, not to speak of a bumptious cable news anchor. But I rarely give him the answers, and this I think is true to the man. The thing about Plato is that he rarely presented himself as giving us the final answers. What he insisted upon was the recalcitrance of the questions in the face of shallow attempts to make them go away. His genius for formulating counter-reductive arguments is at one with the genius that allowed him to raise up the field of philosophy as we know it.

I do make him a quick study, and he has much catching up to do, as much in ethics—what, no slaves?—as in science and technology. This is as it should be if the field he created has made progress. A major contention of this book—and it is a controversial one—is that it has, and that its progress extends beyond the seminar table. In his conversation at the Googleplex, his media escort, a practical-minded woman with little use for the examined life, is able to overwhelm him with the kind of ethical intuitions that she takes for granted and of which he never dreamed—though once she states them he immediately gets the point.

If there is such a thing as philosophical progress, then why—unlike scientific progress—is it so invisible? This is a question that runs throughout the book, in the expository chapters as well as in the dialogues. Ruminating on Plato—the ways in which he's still with us and the ways in which he's been left behind—offers an answer to this question. Philosophical progress is invisible because it is incorporated into our points of view. What was tortuously secured by complex argument becomes widely shared intuition, so obvious that we forget its provenance. We don't see it, because we see with it.

MAN WALKS INTO A SEMINAR ROOM

Plato was born in ancient Athens in the month of Thergelian (May–June) of the first year of the eighty-eighth Olympiad, which would make it the year 428 or 427 B.C.E. by our reckoning, and he died some eighty or eighty-one years later. His antiquity removes him to a time and a sensibility that some have argued are all but irrecoverable to us. And yet, despite the historical distance, Plato could stroll into almost any graduate seminar in philosophy, seat himself at the elliptical table around which abstractions and distinctions would be propagating with abandon, and catch the drift in no time at all.

First off, Plato would have little trouble recognizing the techniques being employed: the laborious constructions and deconstructions of arguments; the intense inspection of intuitions, drawing out their implications and prodding and palpating them for contradictions and other unwelcome consequences; the counterexample tossed in the face of proposed generalizations; the endless attempts to get a grip on slippery terms, to separate out multiple senses that get merged under single expressions.

And then there are the thought-experiments often couched in wildly imaginative terms: Suppose that somewhere out in the universe there's a planet just like ours—let's call it Twin Earth—on which there's a molecule-by-molecule clone of everything and every person, with just one exception. They have something that looks and behaves just like water only it's not H_2O. It's something with an entirely different chemical constitution; let's just call it XYZ. And we're talking a few hundred years ago, so scientists on Earth and Twin Earth can't know about the chemical compositions. Both Earthlings and Twin Earthlings use the word "water," and for all they know, for all that's in their heads when they use the word "water," it means the same thing on Earth and on

Twin Earth. But *does* it mean the same thing, and if it doesn't, then doesn't that prove that meanings are not in the head?*

Or maybe the issue being argued is the ethics of abortion, and someone, wanting to set aside the whole irresolvable question of whether the fetus is a person or not, proposes the following thought-experiment: You wake up in a hospital bed and find yourself surgically attached to a famous violinist. You're told that you, and you alone, being a perfect match for him, can keep him alive for the nine months he requires in order to be viable on his own. There's no question that you're both persons, and he's an important one at that. But still, do you have an ethical obligation to put your life on hold and remain surgically attached to him?†

I mention these famous contemporary thought-experiments not in order to endorse them one way or the other, but simply as examples of what often takes place around philosophy's seminar table. The point I want to make is that, even though the scenarios would be alien to Plato, the techniques employed by the disputants round the table would be largely familiar to him. Plato was himself a master of composing elaborately counterfactual thought-experiments,‡ and we could expect Plato to soon enter the philosophical fray, no doubt dominating the table before the seminar was well under way.

And it wouldn't be the techniques alone that would give Plato the distinct feeling of *been here, done that*. Many of the questions being batted around the table would be owned by Plato. Moral relativism? You mean to tell me you people are still arguing about whether there are any objective facts about right and wrong or rather whether it's all relative to specific cultures, so that in, say, the militaristically regimented city of Sparta, a society I actually admired in many respects, the murder of puny and otherwise unpromising babies, who would only drain the state's resources without reciprocally contributing, is a moral

---

*This thought-experiment was first proposed in a paper by Hilary Putnam, "The Meaning of Meaning" (1975), in *Philosophical Papers*, vol. 2, *Mind, Language and Reality* (Cambridge: Cambridge University Press, 1975). Putnam used the thought-experiment to argue for a thesis called "semantic externalism," meaning that meanings aren't just a matter of what's in the head, a conclusion to which Plato would more than likely be sympathetic.

† This thought-experiment was first proposed by Judith Jarvis Thomson, "A Defense of Abortion," *Philosophy and Public Affairs* 1, no. 1 (Autumn 1971): 47–66.

‡ See, for example, the Ring of Gyges, discussed in chapter γ. Even his famous Myth of the Cave can be seen as a thought-experiment, designed to explore the ethical obligations of the person who has knowledge that would be useful to others even though he has reason to suspect that they will violently reject both it and him. See chapter θ.

obligation, whereas in other societies, perhaps less ruthlessly rational and more prey to sentimentality, infanticide is morally condemned? By Zeus, we were battling that moral relativism rot out with sophists back in the day when Alexander the Great wasn't even a gleam in Philip and Olympias' eyes!*

Or suppose the question on the table is "What is the relevant level of description for explaining a person's action?" And let's say, to make the conversation around the table even more charged, that the action under discussion is of a kind to make the person come under judgment as either guilty or innocent of a crime of some kind. Is the right level of description the state of the brain before and during the time of the action? Or is the relevant description one that displays the action as an expression of the person's character, embedding the action in a more extended narrative of who this person is? Though the physical terms deemed relevant would be new to Plato—the prefrontal lobe and the right temporoparietal junction, the amygdala and dopamine—the general philosophical argument would be familiar to him, as talk round the table focused on the "explanatory gap" between the neural and the narrative descriptions. After all, Plato could lay claim to having first formulated something like this explanatory gap when he considered the explanation for Socrates' decision to stay in prison rather than fleeing to save his life.†

Or suppose the topic of conversation at the seminar table concerns whether abstract entities, such as numbers, truly exist. Mathematicians prove all sorts of truths about numbers, truths that often assert the *existence* of certain numbers (for example, given two rational numbers, there exists a rational number between them) and sometimes the *nonexistence* of certain numbers (for example, there exists no largest prime number). But what does this talk of mathematical existence amount to? Do these proofs really have to do with existence in the same way that tables and chairs, the moon and the sun, and you and me exist? Or is mathematical existence something like saying that a particular move exists in chess—say, when a pawn has moved completely across the

---

* See, for example, Plato's deconstruction of Protagoras' declaration that "man is the measure of all things," in *Theaetetus*, 152a–172d and 177c–179b; and *Protagoras* 320c–327c, 329c–d, and 356c–357b. Protagoras' moral relativism is undermined in the *Protagoras*, while the *Theaetetus* blasts away at the epistemological foundations of relativism.

† *Phaedo* (98c–99b). See chapter 1, "Plato in the Magnet," where Plato discusses such an argument with two neuroscientists.

board to a square on the opponent's back row and can be exchanged for any piece, not just a piece that your opponent has captured, which can result in your having, say, two queens on the board? Is that what mathematical existence amounts to, simply being the logical consequence of stipulated rules? Or is the existence asserted in these proofs something like existence in fictional worlds, where it is no less true that Hamlet was born in Denmark than that Hamlet, being purely fictional, was never born at all?

When the subject is mathematical existence, then Plato would be delighted (or maybe embarrassed) by how central to the argument raging around the seminar table his eponym is. The exact terms of these arguments would be unfamiliar to him—with new mathematical results enlisted pro and con—but the question of "mathematical Platonism" would be front and center. Only last week, an acquaintance sent me an updating email and added this postscript: "This fall I sat in on a seminar on Boolean-valued models forcing extensions of the set-theoretic universe." He then listed the names of the mathematicians and logicians attending, a stellar constellation, and continued, "Very difficult stuff, but utterly beautiful. Arguments over Platonism raged the entire time."

Yes, it's true. A certain percentage of those questions still swirling around philosophy's millennia-spanning seminar, the participants still going at them with everything they've got, were first posed by Plato—and often the "everything they've got" was first gotten to by Plato, too. So comfortable would Plato feel seated at philosophy's seminar table that Alfred North Whitehead could famously write, "The safest general characterization of the European philosophical tradition is that it consists of a series of footnotes to Plato."[*]

Those predisposed to dismiss philosophy—some of my best friends—might hear in Whitehead's kudos to Plato a well-aimed jeer at philosophy's expense. That an ancient Greek could still command contemporary relevance, much less the supremacy that Whitehead claimed for him, does not speak well for the field's rate of progress. Of

---

[*] Alfred North Whitehead (1861–1947) was a prominent British philosopher and mathematician, the collaborator with Bertrand Russell on the monumental *Principia Mathematica*, a work which aimed to explicate the rigorous logical foundations of mathematics, and succeeded so well that it took two weighty volumes to get to the point of being able to prove that $1 + 1 = 2$. The project was abandoned after volume 3, by which time Kurt Gödel had shown that it could not, in principle, be completed. The quote above is from Whitehead's *Process and Reality* (New York: Free Press, 1979), p. 39.

course, not all philosophers would assent to Whitehead's "safest general characterization of the European philosophical tradition." But that lack of agreement in itself bolsters a philosophy-jeerer's charge that philosophy never can establish *anything*.

The reason that some of my best friends are philosophy-jeerers is that many of my best friends are scientists. I do not mean to assert that the majority of scientists are hostile to philosophy. I've known scientists who are philosophically impressive. But there is, in a significant segment of the scientific culture, so ingrained a prejudice against philosophy that, much like other prejudices, people casually express their biases without even realizing they are doing so. To quote from a random example that is fresh in my mind, having read it this morning in a short item in *Science* magazine reporting on the search for "Goldilocks planets," those neither too hot nor too cold to support life: "Just two decades ago, most considered the question of life elsewhere in the universe a fringe topic, more suitable for philosophy than for scientific research."*

The casual equating of philosophy with topics on the fringe, emptily speculated upon, can pass unremarked in scientific circles. Like most prejudices, this one is usually not reasoned out, although sometimes it is. Sometimes a scientist is willing to stand up and bravely defend the claim that philosophy is worthless. "Philosophy used to be a field that had content, but then 'natural philosophy' became physics, and physics has only continued to make inroads," Lawrence Krauss, a cosmologist who writes popular science books, told an interviewer. "Philosophy is a field that, unfortunately, reminds me of that old Woody Allen joke, 'those that can't do, teach, and those that can't teach, teach gym.' And the worst part of philosophy is the philosophy of science; the only people, as far as I can tell, that read work by philosophers of science are other philosophers of science. It has no impact on physics whatsoever, and I doubt that other philosophers read it because it's fairly technical. And so it's really hard to understand what justifies it. And so I'd say that this tension [between philosophy and science] occurs because people in philosophy feel threatened, and they have every right to feel threatened, because science progresses and philosophy doesn't."†

* Yudhijit Bhattacharjee, "A Distant Glimpse of Alien Life?" *Science* 333, no. 6045, (August 19, 2011): 930–932.
† Ross Anderson, "Has Physics Made Philosophy and Religion Obsolete?," interview with Lawrence Krauss, *The Atlantic*, April 23, 2012.

There are many things that one can say in response to this position. For starters, one could point out that the position presupposes that we have a clear criterion for distinguishing between scientific and non-scientific views of the world. When pressed to give the requisite criterion, scientists almost automatically reach for the notion of "falsifiability" first proposed by Karl Popper. His profession? Philosophy. The Kraussian position also presupposes that fields like relativistic quantum field theory (the very theory that, according to Krauss, is helping to render philosophy obsolete) are offering us descriptions of physical reality, even though they employ concepts which refer (*if* they refer) to unobservable states and entities, such as, to take a non-random example, relativistic quantum fields. The view that the strange entities dreamed up in the models of theoretical physics, though unobservable, are nonetheless real (if the theory in question is true) is known as "scientific realism"—a substantive philosophical claim, countered by a view known as "scientific instrumentalism," according to which such theories as relativistic quantum field theory are merely tools for making predictions of observations and are not about any *actual* things that exist in the world. In this view, the success of relativistic quantum field theory offers no reason to believe that there is any such thing as a relativistic quantum field.

Presumably physicists care about the "philosophical" question of whether they are actually talking about anything other than observations when they do their science. And indeed, scientific instrumentalism is by no means a conceptual toy constructed for the extended playtime of philosophers. The view itself was first fully formulated by the physicist Pierre Duhem,[*] and many physicists, including Niels Bohr, a leading formulator of quantum mechanics, have advocated instrumentalism, often motivated by the strangeness of quantum mechanics, which puts up challenging barriers to straightforward realistic interpretations.[†] (A realistic interpretation can give one far more reality than one had bargained for—the so-called *multiverse*.)[‡] Quantum strangeness was

* Pierre Duhem, *The Aim and Structure of Physical Theory* (Princeton, NJ: Princeton University Press, 1954).
† Jan Faye, "Copenhagen Interpretation of Quantum Mechanics," *Stanford Encyclopedia of Philosophy*, ed. Edward N. Zalta (Fall 2008 edition). http://plato.stanford.edu/archives/fall2008/entries/qm-copenhagen/.
‡ See the many-worlds interpretation of quantum mechanics first proposed by Hugh Everett in his 1957 Princeton Ph.D. thesis and now, in many variations, one of the mainstream

Bohr's reason for advocating instrumentalism. Perhaps not surprisingly, other physicists disagree, and when they are disagreeing, they are going beyond the domain of theoretical science and plunging straight into philosophy of science. What they're disagreeing about is the question of what it is, precisely, they are doing when they are doing science. Are they refining their instruments for observation or discovering new aspects of reality?

All of which is to say that one cannot make the claims for science that many philosophy-jeerers make without relying heavily on claims—such as the falsifiability criterion for scientific statements, or the assumption of scientific realism—which belong not only to philosophy, but to that "worst part of philosophy," philosophy of science.* So if philosophy has as little substance as Krauss claims, if there is no way to make progress in philosophical knowledge, then this is as serious a problem for a physicist like Krauss as it is for those who call themselves philosophers.

Krauss mentions the old Woody Allen joke, but I'm reminded of another joke:

After a lifetime of hard work and bad luck, Jake makes a killing in the stock market and buys a villa for himself and his bride of forty years, Mimi, on prime real estate in Miami Beach. The first evening

---

interpretations of quantum mechanics, along with Bohr's. According to the many-worlds interpretation, every possibility represented in the configuration space of quantum events is realized in worlds other than our own. Reality therefore consists of a "multiverse" in which all possibilities are realized. David Deutsch, a proponent, argues that, so far as the multiverse is concerned, the distinction between fact and fiction is illusory. See his *Beginning of Infinity: Explanations That Transform the World* (New York: Penguin, 2012), p. 294. Hugh Everett, who died in this world in 1982, allegedly believed, on the basis of his interpretation of quantum mechanics, in his immortality. Sadly, his daughter, who committed suicide at the age of thirty-nine, wrote in the note she left behind that she was going off to rejoin her father in a parallel universe. See Eugene Shikhovtsev, *Biographical Sketch of Hugh Everett, III* (2003), http://space.mit.edu/home/tegmark/everett/everett.html.

*This is not to claim that physicists, when they confine themselves to doing physics, need know anything about philosophy of science, any more than they need to know history of science. Richard Feynman quipped, "Philosophy of science is about as useful to scientists as ornithology is to birds." And yet, beside Feynman's quip, we might place the contrasting view of Einstein: "I fully agree with you," he wrote to Robert Thornton, a young professor who wanted to introduce "as much of the philosophy of science as possible" into the modern physics course he was teaching and wrote to Einstein for support. "So many people today—and even professional scientists—seem to me like somebody who has seen thousands of trees but has never seen a forest. A knowledge of the historic and philosophical background gives that kind of independence from prejudices of his generation from which most scientists are suffering. This independence created by philosophical insight is—in my opinion—the mark of distinction between a mere artisan or specialist and a real seeker after truth." Einstein to Thornton, 7 December 1944; *Einstein Archive* control index 61-573.

that they're settled in, he and Mimi go out on the patio to enjoy their view of the Atlantic Ocean, and Jake discovers that it's obscured by the trees of their neighbors. It's okay, says Mimi, trying to calm down her excitable husband, but Jake gets right on the phone with the neighbors, and, after extensive bickering, they agree that, if he pays for it, they'll top their trees. The landscaping work is done, and Jake and Mimi take their positions that evening on the veranda. Alas, the topped-off trees still get in the way of the view. Jake calls the neighbors, demanding that the trees will have to go, right down to the roots, but this time the neighbors balk. It's okay, Mimi is heard plaintively begging in the background, but Jake, determined that he and Mimi get the view that their years of scrimping and saving deserve, offers to buy the neighboring villa at an inflated price. The neighbors immediately agree, and, as soon as the papers are signed, Jake has the offending trees cut down. "You know," Jake says to Mimi that evening, as they sit on the veranda drinking in their unobstructed view of the Atlantic Ocean, "there are some things that money just can't buy." Like Jake, some philosophy-jeerers don't take into account all the philosophical cash they have to spend in order to arrive at their view.

But still, even if the most extreme philosopher-jeerer can't altogether avoid relying on a bit of philosophy, Popperian or otherwise, isn't there something to the charge that "science progresses and philosophy doesn't"? After all, if Plato, a man who voiced misgivings about that newfangled technology of *writing things down,** can still find his place at philosophy's seminar table, doesn't that cast the field as a whole in a seriously non-progressive light?

No self-respecting physicist would declare that all of physics consists of a series of footnotes to Democritus, even though that Greek, a bit more ancient even than Plato, managed not only to conceptualize but also to name the atom.† Nor would any biologist describe his field as

---

* See his *Phaedrus*, 274d–276b. Plato worried that writing things down would supplant genuine learning. It was conceptual knowledge on which he was focused, and he worried that writing would undermine the sense of what it really is to have mastered such knowledge. To have mastered it is to have it change the very substance of one's mind; therefore, what need is there to write it down? Some professors have told me that they think of Plato's misgivings about writing whenever a student asks them what's the point of learning some idea when it can be accessed on the Internet whenever it's needed.

† The word "atom" means "indivisible" in ancient Greek. Democritus, who was influenced by his teacher Leucippus to the extent that it is difficult to distinguish between them, said that reality, including us, consists of atoms whirling about in the infinite void. "They said that the first principles were infinite in number, and thought they were indivisible atoms and

mere footnotes to Aristotle, even though Aristotle, with pre-scientific prescience, first laid out the taxonomy of the animal kingdom. Why don't these other ancients have the currency in these scientific fields that Plato still enjoys in philosophy?

The answer, delivered in unison by the chorus of philosophy-jeerers, is that the empirical sciences, so unlike philosophy, make palpable progress. Possessing the self-correcting means to test and dispose, they prod the physical world so that the physical world gets a chance to answer back for itself in the form of experimental evidence. If science oftentimes has charged off in some altogether wrong direction, believing, say, that fire is to be explained by the existence of a fire-stuff, phlogiston, or that life is to be explained by the existence of a life-stuff, the élan vital, then empirical testing will, sooner or later,* disabuse science of such fictions. All mortals are fallible, even the smartest among us, including the scientists. We are prey to cognitive lapses, some of them built into the very machinery of thinking, such as the statistical fallacies we are prone to commit. (Cognitive scientists have recently taken on these cognitive lapses and biases as a subject for scientific explanation.†) Given these cognitive vulnerabilities, it would be convenient to have an arrangement whereby reality can tell us off; and that is precisely what science is. Scientific methodology is the arrangement that allows reality to answer us back. This arrangement was precisely what Karl Popper had in mind when he made falsifiability the criterion of demar-

---

impassable owing to their compactness, and without any void in them; divisibility comes about because of the void in compound bodies." This is from Simplicius' *De caelo* (242, 18). Simplicius lived in the early sixth century c.e. and was one of the last of the pagan philosophers. We owe a great deal of our knowledge of the lost writings of the ancients to Simplicius, all of whose writings are commentaries on these earlier writers, mostly of the classical age.

*The "sooner or later" is meant to remind us that sometimes the scientific revisions take much longer than the empirical evidence warrants. Scientists are just as apt to be bullheaded as other human beings since the fact is that they *are* human beings; and as such they often hang on to wrong theories, on which their reputations and worldviews are staked, long after "the logic of scientific discovery," to quote the title of a Karl Popper book, would have them do otherwise. To complicate the matter, there is also a little business known as the theory-ladenness of observation. A person's holding a theory to be true conditions how the evidence is seen; countervailing evidence will not enter into consciousness, making it all the harder for theories to be falsified. But, though the logic of scientific discovery is not as clear and straightforward as Popper-approving scientists sometimes present it, the elaborate testing and experimentation does provide the means for nature to answer scientists back, sometimes so forcefully and unambivalently that cherished hypotheses and theories are, sooner or later, discarded. The truth of this "sooner or later," with an emphasis on the later, seems to me the crux of what has survived from Thomas Kuhn's incendiary book, *The Structure of Scientific Revolutions.*

† See, for example, A. Tversky and D. Kahneman, "Judgment Under Uncertainty Heuristics and Biases," Science 185 (1974): 1124–31.

cation between the scientific and non-scientific, the very piece of philosophy of science that so many scientists automatically reach for when asked to defend their view that science alone makes progress.[*] Insofar as a claim about reality is scientific, it is, in principle, falsifiable, which means nothing more or less than reality's being afforded the opportunity to answer us back. "Ah, so you think that it's perfectly obvious that two events are either simultaneous or they're not, regardless of which inertial frame of reference they're measured in, do you? Well, we'll just see about that!" Voilà, the theory of special relativity displaces Newtonian mechanics.

Philosophy, in contrast, is like one of those dreaded conversationalists whose idea of engaging with you is to speak endlessly *at* you, not requiring—in fact, actively discouraging—any response on your part, one idea engendering another in a self-perpetuating closed system (as in the classic definition of a "bore": someone who won't change his mind and won't change the subject). In exactly the same way—which is to say, *not at all*—does the actual world get to be involved when it is philosophy that is doing the talking. And it's exactly because philosophy is just such a one-sided conversationalist that its rate of progress is what it is—in a word, null. (Again, still quoting the philosophy-jeerers here.)

And it's not just the empirical sciences that tell so damningly of philosophy's folly of futility. Even mathematics, though just as abstract and non-empirical as philosophy,[†] could hardly be said to consist of a mere series of footnotes to Pythagoras, the number-enchanted seer who died some sixty-odd years before Plato was born but whose mathematically dominated view of the universe had a profound effect on the younger philosopher. Mathematics could not be said to be a mere

---

[*] I have expressed elsewhere my own misgivings about Popper's principle of falsifiability as the absolute criterion for demarcation between science and non-science. See my "The Popperian Sound Bite," in *What Have You Changed Your Mind About? Today's Leading Minds Rethink Everything,* ed. John Brockman (New York: HarperPerennial, 2008).

[†] Mathematics, of course, is utilized within the empirical sciences, but in itself it proceeds by a priori methods of proof, which is why mathematics departments are so much cheaper for universities to maintain than departments of physics, biology, or chemistry, with all their subspecialties and cross-hybrids, all of which require huge outlays of funds in the form of laboratories, observatories, particle colliders, and so on. Mathematicians, on the contrary, require only blackboard, chalk, and erasers, as well as generous quantities of caffeine, a well-worn joke being that a mathematician is a machine for transforming coffee into theorems. Another joke: Philosophers are even cheaper to hire than mathematicians, since you don't need to provide them with erasers.

series of footnotes to any of the Greeks, including Euclid, who was born twenty-two years after Plato died and codified many of the proofs of his predecessors.*

Such ancient thinkers as Democritus, Aristotle, and Pythagoras have been left in the ancient dust by the fields of physics, biology, and mathematics. Democritus, intending to major now in physics, wouldn't get very far with his freewheeling speculative approach and might well be taken aback by the great amount of mathematics—calculus in classical mechanics, for starters—that he would be required to master if he wanted to understand modern conceptions of matter and energy, space and time. The melding of experimental techniques with mathematical description was the great leap forward, accomplished in the seventeenth century, that brought us to the point at which, as Krauss put it, " 'natural philosophy' became physics."† Democritus would also have to put in long hours in the lab, devising experiments under carefully controlled conditions, and taking measurements by means of instruments designed to extract precisely the right information. As for Aristotle, should he intend to major in biology his first assignment would be to master the theory of natural selection, together with genetics, without both of which he could not begin to understand any contemporary explanations for biological structures and functions. And then there is Pythagoras. The legend is that the founder of theoretical mathematics was so outraged when one of his students, the haplessly gifted Hippasus, discovered irrational numbers‡ that he sent the poor fellow out on a raft to drown, initiating a venerable tradition of professors mistreating their graduate students. Pythagoras, should he want

---

*Much of the work that Euclid built on was done at Plato's Academy. Euclid mentions Theaetetus by name.

† For a magnificent discussion of this melding see E. A. Burtt's *The Metaphysical Foundations of Modern Physical Science*. Burtt's book, which was first published in 1924 and was, outrageously, out of print for some decades, was reissued in 2004 by Dover. This too-little-known work remains as rousingly insightful today as when it was first published, even while some of its flashier spin-offs are badly showing their age. Thomas Kuhn seemed to have been unaware that Burtt was influencing him by way of Alexandre Koyré, whom he does credit. For a discussion of Burtt's influence on Koyré, see Diane Davis Villemaire, *E. A. Burtt, Historian and Philosopher: A Study of the Author of "The Metaphysical Foundations of Modern Physical Science"* Boston Sudies in the Philosophy of Science (Book 226) (Dordrecht, The Netherlands: Kluwer Academic Publishers, 2002).

‡ Using the Pythagorean theorem, a square with sides of one unit has a diagonal equal to the square root of 2, which is an irrational number. If one tries to write it as a fraction of two whole numbers, one will be able to derive a contradiction. This, more or less, had been Hippasus' deduction (or so we think).

to continue on for a degree in modern-day mathematics, would have to learn to abide far more counterintuitive results than numbers that cannot be written as ratios between whole numbers. From the square root of -1, to Georg Cantor's revelation of infinite domains infinitely more infinite than other infinite domains, to Kurt Gödel's incompleteness theorems, mathematics has constantly displaced the borders between the conceivable and the inconceivable, and Pythagoras would be in for some long hours of awesome mind-blowing.

And all the while Democritus, Aristotle, and Pythagoras were getting remedial tutoring in their respective fields, our man Plato would be holding forth at philosophy's seminar table. Isn't this ample proof of something seriously awry with the entire field of philosophy?

But wait just a second here. Since Democritus, Aristotle, and Pythagoras are officially classified as philosophers, shouldn't the field of philosophy get some credit, after all, for those progress-achieving, distance-making fields that left those ancients far behind? Not at all, responds the chorus of philosophy-jeerers. Oh, sure, philosophy, by spinning out questions in every direction, like a toddler who has just discovered the exasperating power of mechanically appending "Why?" to every received answer, has managed over the course of its excessively long history to occasionally put forth some good questions, by which is meant questions that have actual answers, instead of variations on those soundless-or-not-trees-in-the-forest non-starters for which no discoverable fact of the matter would count as any solution at all. Philosophers, asking and asking without ever possessing the means of answering, sometimes ask questions that are, so to speak, *protoscientific,* posed before the science yet exists that can pursue them effectively, which is to say empirically. But even though it's the philosophers who ask the questions, it's always the scientists who answer them. Philosophy's role in the whole matter is to send up a signal reading "Science desperately needed here." Or, changing the metaphor, philosophy is a cold storage room in which questions are shelved until the sciences get around to handling them. Or, to change the metaphor yet again, philosophers are premature ejaculators who pose questions too embarrassingly soon, spilling their seminal genius to no effect.

This is the view—pick your metaphor—that Krauss was proposing when he diagnosed why philosophers feel so threatened, as he put it, by the growing power of the sciences, and in particular physics, that they treated his proposed answer to the classic philosophical question *Why is there something rather than nothing?* with less than universal ovation, insisting that, though the cited physics is terrific, it doesn't address the specifically *philosophical* question.* And whether or not the philosophers were correct about Krauss's proposed answer to this specific philosophical question, still there is his larger point that philosophy's main contribution to the growth of knowledge is in providing cold storage. The history of philosophy is, after all, rife with philosophers going after questions that would eventually receive their answers from science.

So take the very first philosophers you will find listed in a history of Western philosophy. Philosophy is said to have begun, toward the latter part of the seventh century B.C.E., not in Greece proper, but on the coast of Asia Minor, in what is now Turkey, in the Greek settlements that constituted Ionia, in rich trading cities that had contacts not only with the rest of Greece but with the older, more established civilizations of Egypt and the Near East. The earliest philosophers—men like Thales and Anaximander, both residents of the Ionian city of Miletus, which is therefore duly recorded as the official birthplace of philosophy—were protoscientists, asking questions, and sometimes even guessing at semi-accurate answers, which Q&A would eventually be taken over by physicists and cosmologists, who minimized the intuitive guesswork, and got to work at experimentally engaging reality to respond.†

These first Ionian philosophers would themselves have made excellent scientists. They were bursting with the right kind of curiosity about the physical world, and their inclinations were thoroughly *materialist*—they intuited that there is some fundamental kind of stuff that's uniform throughout all the myriad phantasmagoria that we perceive—as well as *naturalist*—they intuited that a small number of fundamental laws underlie all the ceaseless changes. Actually, we retrospectively dub it "intuiting" (a verb philosophers call a "success term" and linguists a

---

*Krauss's attack on philosophy was precipitated by a review of his book *A Universe from Nothing: Why There Is Something Rather Than Nothing* (New York: Free Press, 2012) by the philosopher David Albert (who happens also to hold a Ph.D. in theoretical physics) in the *New York Times Book Review*, March 23, 2012.

†The Ionian philosophers didn't bypass observation. Thales' successful prediction of an eclipse, for example, must have been based on careful observation, as well as access to observational records accumulated by the Babylonians. But the full experimental method of setting up conditions to catch nature out had to await the founders of modern science, most notably Galileo.

"factive"), rather than just "imagining," because those Ionians turned out to be right in their intuition that there was some fundamental material principle that constituted everything in the universe ($E=mc^2$ is a materialist principle). And they were right in their intuition that there was an intelligible regularity underlying nature. They were right that physical events are not the outcome of the capricious antics of larger-than-life gods, but rather that they fit into patterns that are law-like, or, as modern philosophers of science put it, *nomological*, from the Greek *nomos*, for law. Of all the conceptions that made science possible, none is more essential than what the physicist and historian of science Gerald Holton called "the Ionian Enchantment": the intuition that nature is governed by a small number of laws which account for all the vast complexity that we observe in the physical universe.* This enchantment, if enchantment it be, ensorcels all of science. Once the Ionians posited this intelligibility, the next question became what is the proper form for conceiving of this intelligibility, and this question continued as a divisive one throughout the Greek classical age. It's this question that forms the crux of the opposition between Plato and Aristotle, with Plato opting for mathematical structure as providing the form of intelligibility and Aristotle opting for teleology.

Science simply cannot subject the Ionian nomological intuition to doubt and still remain science. Should an observation clash with what scientists have heretofore believed was a law of nature, the scientific response is never to consider the possibility that we'd gotten the Ionian intuition wrong; rather, the scientific response is that we got that particular natural law, or cluster of laws, wrong. Scientists may even decide, as they appear to have done, that the laws governing the motions of the subatomic particles of matter are irreducibly statistical. This is a radical rethinking of the nature of natural laws, but not so radical as the negation of the Ionian intuition would be; that possibility is scientifically unthinkable. It is a fundamental condition of doing science that nothing that we could possibly observe would count as a violation of the Ionian Enchantment, at least that part of the Ionian Enchantment that posits the nomological character of physical reality. Nothing would count as evidence that our physical reality is ungoverned by physical

* See Holton's *Einstein, History, and Other Passions: The Rebellion Against Science at the End of the Twentieth Century* (Cambridge, MA: Harvard University Press, 2000), chapter 7, "Einstein and the Goal of Science."

laws. Rather the scientific response would be that we hadn't formulated the laws correctly.*

The Ionians happened on other important aspects of what would eventually become incorporated into the scientific method. Anaximander, who wrote a long and long-lost poem entitled *On Nature*, small fragments of which have come down to us, hypothesized the existence of what contemporary philosophers of science would classify as a theoretical entity or theoretical construct: something that one can't directly observe, as the quantum fields can't be directly observed, but which is conceptualized in the context of an overall theory meant to explain as many observations as possible. Many theoretical constructs have been framed and many have been discarded along the way of scientific progress.† The most abstruse reaches of theoretical physics are still in the business of doing the sort of thing Anaximander first attempted, the big difference being that these theories must somehow be connected with observable consequences, or predictions, by which they might be tested. Genes are a theoretical construct that has allowed the explanatory power of biology to increase by orders of magnitude, a success which should remind us that calling an entity a theoretical construct doesn't mean that we don't know it to exist (at least those of us who are scientific realists). It just explains how we came to know the particular thing in question to exist, which wasn't through direct observation but because of how it functions in a scientific explanation.

Anaximander called his theoretical construct the *apeiron*, or the

---

*David Hume's argument concerning the "principle of the uniformity of nature" comes down to an argument that, since science presumes the lawfulness of nature, science cannot non-circularly provide evidence for it. "For all inferences from experience suppose, as their foundation, that the future will resemble the past and that similar power will be conjoined with similar sensible qualities. If there be any suspicion that the course of nature may change, and the past may be no rule for the future, all experience becomes useless and can give rise to no inference or conclusion. It is impossible, therefore, that any arguments from experience can prove this resemblance of the past to the future, since all these arguments are founded on the supposition of that resemblance. . . . My practice, you say, refutes my doubts. But you mistake the purport of my question. As an agent, I am quite satisfied in the point; but as a philosopher who has some share of curiosity, I will not say skepticism, I want to learn the foundation of this inference." *Inquiry Concerning Human Understanding*, chapter 4, "Skeptical Doubts Concerning the Operations of the Understanding."

†Among the discarded, for example, was phlogiston, the substance of fire; those items that burn have phlogiston in their composition, which is released as fire, which is why one is often left, after a conflagration, with a mere pile of ashes. Phlogiston, as an explanatory hypothesis, was eliminated by the theory of oxidation, which was established by Lavoisier's carefully weighing objects before and after burning. Caloric fluid, which was meant to explain heat, was another theoretical entity that was given up; this elimination was accomplished by the identification of heat with molecular motion.

boundless, a basic something or other which is indefinite in itself, subtending all possible qualities, reconciling in its boundlessness all opposites, out of which precipitates the great abundance of this world. Anaximander's *apeiron* is a first approximation to our modern concept of matter.

Anaximander's conception of the fundamental material principle was a giant leap forward in imaginative theorizing, especially compared to that of his teacher, Thales, who holds the official title of "first Western philosopher." Thales, also proceeding on the first-rate intuition that there is a material unity behind the diversity, had settled for water, though some have argued that Thales' reference to water was a metaphor. If it was, it was lost on Aristotle,[*] as well as on Bertrand Russell, who writes:

> In every history of philosophy for students, the first thing mentioned is that philosophy began with Thales, who said that everything is water. This is discouraging to the beginner who is struggling—perhaps not very hard—to feel that respect for philosophy which the curriculum seems to expect. There is, however, ample reason to feel respect for Thales, though perhaps rather as a man of science than as a philosopher in the modern sense of the word.[†]

I had the good fortune to have Russell's *History of Western Philosophy* assigned by my professor for my first course in philosophy, and my admiration for its verve and clarity has never dissipated. My literary agent once tried to convince me to take Bertrand Russell on and write a new *History of Western Philosophy*, extending it to philosophers who came after John Dewey, Russell's last entry. I dismissed the suggestion for two obvious reasons, both involving comparisons between Lord Russell and me. The first comparison is the obvious one, Lord Russell being one of the preeminent thinkers of his age, and the second is that the long stretch of time that allowed Russell to undertake the tome was

---

[*] Only fragments have come down to us of these earliest philosophers, who appeared to have written short prose pieces or, in some cases, oracular poetry, setting forth their views. Our knowledge of these first philosophers comes largely from the accounts given of them by secondhand commentators, who may or may not have had direct access to their writings. Prominent among these commentators is Aristotle, who writes about other philosophers in his *Metaphysics*.

[†] Bertrand Russell, *A History of Western Philosophy* (New York: Simon & Schuster, 1967), p. 24.

granted him by a stay in prison.* It has been a lifelong goal of mine to stay out of prison. So I offered my agent a counterproposal: a history of western philosophy in limericks, a task for which I might even be better qualified than Lord Russell, and which would in any case be quicker. Here is my first entry, which works best, if it works at all, when read with a New York accent:

> From the beginning philosophy sought for
> The order behind the disorder
>   Thales sipped cheap wine
>   And in this did divine:
> "Why it's nothing at all but pure water!"

The reader will be relieved to learn I abandoned the project.

Anaximander, though demoting water metaphysically, kept the element prominent by proposing that it had once covered the surface of the earth, with all life having originated out of a primordial mud, and with humans developing—or evolving, as we might put it—from fish. (Anaximander might have had recourse to fossils in hypothesizing so happily; we don't really know.)

Another fifth-century philosopher who also fits the mold of a protoscientist in search of an empirical methodology was Empedocles of Acragas, a city not in Ionia but in Greek-settled Sicily.† Empedocles pluralistically listed the basic material elements as four—earth, air, fire, and water—and he speculated that all changes were regulated by two immanent forces, which he named Love and Strife, but which we could advance to scientific respectability by de-anthropomorphizing them into attraction and repulsion. Out of these four elements and these two forces the universe had been generated, including living forms, though not as we know them, but rather in the form of detached organs, which, propelled by the attractive force of Love, merged themselves with other

---

* His crime was distributing pacifist literature during the First World War. Hitler caused him to later renounce his pacifism, to the point that he wished he were younger so that he might don a uniform himself. See Russell, *Autobiography of Bertrand Russell* (London: Routledge, 2000), pp. 438 ff.

† The Greeks colonized all around the Mediterranean basin, and their "Hesperia," or Land of the West, was known as Magna Graecia, Greater Greece. Pythagoras, too, though born on the Aegean island of Samos, eventually settled on the eastern coast of Italy in the city of Crotona. Plato was to spend some time in Italy among the Pythagoreans, and their mathematics-marinated mysticism deeply penetrated his thinking.

organs to form whole organisms, some of which were monstrous and too unfit to survive, a chain of reasoning that brought Empedocles of Acragas intriguingly close to propounding a protoscientific theory of natural selection.*

So Democritus, a philosopher who formulated a theoretical construct (the atom) which was to prove to be the linchpin of modern conceptions of matter,† falls into a deeper tradition in philosophy, of thinkers who asked the kinds of questions that, at a later stage in Europe's history, would be taken up by people like Francis Bacon and Galileo Galilei and Isaac Newton. Only this time around, a methodology of experimental testing under carefully controlled conditions would be brought to bear, supplemented by instruments specifically designed for the task, and this methodology would decisively remove these questions from the domain of speculative philosophy and deliver them into the province of the empirical sciences, that ingenious arrangement whereby reality is afforded the opportunity to answer us back.‡

This mini-history of philosophy's origins can be marshaled as some evidence for the larger point that some philosophy-jeerers are trying to make, which is that the activity of posing scientific questions prematurely is the most useful thing of which philosophy can be accused. But once the appropriate scientific theory develops, which most essentially includes the means for testing itself, then philosophy's usefulness is over, and questions that have been subjected to philosophy's futile gnawings and naggings and nigglings for unconscionable amounts of time, without any progress being made on them, are suddenly propelling us forward into knowledge, the Real Thing at last. Philosophy's inter-

---

*The Greek proscientific idea of organisms attaching to other organisms to create new life forms, some better suited for survival than others, has a counterpart in modern molecular biology. Consider mitochondria, organelles found in the cells of all animals which use glucose to generate ATP, our fuel source. Mitochondria, one of the most essential parts of life forms, used to be free-living organisms. They got swallowed up by the single-cell ancestor of all animals but resisted being digested and maintained their integrity, making complex life possible. The chloroplasts that make plants green and allow them to photosynthesize have a similar history.

† Besides Leucippus and Democritus, ancients who held to a corpuscular theory of matter included Epicurus and Lucretius, who put this philosophy into magnificent poetry in his *De Rerum Natura*, or *On the Nature of Things*. The chance survival of Lucretius' poem was the subject of Stephen Greenblatt's *The Swerve: How the World Became Modern* (New York: W. W. Norton, 2011), which, as the title announces, tries to stake for this poem a pivotal role in Europe's once again picking up the secular-humanist trail that was first laid out in antiquity.

‡ The sciences were known as natural philosophy until well into the nineteenth century, when the word "science," derived from the Latin for knowledge, entered the lexicon.

rogatory irrepressibility means that philosophers regularly pose questions that eventually get appropriated by disciplines of science as they emerge: physics and cosmology and chemistry and biology, and (emerging somewhat later) psychology and logic and linguistics, and (emerging even later) computer science and cognitive science and neuroscience. As scientific disciplines emerge, the number of philosophical questions—the left-behinds—shrinks. If cold storage is *all* that philosophy can provide, then the natural course of scientific progress will eventually empty out the cold storage room until all that is left are those permanent non-starters of the soundless-or-not-falling-trees-in-the-forest ilk.

This prediction can be formulated mathematically (a book centered on Plato ought to have at least one equation):

The Fate-of-Philosophy Equation:

$$\phi_{t \to \infty} = \varnothing$$

which means that as time t approaches infinity ∞, the set of philosophical problems ϕ equals the null set $\varnothing$.

Krauss was, in effect, propounding the Fate-of-Philosophy Equation, though, as the Jake joke suggested, it takes a certain amount of philosophy—philosophy belonging to "the worst part of philosophy," philosophy of science—to make the equation intelligible. But if the philosophy-jeerer can abide that small bit of philosophy, then the Fate-of-Philosophy Equation might just possibly be true.

The question of whether the Fate-of-Philosophy Equation is true is an overriding concern of this book. A millennium and a half have passed since Plato inherited a subset of philosophical questions from an extraordinary character of his acquaintance named Socrates, a man who hung around the agora of Athens and engaged anyone he could—from statesmen to sophists (teachers of rhetoric) to poets to artisans to schoolboys to slaves—in philosophical discussion. Socrates' occupation, as innocuous as it might seem, eventually got him into serious trouble, and he was put on trial, convicted, and executed for the crime of persistently posing his peculiar questions; the formal charges were impiety and corruption of the young. Socrates explained at his trial, at least according to Plato, that he was not interested in the sort of questions posed by Thales and Co.—precisely those questions that we now, looking back, can dub "protoscientific"—but rather was only concerned with questions that helped a person determine what kind of life is worth living.* Socrates called the sphere of this concern *epimeleia heautou,* care of the self.† For Socrates, these were the paramount philosophical questions. And these questions, he maintained, were not to be answered by the inquiries of Thales and Co., although, he affirmed, they also have objective and discoverable answers.

Having received from Socrates a few of these peculiar questions, Plato went on to swell the sphere of philosophical questions beyond those that Socrates posed, formulating questions not just in ethics but in metaphysics, epistemology, political philosophy, philosophy of language, philosophy of mind, philosophy of science, philosophy of mathematics, philosophy of art, philosophy of law, philosophy of religion, philosophy of education, and philosophy of history. Grasping the essential peculiarity of Socrates' peculiar questions, he was able to raise up the entire continent of philosophy, like the lost continent of Atlantis hoisted from the depths, which is an especially apt metaphor given that the first recorded allusion to Atlantis comes from Plato himself.‡

But, as I said, it's been twenty-four hundred years. The inquiries of Thales and Co., now become the mature natural sciences, have ventured into spheres undreamt of by the scientists of fifty years ago, much less a man who spoke the Ionic dialect of ancient Greek. It's not just physics and cosmology in whose name a philosophy-jeerer can claim

---

* For Socrates' spurning of what we can now call protoscientific questions see the *Apology* 19c and 26d and the *Phaedo* 96a–100a. See also Xenophon, *Memorabilia,* I.1.12–16. On our sources for Socrates, who published nothing on his own, see appendix A.
† *Apology* 29d.
‡ Plato described the advanced civilization, destroyed by a natural disaster and swallowed up by the sea, in the *Timaeus.* "Some time later excessively violent earthquakes and floods occurred, and after the onset of an unbearable day and a night, your entire warrior force sank below the earth all at once, and the Isle of Atlantis likewise sank below the sea and disappeared" (25c–d). There is geological and architectural evidence that Plato, in relating what he calls "an old-world story" (21a), was relying on a thousand-year-old cultural memory of the lost Minoan society that had existed on Crete and other islands, including the ancient Thera, whose brilliant civilization (including indoor plumbing!) sat on a volcano that erupted around 1500 B.C.E. The Santorini archipelago, with its massive deposits of pumice, is what remains of what was once the single island of Thera. The tsunamis that were unleashed by the volcano—which is now thought to have been second only to the 1815 volcanic eruption in Tambora, Indonesia—might have been responsible for the destruction of the wealthy and advanced Minoan culture on Crete. See Richard A Lovett, " 'Atlantis' Eruption Twice as Big as Previously Believed, Study Suggests," *National Geographic News,* August 23, 2006. The theme of civilization ending in cataclysmic doom might well have resonated with Plato's historical pessimism, perhaps intensified as he grew older. (*Timaeus* is typically classified as one of his later dialogues.)

to be at last answering age-old philosophical questions with which philosophers have long wrestled. Of perhaps even more pressing relevance are the new sciences of the mind, evolutionary psychology and cognitive and social and affective neuroscience, which have together so ramped up the explanatory powers of how the mind works that both ethics and philosophy of mind have fallen into the sights of science, including the sights of functional magnetic resonance imaging.* And then there is the technology represented by the computer, allowing not only for untold access to information, but also forcing us to rethink the very nature of knowledge, and so of epistemology, and of the entity which knows, namely the mind, and of the philosophical study of that entity, the philosophy of mind. Metaphysics-busting cosmology, ethics-and-philosophy-of-mind-busting neuroscience, epistemology-and-philosophy-of-mind-busting computer technology: What would Plato say about any of this? Would anything he had to say still have philosophical relevance? And if it did, wouldn't that be stunning proof that philosophy—the frozen-hard bits of it still left in cold storage—never makes progress?

Plato's persistence might be all very well for Plato and his reputation, but it doesn't appear to do the case for philosophy any good. To put the point bluntly: If philosophy makes progress, then why doesn't Plato at long last just go away?

There is, however, one aspect of what takes place around philosophy's seminar table that would be different for Plato. Chances are he won't find anybody there writing dialogues. None of the papers presented at the seminar will do what Plato does, which is to enfold philosophical points of view into characters. Why should one waste time on such a project, mere frills around the argument, when it's the argument that counts for everything in philosophy, and the argument is hard enough to get one's head around? Isn't the need to get clear about the argument the very point of Plato and hasn't his point dictated the bare-bones-of-the-argument style of writing that philosophy has adopted, its rigor and impersonality?

Oh, sure there will be plenty of spirited dialogue around the seminar

* See chapter 1 below.

table, a veritable clamor of dialogue. "Several objections come to mind." "There seem to be two possible interpretations of what you have just said. Can you tell me which you mean?" "It's true that if you assume A, then B follows. But doesn't your assumption of A depend on the condition C and can't we imagine circumstances in which C won't hold? For example, consider D." The sort of endless give-and-take—which Plato is at pains to dramatize in his dialogues—is alive and well, just as Plato would have it. But still the style of writing philosophy is quite different, in the following sense: There are no characters to be found—not in the writing, that is. There are characters aplenty sitting around the seminar table. But the voice that is aimed at is impersonal and precise, even in the comments hurtling around the seminar table, and there is good reason for this, again traceable back to Plato and his formative views on the nature of the field.

There will be different points of view sitting around the seminar table, all of them coming at the same arguments, analyzing them, criticizing them, reaching for the grounds good enough to compel acceptance no matter what the personal differences. Progress in philosophy consists, at least in part, in constantly bringing to light the covert presumptions that burrow their way deep down into our thinking, too deep down for us to even be aware of them. Some of these presumptions are societal, spread among us by successful memes. (One of the most successful of recent memes is the notion of memes itself.) Some will veer toward the more personal and eccentric, rooted in one's history and psychology. But whatever the source of these presumptions of which we are oblivious, they must be brought to light and subjected to questioning. Such bringing to light is what philosophical progress often consists of, as Plato himself asserts in what is probably the most famous passage in all his writings, if not in all of Western literature. This is the passage of the *Republic* in which Socrates describes a group of chained prisoners inhabiting a cave, on the back wall of which shadows are being projected by a fire burning behind them. One prisoner frees himself and manages to get out into the light. We'll return to the metaphor or Myth of the Cave (Plato calls it a *muthos*) in a later chapter. Plato presents the journey to the light as a largely solitary one, though some unseen person does yank the prisoner out of the cave; but the format of the dialogues (as well as his having founded the Academy) encourages the view that, on the contrary, Plato conceived of philosophy as necessarily gregarious rather than solitary. The exposure of presumptions is best done in company, the more argumentative the better. This is why discussion round the table is so essential. This is why philosophy must be argumentative. It proceeds by way of arguments, and the arguments are argued over. Everything is aired in the bracing dialectic wind stirred by many clashing viewpoints. Only in this way can intuitions that have their source in societal or personal idiosyncrasies be exposed and questioned. When it came to political democracy, Plato was not a big fan—at least not democracy as he saw it practiced in Athens— but the field he created honors a kind of democracy. It's an epistemic democracy that rules out the appeal to special privilege.* There can be nothing like "Well, that's what I was brought up to believe," or "I just feel that it's right," or "I am privy to an authoritative voice whispering in my ear," or "I'm demonstrably smarter than all of you, so just accept that I know better here." The discussion around the seminar table countenances only the sorts of arguments and considerations that can, in principle, make a claim on everyone who signs on to the project of reason: appealing to, evaluating, and being persuaded by reasons. The whole style of philosophizing has been dictated by Plato's own view about the possibilities for using the project of reason to find our way out of the illusion-haunted cave.

And yet Plato chose to write in a very different style. He wrote in dialogues, lavishing care on idiosyncratic features of his dialogic characters, many based on real people,† and showing us how their entire personalities are brought to bear on their philosophical positions and the way they argue for them. Some of his characters are so alive that some scholars have argued that the dialogues were actually intended by Plato to be acted, and that they were, in his Academy.‡

---

* Josiah Ober uses the notion of "epistemic democracy," but in a different sense: He argues that Athenian democracy was knowledge-based, its principles of political and social organization sensitive to evidence. See his *Democracy and Knowledge: Innovation and Learning in Classical Athens* (Princeton, NJ: Princeton University Press, 2010).

† See Debra Nails, *The People of Plato: A Prosopography of Plato and Other Socratics* (Indianapolis, IN: Hackett, 2002). This book gives thumbnail histories of the many real personages who people Plato's writings.

‡ For the argument that the dialogues were intended for serious performance, see Nikos G. Charalabopoulos, *Platonic Drama and Its Ancient Reception* (Cambridge: Cambridge University Press, 2002). See also the review of Charalabopoulos by Emily Wilson, in the *Bryn Mawr Classical Review* (December 2012), http://www.bmcreview.org/2012/12/20121262.html. And see also Ruby Blondell, *The Play of Character in Plato's Dialogues* (Cambridge: Cambridge University Press, 2002).

His choice of a form that personalizes philosophical positions is remarkable, since he doesn't mean to suggest by his stylistic choice that the truth itself is personal. He's not saying that the most we can say, in confronting an opinion on a philosophical matter, is that that's the way this particular person happens to think, that's *her* "philosophy," end of story. That was the position of many of the sophists of his day, the teachers of rhetoric who taught their art of persuasion without regard to the truth, and Plato despised the sophists. In fact, it's largely through his hostility that the word "sophistry," which derives from Greek for knowledge—sophia, Σοφια—has taken on its pejorative meaning.

For Plato, writing about philosophy itself raised philosophical questions. In the *Seventh Letter* he stunningly asserts that he never committed his own philosophical views to writing: "One statement at any rate I can make in regard to all who have written or who may write with a claim to knowledge of the subjects to which I devote myself—no matter how they pretend to have acquired it, whether from my instruction or from others or by their own discovery. Such writers can in my opinion have no real acquaintance with the subject. I certainly have composed no work in regard to it, nor shall I ever do so in future, for there is no way of putting it in words like other studies. Acquaintance with it must come rather after a long period of attendance in instruction in the subject itself and of close companionship, when suddenly, like a blaze kindled by a leaping spark, it is generated in the soul and at once becomes self-sustaining" (341b–d). Plato didn't think the written word could do justice to what philosophy is supposed to do. And yet he *did* write; he wrote a great deal. And the literary form he invented for his writing should give us an indication of what he thought philosophy was supposed to do.

And what is it, according to Plato, that philosophy is supposed to do? Nothing less than to render violence to our sense of ourselves and our world, our sense of ourselves in the world.

Toward the end of the *Symposium,* Plato has the larger-than-life real historical figure of Alcibiades[*] declare that philosophical questions, once they take hold of one's inner life, exert a frenziedly disorienting power that is akin to the intoxications both of wine and of eros: "I am

---

[*] See chapter ε for more on the extraordinary Alcibiades, who wreaked such havoc on Athens and the greater Greek world.

looking at all the others," Alcibiades declares, and you can feel his dangerously beautiful gaze traveling around the lamp-lit room, the wicks floating in pools of oil to cast their soft glow on the couches drawn into a semicircle, on each of which two men are reclining. They have abstained from drink for the night, at least until this moment when a drunken, laughing Alcibiades crashes in, and have instead gone round the room giving speeches in praise of the god of love, Erōs. They have just heard Socrates give a speech that will spawn the phrase "Platonic love," a speech in which he passionately urges them to transform the erotic longing that tends to fixate on particular boys into an equally passionate longing for abstract truth. Alcibiades lets his gaze wander from one to the other of the symposiasts. "I am looking at Phaedrus, Agathon, Eryximachus, Pausanius, Aristodemus, Aristophanes and all the others—and should one mention Socrates himself? Every one of you has taken part in the madness and Bacchanalian frenzy of philosophy" (218a–b).

Philosophy a Bacchanalian frenzy? This might come as a surprise to readers who have taken a philosophy course or two, finding in the sophistication of the hairsplitting techniques precious little that resembles the sort of reckless abandon that Plato has Alcibiades describing, the violence with which these peculiar questions whip through one's presumptions and certitudes—undermining, overturning, destabilizing, and disorienting. That was how Plato himself experienced the peculiar questions that Socrates had helped him seize upon, and that was how he wanted others to experience them. Their mere internalization is supposed to enact an inner drama, both terrifying and exhilarating, the likes of which can only be compared to the transformations induced by erotic, religious, or artistic inspiration—a comparison that Plato makes in another one of his dialogues devoted to erotic love, the *Phaedrus* (see particularly 244e–245c).

For Plato, this *inner* drama is the essence of philosophy's doing its work, which is perhaps the most important of the reasons Plato had in choosing to present his philosophical ideas in the form of cerebral dramas. Greek drama was, of course, brimming with violence, and there is a kind of quiet violence in philosophy's work. Philosophical thinking that doesn't do violence to one's settled mind is no philosophical thinking at all. Plato himself is always doing violence to his own settled mind, from dialogue to dialogue. (It's instructive to contrast the political stability he thought ideal, though unlikely, with the philosophical turmoil he is constantly inflicting. Keep the state rigid, so that the mind can range free.) And Plato had a contemporary readership, which stayed abreast with what the esteemed founder of the Academy had seemed to argue in his previous dialogues, and could therefore itself witness the constant challenges to philosophical stability that Plato churned up. These attentive readers had perhaps become convinced, by reading his *Republic* and his *Phaedo,* of what he'd urged about the real existence of the Forms, those exemplars that are the referents of abstract universals like Justice, Truth, and Beauty, and maybe even Saltiness, Sleaziness, and Squalor (whether such less-than-lofty universals have referents is one of the worries of the *Parmenides*). And perhaps Plato's contemporary readers felt as if the ground had opened up beneath them when they read his *Parmenides,* which features a time-regressed Socrates unable to answer the challenges to the Theory of Forms posed by the older metaphysician Parmenides. And Plato keeps mum as to what conclusion he means his readers to draw. Should they believe in the Forms, for which he'd argued so well in the *Republic,* or shouldn't they, considering what he's now writing in the *Parmenides?* A reader is left at sea without an author-issued raft. Plato gave great thought to how to inspire the philosophical drama in all of us who will never have the incomparable benefit that he enjoyed and without which he perhaps couldn't have imagined himself becoming the philosopher he became: exposure to the force of Socrates' personality.

His ancient biographer, Olympiodorus,* tells us that Plato had originally set his heart on being a playwright, either tragic or comic. Whether there's any truth in this or not, he did become a dramatist of a sort, creating his dialogues as dramas of philosophical thought. To inspire the inner drama that is philosophical thinking in those of us deprived of the living Socrates, Plato turned his artistry away from writing the kind of stage plays the dramatists wrote, and instead created a new art form, the philosophical drama, which is what his dialogues are.

In some of these dialogues, you might feel that Plato is telling us

---

* Olympiodorus the Younger lived c. 495–570 C.E. and was a Neoplatonist philosopher. Teaching after the emperor Justinian's decree of 529 C.E., which closed Plato's Academy in Athens and all other pagan schools, Olympiodorus was the last to uphold the Platonic tradition in Alexandria. After his death the school of Alexandria converted to Christian Aristotelianism and was moved to Constantinople. Among Olympiodorus' Platonic writings was a *Life of Plato.*

what we ought to think. But in a great many of his dialogues we are decisively not told what to think. Quite often we are led to aporia, an impasse, unable to proceed a step further. Socrates is almost always there, but even he is only a supporting character. The starring role is given to the philosophical question. It is the philosophical question that is supposed to take center stage, cracking us open to an entirely new variety of experience.

Knowing how unsettling this inner drama can be, how disorienting it is to feel our certitudes crumbling beneath us, he seduces us with an abundance of aesthetic delights, with metaphors and allegories and wordplay and wit. (There are other reasons, too, for these aesthetic flourishes, as we'll soon see.) There are characters whose pride and prejudice get in the way of their making progress; their feints can be amusing, but we're never meant to let amusement at others overtake self-criticism. Watching their flailing against the masterful moves of reason, we are supposed to apply the obvious lessons to ourselves. If you read these arguments without internalizing them, turning them uncomfortably against yourself, then you might as well not bother. That's Plato's attitude. Although philosophical argument is personalized by the dialogue form, the characters are shaped by the philosophical work that they must perform. Narrative technique and artistic flourishes are never allowed to get in the way of the all-important philosophical argument. Plato, it is often pointed out, is an artist of consummate skill, despite the hostile words he sometimes casts at artists, and most especially at the dramatists. But unlike in a novel or short story or theatrical play, the characters are not allowed to take on lives of their own. If characters sometimes are flattened and broadened to the point of yes-men or stereotypes, the point to bear in mind is that this is artistic philosophy rather than philosophical art. This is a distinction—and apology—to which I would like to lay claim in chapters β, δ, η, and ι. The characters who will converse with Plato are created to serve the dialogue, rather than, as in genuine fiction, the dialogue being created to serve the characters. The freedom of characters in a philosophical dialogue is constrained. They can never move beyond the arguments, though I hope the reader might sense a certain growth on the part of my characters as they interact with Plato. Perhaps it will seem to the reader that the characters become less one-dimensional and more fully personlike. I hope so, and I think the reader will be able to guess why I hope so. The taking on, the taking in, of the questions that Plato urges on us adds to our internal dimensions.

Another aspect of Plato's dialogues for which, to the extent that I reproduce it, I must beg the reader's indulgence, is their digressiveness. Plato's view of the normativity of reality—that is, that we are morally improved by knowing what is what—has the consequence of merging together fields that we keep resolutely apart. Big questions require answers to other big questions, and the resulting dialogues are not master classes in brevity. Rather his dialogues are assertively discursive, as he himself occasionally points out, appropriating the free style as itself expressive of the freedom of philosophers, that they may take all the time that they need to follow the criss-crossing traceries of questions. If I try to give some mild sense of Plato's expansiveness in the dialogues that follow, I hope it will not overly try the reader's patience. Occasionally in his dialogues Plato will even let loose his bliss-seeking lyricism, though bliss comes in many varieties, and Plato is suspicious of almost all of them (probably because he's susceptible to almost all of them). But when Plato lets loose, he can blast us open with ecstasy. The artistry of the writing is meant to stir the whole of our person, since it's the whole of that person who must feel the force of philosophy and be changed as a consequence.

A few years ago the philosopher Paul Boghossian published an article, "The Maze of Moral Relativism," in the *New York Times,* in its ongoing feature "The Stone." Boghossian attacked moral relativism as internally incoherent.[*] Stanley Fish, a professor of English to whom Boghossian had paid special attention for being, allegedly, incoherently relativist, wrote a rousing reply, called "Does Philosophy Matter?"[†] In arguing that it doesn't, Fish wrote, "[P]hilosophy is not the name of, or the site of, thought generally; it is a special, insular form of thought and its propositions have weight and value only in the precincts of its game. Points are awarded in that game to the player who has the best argument going ("best" is a disciplinary judgment). . . . The conclusions reached in philosophical disquisitions do not travel. They do not travel into contexts that are not explicitly philosophical (as seminars, academic journals, and conferences are), and they do not even make their way into the non-philosophical lives of those who hold them."

---

[*] http://opinionator.blogs.nytimes.com/2011/07/24/the-maze-of-moral-relativism/.
[†] http://opinionator.blogs.nytimes.com/2011/08/01/does-philosophy-matter/.

These lines from Fish might have come straight out of one of Plato's nightmares. Picture Plato waking all of a heart-pounding sudden on an airless Athenian summer night, these words thundering in his head: *Philosophy doesn't travel.* Were these the words of some doom-declaiming oracle or fragments of his own internal doubts? Plato might very well have written with such misgivings because what Stanley Fish claimed was true in the early years of the twenty-first century was precisely what Plato had feared in the fourth century B.C.E. He feared that the conclusions reached around philosophy's seminar table might *stay* around philosophy's seminar table.* He had tried to devise a written form that might prevent this from happening. (His founding the Academy probably resulted from a similar effort.)

It's these philosophical dramas, the dialogues, which he offers up as a substitute for the oracular poetry that many of his predecessors—including Parmenides, whom he held in high esteem, to judge by the dialogue bearing his name—had used to transmit their insights, and the medium is at least partly the message. Truth cannot be transmitted from one mind to another, the pouring out of the full flask of a master into the passive receptacle of a student. Truth-seeing comes from the violent activity of philosophy, a drama enacted deep in the interior of each of us and which manages, in its violence, to deprive us of positions that may be so deeply and constitutively personal that we can't defend them to others. This violent activity is personal even as it leads one in an impersonal direction, where interpersonal agreement is possible.† The dialogues are meant to instigate the strenuous activity of many points of view clashing against one another so that what is personal or cultural—and unable to provide any independent grounds for itself outside of the personal or the cultural—can be extirpated, which is how Plato conceived of philosophy and how philosophy has continued to conceive of itself, though it writes itself so differently now. No written form could take the place of the strenuous activity that ensues when different points of view try to go about convincing one another. This is best pursued in lived conversation, minds in intercourse with minds, a relationship so intimate that sexual relations are a metaphor

---

* In fact, Plato himself presents Adeimantus suggesting something similar to Fish's complaint in the *Republic* 487a–e.

† Of course, a presupposition that stands behind this process is that there is, at least for many questions, such a thing as a true answer. This is a presupposition that Fish, a baton-twirling cheerleader for relativism, spiritedly denies.

for it, rather than, as some Freudians would have it, the other way round.*

But if Plato wrote his dialogues as a way of launching us into philosophy by *not* telling us what to think, then what are we to make of his eponym? If Plato was so deliberately withholding concerning "the subjects to which I devote myself," then how can philosophers hold forth on the content and merits of "Platonism"?

And yet philosophers do speak of a view they call Platonism, with fierce arguments over its claims particularly apt to erupt when the discussion round the seminar table concerns the nature of mathematical truths. There is a position in philosophy of mathematics that needs naming, a position held by many philosophers and perhaps by even more mathematicians, and "Platonism" has historically supplied the name. As the acquaintance I quoted in the prologue had put it: "Arguments over Platonism raged the entire time." This position in the philosophy of mathematics is connected to broader issues that are raised by Plato regarding the status of abstract truth.

Here are what are taken to be three classic statements of the Platonic position in philosophy of mathematics, the first by the mathematician G. H. Hardy, the second by the mathematical logician Kurt Gödel, and the third by the mathematician and physicist Roger Penrose, three famously brilliant thinkers:

> I believe that mathematical reality lies outside us, that our function is to discover or observe it, and that the theorems that we prove, and which we describe grandiloquently as our 'creations,' are simply our notes of our observations. This view has been held, in one form or another, by many philosophers of high reputation from Plato onwards, and I shall use the language which is natural to a man who holds it.[†]

> But, despite their remoteness from sense experience, we do have something like a perception also of the objects of set theory, as is seen from the fact that the axioms force themselves upon us as being true. I don't

---

* Some Freudians, though not necessarily Freud. "What psychoanalysis called sexuality was by no means identical with the impulsion towards a union of the two sexes or towards producing a pleasurable sensation in the genitals; it had far more resemblance to the inclusive and all-preserving Erôs of Plato's *Symposium*." Sigmund Freud, "Resistances to Psychoanalysis," 1925. Reprinted in *Collected Papers: Character and Culture* (New York: Collier Books, 1965), p. 258.

† G. H. Hardy, *A Mathematician's Apology* (1940; Cambridge: Cambridge University Press, 2012), pp. 123–124.

see any reason why we should have less confidence in this kind of perception, i.e., in mathematical intuition, than in sense perception.[*]

I view the mathematical world as having an existence of its own, independent of us. It is timeless. I think, to be a working mathematician, it's difficult to hold any other view. It's not so much that the Platonic world has its own existence, but that the physical world accords with such precision, subtlety, and sophistication with aspects of the Platonic mathematical world. And this, of course, does go back to Plato, who was clear in distinguishing between notions of precise mathematics and the usually inexact ways in which one applies this mathematics to the physical world. It is the shadow of the pure mathematical world that you see in the physical world. This idea is central to the way we do science. Science is always exploring the way the world works in relation to certain proposed models, and these models are mathematical constructions. . . . And it's not just precision. The mathematics one uses has a kind of life of its own.[†]

As these three examples indicate, "Platonism" often expresses itself in the assertion that abstract truths are *out there,* waiting to be discovered, just as scientific truths are *out there,* waiting to be discovered. A Platonist asserts that the abstract is as real as the concrete, the general as realized as the particular. Perhaps the assertion of reality is clarified by contrasting it with the alternatives, what the Platonist is asserting mathematics is *not.* Mathematics is *not* about our own mental ideas, not about the structure of our cognitive equipment, not about our own implicative fictions. We don't do mathematics by introspecting. And mathematics is *not* about axiomatic systems that have been constructed by stipulating a set of formal recursive rules, a kind of higher-order chess. Our systems are tools for discovery, not for creation. As Gottlob Frege, the mathematician who established modern symbolic logic, put it in his own classic statement of Platonism: "The mathematician can no more create anything than the geographer can; he, too can only discover what is there and give it a name."[‡]

[*] Kurt Gödel, "What Is Cantor's Continuum Problem?," *American Mathematical Monthly,* 1947, reprinted in *Philosophy of Mathematics: Selected Readings,* ed. Paul Benacerraf and Hilary Putnam (Englewood Cliffs, NJ: Prentice-Hall, 1964), p. 271.
[†] Karl Giberson, "The Man Who Fell to Earth: An Interview with Roger Penrose," *Science and Spirit Magazine* (March–April 2003).
[‡] *Die Grundlagen der Arithmetic* (Breslau: W. Koebner, 1884). *Foundations of Arithmetic,* trans. J. L. Austin (Evanston, IL: Northwestern University Press, 1968), section 96.

Platonism reifies the abstract—but there is reification and there is reification. Talk of the "world" of Platonic entities suggests a picture of some sort of separate *place,* sometimes lampooned as "Plato's heaven." Here in the perfection of eternity, beyond the reach of the corrosive tides of time, such things as numbers and non-numerical abstract universals shine forth. They are to be glimpsed not by the crude organs of the body but by the far more refined—and inequitably distributed—faculties of mind. Such are the eternal exemplars that "virtuous logicians" may hope to meet in the "hereafter," in the derisory words of Bertrand Russell, describing the view of the "unadulterated Platonist," Kurt Gödel.* "Plato's heaven" may be invoked—or mocked—as the place in which all concepts, not just those having to do with mathematics, reside. Such talk of a world of abstract things, parallel to our sensed world of concrete things, a kind of space beyond space, is one way of presenting Platonism, though it isn't the only way, and to my mind it doesn't do justice to the subtlety of contemporary Platonist views.

Nor, to my mind, does it do justice to the subtlety of Plato's Platonism—that is, his reification of the abstract—which kept evolving throughout his long philosophical life. Perhaps Plato once did have something like the view that Russell mocks; he himself subjects some such view to his barrage of criticism in the *Parmenides,* propounding such difficulties as convinced at least one of his students, Aristotle, to give up on Platonism altogether and start all over again on the problem of abstract universals. But the ways in which Plato continues to reify the abstract don't fit this lampooned picture. Yes, he continues to assert, in such works as the *Timaeus,* that the intelligible forms can't be reduced to the "stuff" of the spatiotemporal world, to the world of appearance that we sense. But the abstract doesn't transcend the spatiotemporal world of stuff either; it can neither be *reduced* to it nor exist in isolation from it. Abstraction—most especially mathematical abstraction—is the permanence within the flux, the very permanence that provides the explanation for the flux, that provides the *right form* for rendering the intelligibility of nature that the Greek thinkers had been chasing ever since the protoscientists of the Ionian Enchantment intuited that there was intelligibility out there. But the *out there* of the rationally apprehended is immanent within the *out there* of the empiri-

* Russell, *The Autobiography of Bertrand Russell* (London: Routledge, 2000), p. 466.

cally given. It inheres in the structural features of the given, and these features are captured in mathematics. This is the far subtler view that Plato suggests clearly enough so that such thinkers as Galileo can, millennia later, pick up the thread again.

So, at least under some interpretation, Plato appears to have held firm throughout his life to the "reification of the abstract." Evidence for this comes not only from the dialogues but from the Academy he established. To his Academy, Plato gathered all the best mathematicians of his day and put them to work on what the eminent philosopher Myles Burnyeat has called his "research program," which was to discover the mathematical structures immanent in nature. Plato's assertion of the reality of mathematical structures found its practical realization in the study of plane and solid geometry, of astronomy, harmonics, and optics—all of which were pursued in his Academy. His search for mathematical proportions and "harmonies" even lent itself to medical theories, premised on the supposition that health is a matter of the correct mathematical proportions between the "opposing" constituents of the body, which in those early days were thought of in terms of the hot and the cold, the moist and the dry.

Was Plato a Platonist? The question sounds as dopey as asking who's buried in Grant's tomb. But the non-dopey answer is "It depends on what you mean by Platonism." Some version of maintaining the primacy of the abstract, including, most essentially, the abstraction that finds expression in mathematics, seems to be a view we can pin on Plato.* It's a commitment that seems to have persisted relentlessly, if restlessly, throughout his philosophical life. In that sense, we can, with some relief, affirm that Plato was a Platonist. But no matter what his precise attitude toward the issue of the existence of the abstract, there's no question that it was he who raised the issue, and that, according to Aristotle, it was a topic of fierce debate within the Academy—and it is an issue that remains with us still, robustly philosophical and scientifically unresolvable. Do mathematicians discover mathematics, construct mathematics, introspect mathematics, imagine mathematics?

---

* Myles Burnyeat argues that Plato raises the question of the precise ontological status of mathematical objects in the *Republic*, only to decide to leave it unresolved. See his "Plato on Why Mathematics Is Good for the Soul," *Proceedings of the British Academy* 103 (2000): 1–81, especially pp. 33–35. Plato also raises the question explicitly in the *Timaeus*, again leaving it largely unresolved. See especially 51c–52c.

Science makes use of mathematics, but it doesn't tell us what mathematics *is.*

Another doctrine (although closely connected to this one) to which Plato seems to have held firm through all the philosophical twists and turns with which he presents us is the intertwining of truth, beauty, and goodness. Call it the Sublime Braid: truth, beauty, and goodness are all bound up with one another, sublimely. This assertion appears, at first blush, like the worst kind of metaphysics, like a positivist's parody of metaphysics. Truth! Beauty! Goodness! Together again! (Well, actually since forever.) And the metaphysics doesn't end here. Entailed in the Sublime Braid are other doctrinal strands. For starters, beauty and goodness are as objective as truth itself is. "Beauty—be not caused—It Is—," said the poet. Yes, Emily, Plato agrees. Beauty is. And because beauty is, the world is the way it is. If the world really is shot through with intelligibility, as the Ionians first supposed, then this intelligibility is itself beautiful, and the more intelligible it is, the more beautiful it is; and the more beautiful it is, the more intelligible it is. Mathematics provides, in itself, the most perfect intelligibility. When we understand a mathematical truth, we understand that it will always be so: no changes of perspectives or of contexts will render it untrue.[*] This invulnerability to perspectival distortions makes it unqualifiedly what *is,* and thus unqualifiedly knowable or intelligible (*Republic* 477a). So mathematics, being maximally intelligible, is maximally beautiful. And this is why mathematics supplies the *right form* for explaining the world, and it is how it is that our sense of beauty becomes our most sure-footed guide on the vertiginously steep path to truth. Given two empirically adequate scientific explanations of the same phenomenon, go for the more mathematically beautiful one and you'll go for the truth.

Is Plato's metaphysics sounding a little more congenial to the scientifically oriented philosophy-jeerer? After all, Copernicus, Galileo,

---

[*] According to Burnyeat, Plato doesn't present the specialness of mathematical truths in terms of their necessity, but rather of their context-invariance: "Regardless of context, the sum of two odd numbers is an even number. It is not the case that in some circumstances the square on the hypotenuse of a right-angled triangle is equal, while in other circumstances it is unequal, to the sum of the squares on the other two sides. Pythagoras' theorem, whoever discovered it, is context-invariant. It is important here that Plato does not have the concept of necessary truth. Unlike Aristotle, he never speaks of mathematical truths as necessary; he never contrasts them with contingent states of affairs. Invariance across context is the feature he emphasizes, and this is a weaker requirement than necessity; or at least, it is weaker than the necessity which modern philosophers associate with mathematical truth." Burnyeat, "Plato on Why Mathematics Is Good for the Soul," pp. 20–21.

and Kepler all appealed to Platonic doctrines—Galileo and Kepler both referring to "the divine Plato"—in order to argue the superiority of Copernican heliocentrism over Ptolemaic geocentrism. Even though the Ptolemaic view was itself a product of the mathematically oriented doctrines of the Academy, switching the point of orientation from the earth to the sun made the mathematics so much more beautiful. Being led by the beauty of the mathematics was quite an important aspect of that evolution of "natural philosophy" into science applauded by certain philosophy-jeerers.

Plato's intuition—of the intertwining of (mathematical) beauty and truth—is unabashedly echoed by many modern-day physicists of unassailable caliber. The Nobel laureate Paul Dirac, for example, said, "It is more important to have beauty in one's equations than to have them fit experiment." Einstein, too, often made similar remarks, for example telling the philosopher and physicist Hans Reichenbach that he had been convinced that his theory of relativity was true even before the 1919 solar eclipse, which delivered the first confirming evidence, because of its mathematical beauty and elegance. In our day, the sovereignty of beauty—of the mathematical variety—has often been most vociferously proclaimed by champions of string theory, which has so far been unable to produce any testable predictions. "I don't think it's ever happened that a theory that has the kind of mathematical appeal that string theory has has turned out to be entirely wrong," Steven Weinberg—the third Nobel laureate quoted in this paragraph—has said. "There have been theories that turned out to be right in a different context than the context for which they were invented. But I would find it hard to believe that that much elegance and mathematical beauty would simply be wasted."[*]

Physicists have long been helping themselves to Plato's metaphysics, without going through any of the steps he took to arrive at it, rather like people who consume hot dogs and would rather not know how they are made.

All of this metaphysics comes spilling out of Plato's Sublime Braid, and we haven't even considered goodness yet. We'll be considering goodness all through this book. It's always Plato's major concern, no

---

[*] Quoted on *Nova, The Elegant Universe,* "Viewpoints on String Theory," http://www.pbs.org /wgbh/nova/elegant/view-weinberg.html.

matter whether he's doing moral philosophy, political philosophy, epistemology, metaphysics, or cosmology. It turns out, on Plato's view, that our sense of beauty is more reliable than our sense of goodness. It's our sense of beauty that is enlisted to lead us to the truth, whereas our sense of goodness has to undergo a major revision in the light of the truth.

But what does Plato mean by goodness, and how does he entwine it with truth and beauty?

Plato's truth-entwined goodness can best be gotten to by way of "the best reason" that he sees lurking inside truth. The truth is as it is because "the best reason" is determining it to be so.* His language is, at first blush, suspiciously teleological, even suggestive of intentionality. Did someone—Some One—implement this best reason, designing the world accordingly? Or is it rather that the best reason works all on its own, a self-starter, with nothing external to it required to put it into action? It was the latter possibility that Plato had in mind. If there is "mind" determining the truth, an idea put forth in the *Phaedo* and explored in greater depth in the *Timaeus,* the existence of this mind amounts to nothing over and above the assertion that the truth is determined by "the best reason." In other words, the best and final scientific theory would work all on its own to create the world in accordance with itself. In the *Timaeus* he presents a creation myth, in which a demiurge, or divine Craftsman, is implementing "the best reason," but his using a myth to dramatize the point is in itself an indication that it's a more abstract metaphysical principle he has in mind: the best reason is, in itself, a self-starter, an explanation that explains itself, a *causa sui,* as Spinoza—who picked up this Platonic intuition and ran all the way with it—was to put it.

The determining role of "the best reason" in making the world what it is is what the goodness in Truth-Beauty-Goodness consists in. Goodness is interwoven with truth because the explanation for the truth is that the truth is determined by the best reason, and the best reason works all on its own—which is as good as it gets. The truth, being determined by the best reason, is ultimately capable of explaining itself. This makes reality as intelligible as it could possibly be. It's its very intelligi-

---

* Cf. *Phaedo* 97b–d, *Timaeus* (passim), *Philebus* 27–30, and *Laws* X. Leibniz is often credited with first formulating the question of *why is there something rather than nothing?* But here, too, Plato beat everyone to it—including Spinoza, who also wrestled with the question, a precedent that I imagine both Spinoza and Leibniz readily acknowledging.

bility that provides the reason for its existence. For intelligibility-craving minds, what could possibly be more sublime?

And once again, as it was with beauty so it is with goodness: it is mathematics that largely foots the bill. The best reason is the reason that is thoroughly intelligible, that presents its own justification transparently to the mind, which is what mathematics does (*Republic* 511d, *Timaeus* passim). In the creation myth of the *Timaeus,* the divine Craftsman imposes as much mathematics on the material world as it can possibly hold, because mathematics is the most perfect expression of the good intentions—the best reasons—by which the mythical Craftsman works (29d–e). The mythical Craftsman doesn't make the forms he imposes on the world the best by virtue of choosing them; rather he chooses them because they are, independent of him, the best of forms, and their being the best of forms in itself explains why they must be realized.

The talk of "the best reason," which sounds deceptively teleological, is not teleological at all. The causality is fueled by the mathematics. The causality is at one with the intelligibility. In fact, it was the return to this version of Platonism that managed to get the teleology *out* of physics, by displacing Aristotle's final causes with Plato's mathematical conception of causality. Spinoza, who, like other seminal thinkers of the seventeenth century, was rebelling against the Aristotelian-scholastic teleology that held sway, put the point this way: "Such a doctrine (teleology) might well have sufficed to conceal the truth from the human race for all eternity, if mathematics had not furnished another standard of truth . . . without regard to . . . final causes."*

So there you have it: truth, beauty, and goodness, all bound up with one another, providing the ontological structure of reality. Such a confluence of truth, beauty, and goodness suggests a notion like the sublime—not identical with truth or beauty or goodness but rather with the confluence of all three. Reality is shot through with a sublimity so sublime that it simply *had* to exist. Existence explodes out of the sublime.

Notice that the goodness that we're speaking about here isn't a specifically human goodness. The point is *not* that the world has been created with *our* good in mind. I can't think of a single place in the corpus where Plato even floats this idea. It's entirely foreign to his conception

---

* *Ethics* I, Appendix. Trans. R. H. M. Elwes, 1883. Revised edition (London: George Bell and Sons, 1901).

of the world. (It's pretty foreign to the entire Greek conception of the world, even non-philosophically speaking. Those gods and goddesses pursue their own ends and pleasures. We mortals are, at best, incidental to their purposes.) The goodness that's woven into the Sublime Braid has no more of the human element in it than $E=mc^2$ does.

But Plato also seems to suggest, all through his dialogues, in one form or another, that there is also some goodness—in the way that we humans understand goodness, as it applies specifically to *people*, the lives we live, the actions we perform—to be gained from knowledge of the way the world is, the way it *has* to be because of the Sublime Braid that furnishes its structure. Knowledge is not only *of* the good, but also makes us good, reforming us so that we become more virtuous—more inclined, because of our knowledge, toward justice, temperance, courage, and reverence. Metaphysics—understanding how the world *is* by understanding how it *must* be, understanding, for example, that it must be maximally intelligible[*]—is ethically reforming.

The term "goodness" is a placeholder. It needs filling in. Yes, indeed, we ought to be good; so much is trivially true. But tell us what we must be—or do—in order to be good. For Plato, it's knowledge that does the filling in of the placeholder "good." Knowledge is *ethically active*, even when it's knowledge of the most impersonal kind, as indifferent to the world of humans as pure mathematics.

In fact, it's the very *im*personality of impersonal knowledge that renders such knowledge the most ethically potent of all. Simply to care enough about the impersonal truth, devote one's life to trying to know it, requires disciplining one's rebellious nature, which is always intent on having things its own way, on seeing the world in whatever light does most justice to one's own petty ego so that the truth-as-one-sees-it will push one's own self-serving, power-centric agenda along. So simply to allow oneself to be *overtaken* by the reality of Truth-Beauty-Goodness—to become embraided oneself in the Sublime Braid—is to exert discipline over one's unruly nature, to call a halt to its self-enhancing fantasizing.

But that is only the beginning. Reality is of such a kind as to do us ultimate good, and that because of the principles by which it has

---

[*] "Intelligible" is no more meant to entail "intelligible to us humans" than "good" is meant to entail "good for us humans."

been fashioned. As we take in the Truth–Beauty–Goodness that structures reality, its rational order is replicated within our own minds in the act of knowing it—and we are made better for this replication. We are *rationalized* by nature's own rational order, our minds' constituents reconfigured in their ideal proportions to one another, just as in health the constituents of the body are configured in their ideal proportions to each other. We become structurally isomorphic to reality itself, and in that way our natural affinity to it is strengthened. We become more *like* it (*Timaeus* 47b–c). This, too, further removes us from the smallness of our own lives, the strengthened kinship with the cosmos expanding us outward to take it in. Our reality-enhanced minds can't help but see their own small place in the grand scheme of things and will be appropriately humbled in the process, which is what this secular kind of piety consists in (as Spinoza thought: piety is humility before reality). Knowledge of impersonal truth drives all personal thoughts from the mind [*Timaeus* 90a–c]). Plato would say, about a physicist avidly awaiting that call from Stockholm, or thinking only of the fame she can acquire by writing one of those scientific blockbusters, that she never was earnestly in love with the beautiful, not so that it overtook her own love of herself. Such a scientist has been fueled by intelligence but not by wisdom, which must include an overwhelming love for that which isn't oneself. The appropriate reaction to the beauty of the Sublime Braid can only be love.

The historical Socrates had perhaps taught that human virtue is a kind of knowledge, a view that Plato took sufficiently seriously throughout his life to be constantly probing it. Sometimes he endorses it (as in the *Protagoras*), and sometimes he challenges it (the tripartite theory of the soul he puts forth in the *Republic* amounts to a challenge of it). But that knowledge is the most potent form of ethical transformation that we have, does seem to be, in one way or another, a continuous aspect of Plato's thinking, another strand of the Sublime Braid. Ethical progress requires knowledge, even if that progress may require something in addition to knowledge, a kind of surrender to that knowledge that is a kind of love. The best among us are those who have allowed the abstract knowledge of the True-the Beautiful-the Good to subdue what is mean-spirited in us, banished from our thoughts what is unworthy of minds privileged enough to behold what it is they behold. And though this doesn't mean that the very intelligent are necessarily good—an easily falsifiable proposition—it does seem to suggest that the very good must be very intelligent. Knowledge, though perhaps not sufficient for virtue, is necessary.

And in this last proposition, Plato might already have hit a live nerve in your moral fiber. I hope so. I hope you're thinking something like this: How dare Plato suggest—or this author suggest that Plato had suggested—that goodness requires an intelligence for abstractions? Ridiculous! People can't help the degree of intelligence with which they were born. That obviously doesn't mean that they can't be good people, often far better than the arrogantly smirking specimens strutting their stuff at the far end of the bell curve. Perhaps that was Plato's problem! In any case, there's obviously something abominably wrong with either Plato's reasoning or with this author's interpretation of Plato's reasoning, to have allotted any attention at all to a conclusion so morally repellant. If this is how the truth is supposed to reform our sense of goodness, then I'll stick to my own unreformed sense, thank you very much. I have far more faith in my own moral sense than in these admittedly *metaphysical* intuitions.

If you are reacting in some such way, perhaps even at this moment considering why you have so much more faith in your own sense of goodness (quite different, of course, than the sense of your own goodness), than in Plato's claims about how knowledge might better reform that sense of goodness—then Plato has succeeded in his larger aim, which is to engage us in just these kinds of questions, as rigorously as we know how and always on the lookout for the unexamined preconceptions that are in need of vigorous rattling. His belief that we can make progress of this sort was a kind of prediction that itself comes out of the tangle of views—metaphysical, epistemological, aesthetic, and ethical—of the Sublime Braid. If Plato is correct in a big-picture sort of way, then we should be able to look back at him and see ways that we've left him behind, not only scientifically but philosophically and ethically. Can we? That's one of the questions that Plato bequeaths to us. And there are many more.

It's Plato's questions, or successive iterations of them as they have arisen in response to changing circumstances and growing knowledge, that subtend many of our most raucous contemporary disagreements. Here are just a few:

When we disagree over whether the 1 percent really contribute more to society than the 99 percent and whether, if they do, their contributions should be recognized in the form of increased privileges or increased obligations, then Plato is there.

When we argue over what the role of the state is, whether it is there to protect us or to perfect us, then there is Plato.

When we worry about the susceptibility of voters to demagoguery and the dangers of mixing entertainment values with politics, then there is Plato.

When we wonder whether professional thinkers who come out of our universities and our think tanks should have a role to play in statesmanship, or whether their expertise is useless or worse in the practical political sphere, then there is Plato.

When we argue over whether ethical truths are inextricably tied to religious truths, then there is Plato.

When we wonder whether all truths—even the scientific—are no more than cultural artifacts, then there is Plato.

When we wonder whether reason is sufficient—or even necessary—to guide us through life, or whether there are occasions when we should abandon reason and go with our hearts, then Plato is there.

When we ponder the nature of romantic love and whether there is something redemptive or rather wasteful about the amount of attention and energy we're prepared to sacrifice to it, then Plato is there.

When we wonder over the nature of great art and whether it is able to teach us truths we can't otherwise know, then there is Plato.

When we wonder whether we should instill in our children a discontent with the ordinary so that they will be inspired to be extraordinary, then there is Plato.

When we wonder whether there is a real difference between right and wrong, or whether we're only making it up as we go along, then there is Plato.

When we wonder how, if we do know the difference between right and wrong, we come to know it, then there is Plato.

When we wonder how we can teach the difference between right and wrong to our children, whether it is through storytelling or reason or threats or love, then there is Plato.

When we wonder why virtue so often seems to go unrewarded, with good people suffering while bad people prosper and get tenure, then Plato is there.

When we wonder whether the scientific image of the human—as subject to the laws of nature as the computer on which I write—has rendered the grander humanist image of us quaintly obsolete, then there is Plato.

When we ponder the moral shape of history, whether mankind is making moral progress or only finding more efficient ways of expressing savagery and ruthless self-regard, then there Plato is.

And when we wonder whether we have at last grasped the truth or ought rather to hear further arguments from the other side, then there, too—always—is Plato.

β

PLATO AT THE GOOGLEPLEX*

*The Googleplex is the corporate headquarters complex of Google, Inc., located at 1600 Amphitheatre Parkway in Mountain View, Santa Clara, California. The word "Googleplex" is a portmanteau of "Google" and "complex," but it is also a pun on googolplex, which is an enormous number. First, start with a googol, which is 10 raised to the power of 100, or a 1 followed by a hundred zeroes. A googolplex is 10 raised to the power of googol, or a 1 followed by a googol zeroes. Google, Inc., has always thought big.

Dramatis Personae

Cheryl, media escort
Marcus, software engineer
Rhonda, narrator and Cheryl's friend

The other day, I came into the city to meet my friend Cheryl for a drink and—her expression—a little tête-à-tête-ing. Cheryl and I are both New Yorkers transplanted to the West Coast. That's one of the ties between us. It might be the only tie between us, but somehow we've fallen into the habit of being friends. We met at a pricey hotel bar on Nob Hill that's decorated like an Italian bordello, with heavy red velvet drapery and gilded statuary. But it is—again Cheryl's expression—quiet as a vault, which means you can hear yourself talk, even though, as usual, Cheryl did most of the talking. You can't altogether blame her, given the interesting people she's constantly meeting. She's my own personal version of Gawker, a way of my getting a glimpse into the lives of the famous, the near-famous, and the willing-to-do-anything-short-of-landing-themselves-on-death-row-in-the-hopes-of-someday-being-famous.

She was late, which was my first tip-off that something was up with her. Cheryl is super-organized, which is something you have to be in her line of work. Here's how organized she is: while she was parking her Lexus, she called me and told me to order her a Long Island Iced Tea, which is a far stronger mixed drink than our usual Chardonnay.* The drinks were just being brought to the table when Cheryl arrived, amid

* A Long Island Iced Tea, tasting innocuously like the kind served on the back porch by your maiden aunt, is typically made with equal parts vodka, gin, tequila, rum, and triple sec, with one and a half parts sour mix and a splash of cola.

all the jangling of the large silver bangles she was wearing. Cheryl is always in full Tiffany armor.

After she'd made her little joke about the waiters, who all act as if there were stiff entrance requirements enforced to get in here, including letters of recommendation from your high school math and English teachers, she settled down to tell me about her latest adventures escorting authors from one media event to another. Since everybody's writing books these days, Cheryl gets to meet politicians, movie stars, all sorts of has-beens, alcoholics, and junkies, and even some authors who do nothing but write books. She's got the knack, she says, so that people open up to her, and if she ever retires and writes a tell-all memoir she'll need her own media escort as well as a good lawyer.

Boy, did I have an experience today, she launched in with little preamble. My author was a philosopher, which I just figured was going to be awkward and tedious. And he uses just the one name Plato, which struck me as not a little off-putting, as if he were on a par with a Cher or a Madonna. From the start I figured it was going to be one very long day, but I had no idea.

She took a long sip of her drink.

No idea at all, she continued. Plus his event was one of those Authors@Google things and that place always puts me on edge. It's hard to breathe in the congested self-congratulation up there at the Googleplex. When somebody tells me that they work hard and play just as hard, which I hear every frigging time I go there, then I make it a point to roll my eyes . . . hard.

Cheryl rolled her eyes as she said this. Her coming down so hard on the Googlers for their high self-esteem is funny, in its way. If I had to escort the high-and-mighty the way Cheryl does, I'd be so intimidated I wouldn't open my mouth unless absolutely necessary. I'm intimidated at one remove, just hearing about Cheryl's authors. But no matter who Cheryl is escorting, she doesn't know from awe. On the contrary, if you know what I mean. So it's funny how irked she is by other people's little gestures of self-importance.

Of course, there *is* the food there, she was saying. I always make it a point to take my authors to lunch there first. I've told you about the food there, right? I mean it's gorgeous. Yoscha's Café is my favorite. It's huge and airy, and they've got dozens of food stations with different gourmet food so lovingly prepared you can just imagine the doting caretakers who sent their darlings out into the world. And of course it's all free, as I explained to Plato. That's the first thing to know about the food here, I said to him. They get breakfast, lunch, dinner, whatever, absolutely free. It's feeding on demand.

I'd hate that, I told Cheryl. I'd gain ten pounds in a week.

Yeah, well, apparently that's a "problem"—she air-quoted—which they complain about in their bragging sort of way. We work hard, play hard, *and* eat hard, which makes us *exercise* hard. Oh, my goodness, can you possibly grasp what a bunch of superior people we are? Cheryl was rolling her eyes again. Anyway, she went on, Plato was listening to me very intently—it's almost disconcerting how intently he listens—even though I was just rambling on, kind of free-associating, just trying to make conversation because I could tell this guy's skills at small talk were not the highest. You know, very ivory tower, though with extremely good manners, almost something aristocratic about him. Also he makes eye contact, unlike a lot of these types. In fact, he makes *serious* eye contact. His stare is penetrating to the point of aggravating. Anyway, when I finally stopped to take a breath, he asked me: And what is the second thing to know about the food here? You see, he's got this very logical mind. If you say to him, here's the first thing to know about something, then you've also got to give him a second thing to know about it. So I said, well, I guess the second thing is that it's yummy. And of course it's local and organic and all those other kinds of things that people around here are into.

And he asked me, have you ever heard of the Prytaneum?

No, I answered, what's that, some hot new restaurant?

He sort of smiled, which he tends to do more with his eyes than his mouth, and said, in a manner of speaking, yes, it is hot. The sacred fire of the city is kept going there at all times, its flame carried to any new colony established by the metropolis.

Well, of course, I had no idea what he was talking about, though I vaguely sensed he was making some kind of a joke. He comes from Athens, I forgot to tell you that, and even though I'd been to Greece on that cruise with Michael before the kids were born, the more Plato spoke, the more I realized that Michael and I hadn't seen the real Greece. I mean, you have no idea of how different they do things over there, at least to listen to Plato describe it. Anyway, he told me, the Prytaneum also serves free meals.

So I said to him, no kidding! That's quite a deal. How can they afford to stay in business?

It is run by the city, he answered, and the meals are mainly for those who have rendered extraordinary service to the city.[*] I had a friend who got into some very unfortunate legal trouble. Socrates was charged on two counts, impiety and corruption of the youth.

Corruption of the youth? That sounds pretty dark. Was he some sort of pedophile? I asked him.

Not in the sense that you are most likely thinking, he said, though he loved youth.

Well, I hope not in the sense that I'm thinking! I said right back at him, which made him kind of wince.

The charge was more a matter of his not accepting the moral values of his society and his encouraging the young to question them as well. And he was right to question them and to get us younger men to question them. As proof of how corrupt the society was, the jury ended up convicting him.[†]

And you should have seen his face when he said that, Rhonda. This was the first inkling I got that there was a lot going on behind his façade. He's a restrained kind of person—very, I don't know, formal.

And it's true that every time Cheryl spoke Plato's words she took on a formality, speaking slowly and precisely, as if every word had been carefully considered. She's a natural-born actress who just automatically slips into impersonations.

In fact, the longer the conversation went on, she continued, the more I could see glimmers of genuine human feeling going on behind his marble façade. I could tell from the tightening of his jaw and from the way his voice, which is very soft to begin with,[‡] went even softer, how traumatic this whole business with his friend Socrates must have been for him.

---

[*] The grant of being fed at public expense in the Prytaneum or the tholos, a round building, was known as *sitēsis* (food grant). The prytaneis (the fifty citizens who led the Boulē, the Council of 500, for each of the ten months of the Athenian calendar) were also fed at public expense, usually in the tholos.

[†] See Myles Burnyeat, "The Impiety of Socrates," *Ancient Philosophy* 17, no. 1 (1997): 1–14 for the now widely accepted view that Plato represents Socrates as guilty as charged. The specific charge brought against Socrates was that he didn't believe in the gods of the city, and Burnyeat argues, on the basis of the *Apology*, the *Euthyphro*, and parts of the *Republic* and the *Laws*, that that's exactly how Plato represents Socrates. I will argue that his skepticism was such as to undermine the very identity of his Athenian society. Cf. chapter ζ below.

[‡] There are, in fact, reports that Plato lectured at the Academy in a very soft voice, perhaps a ploy to get his students to lean in closer to catch every word.

So I asked him: How long ago did this happen to your friend?

Oh, it's ancient history, he said. I was a young man, not yet out of my twenties.

That's interesting, I said, breaking into Cheryl's narrative, which she doesn't exactly encourage. It's rare for a man to care so much for a friend, I said. Are you sure that Socrates was just a friend and not something, you know, more?

Well, of course the thought occurred to me, too, Cheryl said. But you don't just come out and ask someone about that, especially not someone like Plato. You know, my trick to getting my authors to tell me so much? It's asking the question just to the side of the one that I really want to ask. So I just said, what a terrible story. Didn't he have a good lawyer?

Lawyers, said Plato and smiled. I have heard of such people.

Well, of *course* you have, I said to him, again wondering if this was an example of some kind of humor, you know a *lawyer* joke, especially since he said it with a slight smile. He has a pretty stiff face, with very strong bone structure, kind of broad around the forehead, and he doesn't make any sudden motions, facial or otherwise. You can see what a powerful physique he must have had when he was younger, and he still holds himself ramrod straight.*

We have no such people in Athens, Plato said. Accusers accuse and defendants defend. Everybody acts as his own lawyer. Those who can afford to usually hire a logographer to write their speeches.

No lawyers, I interrupted Cheryl. He's got to be putting you on. Whoever heard of Greece having no lawyers?

No, that's what I meant about Greece being so unbelievably different, Rhonda. It's kind of mind-boggling.

Are you sure this Plato isn't one of your fiction writers? I asked her.

Well, if he is, he's more convincing than any of them. I'll never hear

---

*According to Alexander of Miletus, who is quoted by Diogenes Laertius in his *Lives and Doctrines of Eminent Philosophers* (Book III, *Life of Plato*, chapter 4), which was probably written toward the beginning of the third century c.e., Plato's real name was Aristocles, son of Ariston, of the deme Collytus, and later sources confirm his name as Aristocles, though perhaps they are simply depending on Diogenes as their source. Aristocles, we know, was the name of his grandfather, and it was customary to give a boy the name of his grandfather, so that is some (slight) independent evidence as to his original name. Diogenes goes on to say that the name Plato was given him by one of his teachers in gymnastics, Ariston of Argos, "because of his robust figure," adding "but others affirm that he got the name Plato from the breadth (*platutèta*) of his style, or from the breadth (*platus*) of his forehead." Few scholars continue to hold this view of Plato's name having been other than "Plato," ever since James A. Notopoulos disputed it in "The Name of Plato," *Classical Philology* 34 (1939): 135–145.

the word "gravitas" again without thinking of him. This guy is like *hewn* from gravitas. The procedure in our city, he said, is that if you are found guilty you get to propose the penalty that you think would be fair. Then the accusers pose another penalty, harsher of course, and then the jury votes on the penalty, often aiming for the mean. This procedure worked to Socrates' detriment. My friend was famous for his irony, and he was not inclined to abandon it, not even with his life hanging in the balance. I should say *especially* when his life hung in the balance, since to cower before death, showing a readiness to do anything, throw overboard any principle, in order to stave off death just a few moments longer—for it is only a few moments from the standpoint of eternity—is unmanly.

That's an interesting perspective you've got there on death, I told him, but just one helpful hint. I'd avoid the use of adjectives like "unmanly." They can come off sounding sexist, as if you think maybe men are superior to women.

How'd he take that? I asked Cheryl.

Surprisingly well, Cheryl said, especially for someone so old-school. He thanked me for my advice, promising that he'd try to remember to avoid sexist words in the future. I have not failed to notice, he said, how differently women are regarded in your society compared to mine. It had always struck me as an unreasonable waste of human resources to keep talented women secluded in their homes, which is what our practice is.* Yours is a much more rational way of utilizing human potential. So let me amend my last statement and say rather that Socrates held it to be *ignoble* for a person to undertake an action with the only aim of postponing death, especially since the proposition that death is an evil turns out to be non-trivial to justify.† During his sentencing, Socrates made a point of mentioning Achilles,‡ who is considered throughout Greece to have been the greatest legendary hero. Achilles

---

* In fact, Plato showed astonishing gender egalitarianism. See *Republic* 451–457b, for his general discussion of why exceptional girls should receive exactly the same education as exceptional boys, designed to train them to be the guardians of his utopian city, his *kallipolis*. His discussion includes such statements as this: "Then there is no way of life concerned with the management of the city that belongs to a woman because she's a woman or to a man because he's a man, but the various natures are distributed in the same way in both creatures" (*Republic* 455d). It is, however, important to point out that his gender egalitarianism derives more from considerations of what will make for the best-run state. The considerations focus less, if at all, on the unfairness to women in depriving them of equal opportunity but rather on the unfairness to the state in depriving it of all its talented individuals.

† *Apology* 28b–29c.

‡ Ibid. 28b–d.

had been given the choice of either a brief but glorious life or a prolonged but less exceptional life. Of course, Achilles made the heroic
choice, and so did Socrates, though I should mention that my friend
had already reached his seventieth year, so the option of a short life
was foreclosed.* Nevertheless, he would not succumb to the indignity
of acting only to eschew imminent death, especially when doing so
required violation of the principles on which he had lived out his life.
So when asked to propose a penalty that would accurately reflect his
culpability Socrates responded that since he had performed an invaluable service to his city, trying to wake its citizenry from its sleep of
complacency, and had never asked for any recompense for his services,
the city, if it truly wished to show justice toward him, should vote him
free meals for life at the Prytaneum. That was the penalty he proposed
after he'd already been voted guilty of a capital offense (*Apology* 36c–d).

That's some chutzpah your friend had there, I said to him.

Chutzpah? he asked me. This word I do not know.

Audacity, I explained. I was going to say "balls," but the word froze
on my lips given that hunk of graven gravitas I already mentioned.

So how did that work out for your friend? I asked him.

Not so well, he answered, looking down at his folded hands. The
jurors were so outraged at Socrates' chutzpah—he pronounced it perfectly, Rhonda—that more people voted for him to be executed than
had voted for a verdict of guilty in the first place. A perfect display of
Athenian irrationality and its cultish valorization of the crowd. It would
have been funny had it not been so tragic.

What a sad story, I said, and really, judging from his expression you'd
have thought his friend's death had happened just yesterday. I could
almost see the emotions bleeding through the marble. I can see how
it still really affects you, I said to him. Your whole demeanor changes
when you talk about him. You might think about talking about him

---

*Xenophon asserts that he believed that Socrates wanted to die, having reached an age when
life would progressively deteriorate: "Socrates was already so far advanced in years that had he
not died his life would have reached its natural term soon afterwards; and secondly, as matters went, he escaped life's bitterest load in escaping those years which bring a diminution
of intellectual force to all—instead of which he was called upon to exhibit the full robustness
of his soul and acquire glory in addition partly by the style of his defense—felicitous alike in
its truthfulness, its freedom, and its rectitude—and partly by the manner in which he bore
the sentence of condemnation with infinite gentleness and manliness. Since no one within
the memory of man, it is admitted, ever bowed his head to death more nobly." *Memorabilia*,
Chapter VIII, trans. H. G. Dakyns (Macmillan, 1897).

more. Audiences will eat it up. Of course, I was trying to buck him up, since he had to go on in less than an hour. And I've heard from so many of my authors that the one consolation for the bad episodes they've lived through is that they can always use them in their writing.

He was the best man of his time, he responded (*Seventh Letter* 324e).

Do you write about your special relationship with him in your book? I asked him.

I've written about him in many of my writings, he said.

But about how you feel about him, the effect he had on you?

No, he answered, I've never written specifically about that.

Well, you should, I told him. You'd have a best seller right there. *Tuesdays with Morrie* meets *Dead Man Walking*.

He just looked at me and smiled, and I had the sense that he had no idea what I was talking about. It wasn't just a matter of his being Greek, Rhonda, but of his being a foreigner in a weirder way, which became increasingly clear as the day wore on. He's the first philosopher I've ever escorted. I mean the first *professional* philosopher. I've had guys like William Bennett and Dinesh D'Souza. And then that time when I had Bono.

Anyway, all this time we had been wandering around, gathering food from the different stations, with me urging Plato to try a little of this and a little of that. He was being very finicky, like one of those kids who don't want to try anything new—my Jason was like that and it drove me crazy. I was thinking of Jason while I urged Plato to try the sushi and he looked at it like it was boiled cockroaches or pie à la mud, which is what I used to say to Jason. They have these long communal tables at the café, kind of the way it was in elementary school, which is probably no accident since most of these Googlers are barely out of elementary school, still running around with their backpacks and T-shirts and jeans and bringing home fat paychecks to splurge on their toys. I spotted a fairly empty table and shepherded him over, spreading my bag and scarf and sweater around to discourage nudniks. I explained to him that we didn't have all that much time because a delegation was going to be arriving in about forty minutes to give him a tour of the Googleplex, which I told him I could get him out of if he wanted.

Why should I want to forgo seeing the Googleplex? he asked.

Well, you know, I answered, trying to be as delicate as possible, I was just thinking that this is a pretty demanding book tour. What are you doing, like twelve cities in three weeks? You're in great shape, don't get me wrong, I know how you Greeks love to work out and all, but still, you have a pretty hectic schedule ahead of you, just for today. They put you in their largest room here, which means they're expecting a crowd, so I'd like to keep you perky for that. And then there's the book signing afterward, which is the point of this whole thing and let's hope your signing hand will get a good workout. I think maybe you should rest a little instead of taking the Google tour. I'm sure they have nap rooms here, since they have every other sort of creature comfort. Including, Rhonda, which I usually make a point of mentioning to my authors, heated toilet seats that wash you and dry you and do all but burp you. Somehow it seemed inappropriate to mention the toilet seats to Plato.

But anyway he really was eager for the tour. I don't want to squander my opportunities for learning as much about your *polis* as I can, was the way he put it.

Police? I said to him. I don't get it. What do the police have to do with anything?

Sorry, he said. I meant the city. I want to learn as much about this city as I can.

Mountain View? I asked him dubiously.

I mean, perhaps, more this city of the Googleplex, he answered.

Okay, I said. I guess it sort of is a self-contained city. But why are you so eager to learn about it?

Is not Google the most powerful way of acquiring knowledge?

Yes, I said, it's a powerful search engine, but you don't have to understand how they actually do it in order to use it. Everybody in the world googles, but nobody understands how it works. It's techno-magic.

If we don't understand our tools, then there is a danger that we will become the tool of our tools, Plato said, which I thought was a very astute observation, especially considering how little it turned out that he actually knew about Google or really anything about the Internet.

I know nobody's asking me for my opinion, I said, but for my money we could have just skipped this whole Authors@Google thing. They never buy that many books anyway. I usually get very lackluster audiences for my authors, except when Google buys the books and gives them out as goodies, which happens far too infrequently, considering the resources of this place. And I'm going to warn you right now. They'll have their computers open the whole time that you're speaking, their eyes glued on their screens instead of on you. That's off-putting for a lot of my authors, because you know, Rhonda, I may be repeating myself here, but authors are probably the most insecure people in the world. And the more self-infatuated, the more insecure. I don't know whether it's the insecurity that drives them to write books in the first place, or it's writing the books that makes them insecure. All I know is that most of them are as neurotic as a love child of Lindsay Lohan and Woody Allen. Now there's a thought! Anyway, just don't let their staring at their computers throw you, I told Plato. The most important thing to remember is that your talk gets thrown up on the Web, and that's where you'll sell your books. Just keep looking at the cam and forget that everybody in front of you isn't listening to a word you're saying.

Will they then be watching my image on their screens rather than watching me as I speak? That is very much like a certain scenario I had once imagined.[*] In fact, I had thought about speaking of that scenario to the audience here, since there is a way in which what I was imagining there relates to the idea of codes of information from which the whole picture can be generated, which seems to me relevant to ideas which are pursued here, at least from the little I have been able to understand.

Well, I said, treading carefully around his exposed ego, which is, of course, the first thing you learn in my line of business, they *could* be googling information about you as you're talking. That's entirely possible. And as far as what you should talk about, that's up to you, of course, but it should definitely be related to your most recent book, since that's the one that you're here promoting.

Plato has this strange combination of sophistication and utter cluelessness. I had the sense that any pimply seventeen-year-old who's just published his first memoir has a better grasp of what it takes to sell a book than Plato has.

So they can, as you say, *google* about me, or about other topics as well, he said.

Yes, you know "google" is a verb now. As in, you google whenever you want to know anything at all, any subject, big or small, I explained,

---

[*] Plato is, I think, referring to his Myth of the Cave. See *Republic* 514a–518d. And see chapter θ, in which the Myth of the Cave is discussed in more detail.

wondering if it was just the language thing or something else going on here. It's not as if they don't have Google in Greece, or at least I assume they do. Of course, if they don't have lawyers then who knows what else they're missing?

You google whenever you want to know anything at all, he repeated, any subject, big or small. So all of knowledge is concentrated right here at the Googleplex, and those who work here are privy to all knowledge. This is so extraordinary it almost strains credulity, that knowledge could be localized in this way.

Well, as you'll learn on your tour, the knowledge isn't actually at the Googleplex. This is just corporate headquarters.

Where is it then? he asked me. Where's the knowledge?

It isn't anywhere in particular, I said. It's in the cloud.

At this he became extremely excited,* and started asking me all these questions that I couldn't answer. I mean, I've been on the tour too many times to count but I never listen. But your other points, I said to him, about Google being privy to all knowledge? I think you put your finger on something there. Frankly, I don't even want to think about all the information these Googlers are privy to.

But it can only be a good thing to have knowledge,† Plato said, which is an example of the kind of cluelessness that I'm talking about here which seems to go way beyond English-as-a-second-language cluelessness.

Well, I wouldn't be so sure of that, I told him. In fact, just last week I was escorting an author who had published a book about all the information that Google is gathering about each of us, which, since Google is a corporation, and the purpose of corporations is to make money, it's probably selling to advertisers, so that they can tailor their ads specifi-

---

* Plato's excitement is understandable, considering that the non-localized cloud has something Platonic about it.

† In some dialogues Plato goes so far as to connect goodness and knowledge so strongly as to suggest that knowledge of what is good is both necessary and sufficient for doing the good. This strong connection between goodness and knowledge renders the notion of *akrasia*—or weakness of the will—problematic, since weakness of the will consists in knowing what is good but nevertheless not doing it, presumably because one doesn't want to. Claiming knowledge is sufficient for goodness depends not only on a theory of the good but also on a theory of the human will, according to which it is in the nature of this will to want the good. See, for example, *Protagoras* 358d: "No one goes willingly toward the bad." It is possible that this view corresponds more closely to the historical Socrates' view than to the mature Plato's. Plato's theory of the tripartite soul, worked out in the *Republic*, is a reformulation of the theory of will that would leave room for *akrasia*.

cally to us. This guy, Siva Vaidhyanathan, who's also a professor like you, said that we think of ourselves as Google's customers, but really we're its products. We—all of our secret desires and whatnot, which Google keeps track of by following what we click on—are what Google sells to advertisers.* I'm not sure that everything he says is right—I mean, I take all my authors with multi-grains of salt, not that I told Plato that—but I do think there's something creepy about how much Google knows and how it's always trying to learn more, just gobbling up every fact in the multiverse, which, Rhonda, is not only *this* universe but all of them and was a VOOM that a scientist I had last week was peddling pretty hard.

VOOM stands for Vision of Outstanding Moment. All the big thinkers that Cheryl escorts have VOOMs. I can't remember whether Cheryl made up the word or whether the word "VOOM" was one of her authors' VOOMs.

So you're telling me that the purpose of all this knowledge is merely to make money? Greed is driving the great search engine for knowledge? This bewilders me more than anything else I've gathered about this place. How can those who possess all knowledge, which must include knowledge of the life that is worth living, be interested in using knowledge only for the insignificant aim of making money?

Well, what do you do when you're faced with monumental cluelessness of this sort?

Plato, I said, I think you have a somewhat exalted view of Google and the nerds who work here.

Nerds? he said. Another word I do not know.

Well, again I was in a somewhat awkward position, since I didn't want to offend Plato, who struck me, despite his excellent manners and eye contact, as a nerd par excellence. So I fell back on something I'd once heard from one of my authors, who said that the word was originally "knurd," which is "drunk" spelled backwards, and was used for students who would rather study than party. Anyway, that's the explanation I gave Plato.

And the people who work here at Google are all nerds? he asked me.

---

* *The Googlization of Everything (And Why We Should Worry)* (Berkeley: University of California Press, 2011). Cheryl is not quoting Vaidhyanathan verbatim. Here is what he writes: "We are not Google's customers. We are its product. We—our fancies, fetishes, predilections and preferences—are what Google sells to advertisers." This specific claim of Vaidhyanathan's has been challenged. At this point, Google does not sell specific information about each of us to advertisers, though Facebook does. For an excellent discussion of Google and its omniscience, see Daniel Soar, "It Knows," in the *London Review of Books*, October 6, 2011.

I would say each and every one, I told him.

He smiled and looked around the café as if he had died and gone to philosophers' heaven. Apparently, I hadn't done a thing to dislodge his crazy idealized view of the Googleplex. My chosen term for nerd, he said to me, still smiling, is "philosopher-king."

Was he joking or what? Cheryl asked me. What do you say?

I'm not sure, I said to Cheryl, but I think not.

He wasn't, she said, only at that point in the conversation I wasn't sure, so I decided to just pass it off as a joke. These guys, *kings?* I said to him. I doubt any of them even owns a piece of clothing that isn't a pair of jeans or a T-shirt.

You know, Plato said, still surveying the room, I spent the better part of my life trying to figure out how to ensure that those who are most fit to rule are the ones who end up ruling. I gave much thought to the question of how to educate rulers so that they wouldn't fall in love with their own power.[*] But I confess I never once considered what these rulers should wear to work.

I can see how you don't give too much thought to clothes, I answered. You're my first author in a toga.[†]

Wait a minute, Cheryl, I interrupted her again. Are you telling me that your author was running around Mountain View in a toga?

Oh, didn't I mention that? she said airily, so that I knew she'd just been waiting for the right moment to drop it in. Yes, he was in a cute little white toga, together with some very authentic-looking sandals, like something they might have worn to watch triple-A-grade Christians being fed to the lions.

I think that was the Romans, I said to Cheryl.

Whatever, she said. Anyway, that was his getup. And I think it's a pretty clever piece of marketing. It was probably his publisher's idea, but I'm surprised that they were able to get Plato to go along with it. They're clearly trying to brand him.

Lucky thing we're so near San Francisco, I told him. You can get away with anything here. Look at that guy over there, the one with the dreads. There's a philosopher-king for you.

---

[*] "But what we require, I said, is that those who take office should not be lovers of rule" (*Republic* 521).

[†] Cheryl is not an entirely reliable narrator on the details of Greek clothing. As a Greek male, Plato would likely have been wearing a *chitōn* (tunic) with a *himation* (cloak) if an extra layer was needed.

Perhaps, Plato said. I would have to question him.

The guy we were looking at not only had dreads but was wearing a super-sized Grateful Dead T-shirt, and of course jeans. He was probably a decade or two older than your average Googler, but it was obvious he hadn't used the extra years to put any maturity between himself and the young ones. Of course, I'm not one to judge people by their appearances, Rhonda, but from how this guy looked I would have said he had graduated high school with three friends tops, all of them in the computer club with him, and that he had some super-obscure hobby he was obsessed with, like collecting ancient musical instruments or making origami rocket ships that could break the sound barrier, and that, if he noticed women at all, he tried to impress them with how many decimal places of pi he had memorized.

Does his unusual hair arrangement indicate adherence to a particular religious code? Plato asked.

You mean like Rastafarian? I doubt it. Probably more like a religious adherence to not caring how he looks.

Ah, then a philosopher! he said,[*] and again I couldn't for the life of me tell if he was being serious or hilarious. All I know was that he was staring at the guy as if actually contemplating whether he was a candidate for the position of philosopher-king.

Don't make eye contact! I warned him, but it was too late. The potential philosopher-king was already making a beeline straight for us, pumping his arms and moving really fast. Don't worry, I said to Plato, I'll get rid of him for you.

But when the guy asked if he could join us, Plato said that he would be delighted, and he proceeded to make room for this guy by taking my bag off of the seat, where I had purposely put it. I'm supposed to protect my authors, but what can you do when they sabotage you like that?

I'm Marcus, this guy introduced himself. He had a pile of sushi on his plate that could have fed Tokyo and its environs.

Well, Marcus, this is Plato, I said, who's a writer and a philosopher and who is going to be speaking to you all very soon, and is just relaxing

---

[*] Long hair as a sign of aristocracy had passed out of fashion by the fifth century B.C.E., but it's possible that philosophers wore their hair long to indicate that they couldn't be bothered with appearances. Aristophanes, in any case, makes such a claim about Socrates and his followers.

now, so we shouldn't tire him out with any needless chitchat. As you can see, he's a foreigner.

Sure, Marcus said. I'm really looking forward to your talk, Plato. I've read everything you've written.

Really? I said. So did you read the one about philosopher-kings?

Well, of course, Marcus responded, looking at me like I was yesterday's sushi. Who hasn't read the *Republic*?

Naturally, I was embarrassed, and looked daggers at Marcus. I have so many authors to take care of, I explained to Plato, that if I were going to read everything they write I wouldn't have time to take care of them.

That stands to reason, he said, smiling at me with not the slightest hint of taking offense. I already told you he has impeccable manners. And I think Marcus is exaggerating my readership, he went on. I do not write for everyone. Sometimes I wonder whether I write for anyone.

Of course, I hear that sort of thing from a lot of my writers who can get a bit despairing about their readership, especially nowadays when they can actually read the reader reviews on Amazon. That's my number one piece of advice I give to authors. Whatever you do, don't read your Amazon reviews. Second piece of advice, after they ignore my first piece of advice, just remember that anonymity brings out the nasty in people, especially people who haven't gotten out of their pajamas and slippers in several days, which is how I tell my authors to picture their meanest reviewers.

Just give me the gist of it, Plato, I said. You know, like how you would on *The Colbert Report*. Five words or less.

He thought for a few moments and then he said, counting off on his fingers for each word: The. Perfect. State. Defines. Justice.

Okay. I'll bite, I said, you know the way Colbert says. Which is the perfect state? California? I guess a case can be made for Hawaii, too.

I heard Marcus snickering, but Plato wasn't paying any attention to him, which I thought showed good sense. The perfect state, he said, is the one that is ruled by those who have the knowledge to rule, and that knowledge is philosophical knowledge, just as the fundamental question of what constitutes justice is philosophical. Since these most abstract questions are philosophical questions, requiring a philosopher's insight, and the person who has this insight into the nature of justice will not allow himself to be corrupted by the perquisites of power, I arrived at the view that until a philosopher acquires political power, or, alternatively, someone who already has political power can be made into a philosopher, there can be no justice in the state (*Republic* 610c).

Well, that's a new idea, I said.

When I first proposed it, it was new, he answered. Now, not so new.[*]

And how did that work out for you? I asked him.

Not so well, he said.

Book didn't sell? I asked sympathetically.

Oh no, far worse, he answered. It led to the greatest fiasco of my life.

Well, somebody can't just come out and make a dramatic statement like that and not elaborate, so I asked him what had happened.

An opportunity opened up, he said, which allowed me to try and put my views about justice into practice. I had a friend whose name was Dion, who had studied with me at the Academy,[†] the quickest of all the students I have ever taught.[‡] He came from the city of Syracuse, and he was very well connected there. In fact, his wife's brother was the tyrant of Syracuse.

This, of course, took me by surprise. Syracuse? I said. They have tyrants running Syracuse?

At this Marcus couldn't contain himself. It's another Syracuse,[§] he said, with a smirk nasty enough to rot the sushi on his plate, of which,

---

[*] The ruler who came closest to realizing Plato's ideal of a philosopher-king was, most Plato scholars would say, Archytas of Tarentum (428–347 B.C.E.). Later in history, there was Marcus Aurelius (121–180 A.D.).

[†] Plato founded his Academy around 387 B.C.E., after he returned from the years he'd spent away from Athens, prompted by the execution of Socrates. It was located in what was originally a public garden that had been left to the Athenian citizens by the Attic hero Academus. The Academy has a claim to being the first European university, although there were other schools in existence, specifically one that was founded by Isocrates, but these were confined to the teaching of rhetoric. The most famous student to have studied at Plato's Academy was Aristotle, who came to Athens from Stagira, in northern Greece, where his father was the personal physician to the Macedonian royal family. Aristotle stayed for twenty years, leaving after Plato's death to found his own school in Athens, the Lyceum. The Academy continued to exist throughout the Hellenistic period, terminating, for the first time, after the death of Philo of Larissa in 83 B.C.E. During the Roman period, philosophers continued to teach Plato's ideas, but it wasn't until 410 C.E. that a revived Academy was reopened as a center for Neoplatonism. The Academy lasted then for a little more than a hundred years. Its doors were finally shut in 529 by the Byzantine emperor Justinian I, a stickler for Christian orthodoxy.

[‡] "At any rate Dion, who was very quick of apprehension and especially so in regard to my instruction on this occasion, responded to it more keenly and more enthusiastically than any other young man I had ever met, and resolved to live for the remainder of his life differently from most of the Greeks in Italy and Sicily, holding virtue dearer than pleasure or than luxury" (*Seventh Letter* 327 a–b). If the *Seventh Letter* is authentic, it would have been written sometime after Dion's death, round about 352 B.C.E.

[§] Syracuse was a *polis* in what is now Sicily, and was first founded by Corinth.

by the way, there wasn't much left, since he'd been scarfing it down the whole time he was listening. My sense was that he was resenting all the attention that Plato was paying to me, a mere media escort and him a potential philosopher-king.

Oh, okay, I said. You know, my brother-in-law is in politics, too. Nothing as important as a tyrant. He got himself elected to the school board in Freemont. Although I have to say Leon can be a bit of a tyrant at home. My sister doesn't have an easy time of it.

I could hear Marcus snickering again, but Plato simply explained to me that where he comes from they use the word "tyrant" somewhat differently. For us, he said, "tyrant" strictly means someone who seizes control through irregular means, arrogating the legitimate transfer of power.* Tyrants, at least so far as the semantics of our term goes, need not behave as oppressors and tormentors who abuse their power. In fact, when tyrants first appear, at least as I have observed, they more often than not present themselves—and may even initially be—a spokesperson on behalf of the people against the abuses of the powerful few, who have accumulated a disproportionate amount of capital and power.

You mean like the 1 percent versus the 99 percent, I said.

Yes, exactly, Plato said. The 99 percent have often some champion whom they set over them and nurse into greatness (*Republic* 656c–d), Plato said, going into lecturing mode. These professor types can be counted on to do that. You ask them a question and out pops a lecture. Anyway, he went on for a bit explaining that tyrants are first okay and actually protectors of the people but gradually get rid of all those who can restrain them, and so end up demonstrating that tyranny is the worst possible form of government (*Republic* 656c–d), as this particular tyrant of Syracuse apparently did.

Nasty piece of work? I asked him.

As all must turn nasty if there is nothing to restrain their insatiable appetite for power. He was so jealous of his power and suspicious that someone might possibly usurp his supremacy that he kept his son,

---

*In later years the Greeks condemned tyranny, but at first it was regarded as an irregularity that wasn't always objectionable. In Attic tragedy the word *tyrannos* is often used more in the sense of "king." Plato and Aristotle both condemned tyranny as the worst possible form of government, but by their time tyranny had outlived its original usefulness and developed vices with which we're all too familiar. See Sian Lewis, *Greek Tyranny* (Liverpool: Bristol Phoenix Press, 2009).

Dionysius II, completely uneducated so that he could never pose as a rival (*Seventh Letter* 332c–d). Then the elder Dionysius died quite suddenly, whether of natural causes or not, and the son, as unprepared as a slave for leadership, assumed his place. My friend Dion, who at that time felt that he exerted some beneficial influence over his nephew, saw this as an opportunity to put the ideas of the Academy into effect. He implored me to come to Syracuse to take charge of the education of the younger Dionysius (ibid. 328c).

You were going to turn the one who already had political power into a philosopher, Marcus said, entirely unnecessarily, as if only to get himself back into the conversation.

That was the general idea, Plato said, his soft voice managing to convey a strong dose of dryness. Or failing that, at least to convince the one who had political power to consult philosophers for advice. Though I must say it took some convincing for Dion to get me to take on the project. First of all, the education of the younger Dionysius ought to have been undertaken long before. He was eighteen when he assumed power, and that is entering rather late in the project of molding character. Young men of that age have sudden impulses and often quite contradictory ones (*Seventh Letter* 328b). And the young man's character was, to begin with, a weak one. To use a metaphor of mine I had once thought to put to political use, the ore of his soul was not of the gold required for true leadership, nor of the silver that makes for a good soldier, but was of the less precious metal, bronze (*Republic* 414c–415d).*
I also knew that there was much intrigue in the court, schemers who wanted to exploit the weakness and lack of self-discipline and self-

---

* Plato is here referring to his "noble lie." He arrives in the *Republic* at the conclusion that the just state is one in which a person's role is determined by their fitness for that role. His utopian state would consist of three classes. On the top, making the decisions, is the ruling class, whose members would have, both by their intrinsic natures and their training, the self-discipline to act strictly in the interests of all the citizens. Next come the soldiers, charged with implementing the decisions of the rulers. Last come the farmers and the craftsmen, who live by those decisions, pursuing the necessary tasks of keeping the material support of the city flourishing. In order to ensure that members of each class would obligingly perform their respective tasks, with no thought of destabilizing the society by trying to jump into a class for which their natures don't suit them, Plato proposed that the people all be told the myth that there are different metals mixed into their constitution. The rulers have an admixture of gold, the soldiers silver, and the farmers and craftsmen bronze. He himself puts the suggestion for the noble lie forward with some embarrassment. Noble lies, of one sort or another, have gone on to have a long history in governments, both democratic and not, such as to vindicate Plato's embarrassment. Still, it is an important and thorny question, both moral and political, whether governments—or individuals, for that matter—are ever justified in lying, and if so under which conditions.

knowledge of the new ruler. Still, there was Dion, a man of the utmost probity and noblest disposition, imploring me to come to Syracuse and undertake the great experiment about which we had so often spoken, shrinking the long hours between dusk and dawn to what seemed mere moments, carried away out of time by our vision of what could be, so that we watched together in amazement as the sun rose over honey-yielding Mount Hymettus. And it seemed to me that for the sake of my own self-respect, I ought to see whether some of my ideas could be realized (*Seventh Letter* 328c). But I was a soul divided. I had written that, on the one hand, the philosopher keeps quiet and minds his own business, content if he himself shall live his life in purity and free from injustice and take his departure finally with fair hope and in a spirit of graciousness and kindness. In that case he will have performed not the least of achievements before he departs (*Republic* 496d–e).

Were you thinking of your other friend there? I asked him gently. You know, the one who didn't get to have his free meals in the what-do-you-call-it?

The Prytaneum, he said. And I was. I was thinking very much of Socrates when I wrote those words. But then, on the other hand, I had also written that such a philosopher will not have performed the greatest achievement either, if he doesn't find a state that fits him. For in the state that fits him he himself will attain greater proportions and, along with his private salvation, will save the community as well (*Republic* 497a). Dion's summons, therefore, fell heavily on me. I was urged by a sense of shame in my own eyes that I should not always seem to myself a kind of argument pure and simple, never willing to set my hand to anything that was an action (*Seventh Letter* 328c).

So how did that work out for you, I asked him, to prod him back to the story line.

I barely escaped with my life, he said. It was a grim statement, but he didn't say it grimly. He only raised his eyebrows slightly.

Well, look, don't blame yourself too much, I said to him. You know, the best-laid plans and all that. And the royals can be a handful. Believe me, I know. I had Fergie for her last two books. Anyway, at least you didn't end up getting executed like your poor friend, I said, and then immediately felt bad that I had, since he got that stricken look again. And I've got to give you credit, too, I said. At least you got yourself outside of your ivory tower for a little bit, and from the sound of it you got yourself a real adventure. I really hope you worked all this into your new book. The thing I don't understand, though, is why you harp so much on philosophers, as if they possess the secret of saving the whole world. You sort of remind me of an orthodontist who thinks that the whole secret of living a good life is having perfectly aligned teeth. No offense, Marcus, I added, since on top of all of his other charms he has a perfectly hideous set of teeth. In fact, come to think of it, that's probably unconsciously what made me think of orthodontists.

What do you mean? Marcus said.

Never mind, I said. Forget it.

But for some reason it was Plato who grabbed hold of the orthodontia thing and wouldn't let it go.[*]

But if you were to want your teeth to be perfectly aligned, he said, to whom would you go?

There's nothing wrong with *my* alignment, I answered him.

No, of course not, he said, but just bear with me a moment. I'm trying to see if I might offer you an answer as to why I, as you say, harp on philosophers.

Oh, okay, I said.

So let's say, just hypothetically speaking, he said, that there were something the matter with the way your teeth lined up with one another. To whom would you go to get this problem corrected?

I'd probably go to Dr. Kolodny, I answered him.

And what is it that makes Dr. Kolodny the right person to correct this problem? he asked.

He's one of the best orthodontists in the Bay Area, I answered. Both my kids had their braces put on by him, and he did a beautiful job. Jason and Valerie both have smiles to die for. I took out some pictures of the two of them—and you know how gorgeous they both are—and Plato looked at them and complimented their teeth, while Marcus didn't even deign to look.

---

[*] In Plato's dialogues, Socrates is very fond of using homely examples, and Plato was probably being faithful to the historic Socrates in having his character reach often for plebeian analogies and dwell on them at length, as a passage from Xenophon's *Memorabilia* bears out. When Socrates was threatened by the oligarchs who briefly took control of Athens after the Athenian defeat in the Peloponnesian War (see chapter ζ), Critias, forbidding Socrates to speak philosophy publicly, said, "But at the same time you had better have done with your shoemakers, carpenters, and coppersmiths. These must be pretty well trodden out at heel by this time, considering the circulation you have given them" (Book IV).

And is it because of what Dr. Kolodny knows that you brought your children to him? he asked me. I didn't exactly see where he was going with all this, but he seemed so intent on his questions that I was willing to answer what seemed brain-dead obvious.

Well, yeah, I answered.

And what is it that Dr. Kolodny knows? he asked.

He knows bites, I answered. Meanwhile Marcus was grinning like the conversation wasn't at all about his bad teeth but about something entirely different about which he, Mr. Philosopher-King, was in the know and I wasn't.

So when it is a matter of fixing the alignment of teeth, Plato continued, then you seek the person who has the right sort of knowledge to remedy the situation. The person who knows bites—who knows, in other words, what a good bite is and what must be done to a less-than-good bite to change it into a good bite.

I was figuring that Plato, with his good breeding, was simply trying to give a very broad hint to Marcus, so I said, If you want, Marcus, I can give you Dr. Kolodny's phone number.

Marcus reared up at this. He didn't mind Plato hinting around, but he wasn't going to take it from me. Listen, he said, are you trying to imply that there's something wrong with my teeth?

Nothing that Dr. Kolodny couldn't fix, I said soothingly.

Because when it comes to teeth and the matter of their alignment, Plato jumped right in, there is a right way and there is a wrong way, and the person who is the expert is the one who not only knows the right way but knows how to change the wrong way into the right way.

My teeth work just fine, Marcus said, and in some sense he was right, since he'd managed to get all that food down in record time, while Plato had barely touched his fruit salad and Greek yogurt.

But you are not, like Dr. Kolodny, an expert on teeth alignment, Plato said softly.

Look, they're my teeth, and they feel just fine to me, said Marcus.

I was pretty sure that Plato and I were both on the same page here, so I said, Maybe they just feel fine to you because they're the only teeth you've ever had. You don't even know what it feels like to have a perfect bite.

A perfect bite? Marcus spit out. Maybe it's because I'm just a software engineer and not a mathematician like you, Plato, not to speak of whatever the hell it is that you are, Cheryl, but I have to say I'm dubious of any claims about perfection, including when it comes to bites. When you've got teeth that are fully functional, which my teeth happen to be, then the only reason to mess with them is for some trivial matter of aesthetics.

To answer your first question, I said to Marcus, I'm a media escort, hired by an author's publicist to make the author's book tour, in any given city, go as smoothly as possible no matter what unforeseen obstacles there are. And to answer your second question, I wouldn't poohpooh aesthetics so much. Out in the real world, there's nothing trivial about aesthetics. And I think Plato here agrees with me.

If I'm not mistaken, Marcus said in his gotcha sort of way, Plato once wrote that beauty is a short-lived tyranny.

At this Plato looked almost as stricken as when he was speaking about his friend the felon. When people quote things back to me that I have supposedly written, he said, looking down at his beautiful hands folded on the table in front of him, then I regret ever having taken stylus to papyrus. That's what he said, Rhonda, and I figured it was just part of the whole toga shtick. I have never committed my true philosophical views to writing (*Seventh Letter* 241c–d), he continued, which I have to say, considering all the money that his publisher seems to have committed to him, sending him on a twelve-city tour, not to speak of the whole branding thing, was pretty bizarre.[*] As for that particular line, I don't remember ever having written anything like it in any of my philosophical works, he said.[†]

---

[*] "One statement at any rate I can make in regard to all who have written or who may write with a claim to knowledge of the subjects to which I devote myself—no matter how they pretend to have acquired it, whether from my instruction or from others or by their own discovery. Such writers can in my opinion have no real acquaintance with the subject. I certainly have composed no work in regard to it, nor shall I ever do so in future, for there is no way of putting it in words like other studies. Acquaintance with it must come rather after a long period of attendance on instruction in the subject itself, and of close companionship, when, suddenly, like a blaze kindled by a leaping spark, it is generated in the soul and at once becomes self-sustaining" (*Seventh Letter* 341c–d). This is rather an extraordinary admission by Plato, and has, in fact, led some to argue that the *Seventh Letter* must be authentic. A similar claim that he had never committed his philosophy to writing is made in *Letter II* (314b).

[†] Plato is being cagey here. The line doesn't come from any of his dialogues but rather from a fragment of lyrical poetry attributed to him: "I throw the apple at you, and if you are willing to love me, take it and share your girlhead with me; but if your thoughts are what I pray they are not, even then take it and consider how short-lived is beauty." *Epigrams*, translated by J. M. Edmonds, in *Elegy and Iambus* (Cambridge, MA: Harvard University Press, Loeb Classical Library, 1931), vol. 2. Revised by John M. Cooper and reprinted in *Plato Collected Works*, edited by John M. Cooper and D. S. Hutchinson (Indianapolis, IN: Hackett, 1997), p. 1, 744.

And does that include your philosophical views on orthodontia? Marcus asked snidely. I was in an awkward position here. I'm supposed to run interference for my authors and protect them from nudniks like Marcus, but Plato had invited the guy to sit down and now seemed oblivious to his obnoxiousness.

You mentioned mathematics, Marcus, Plato said. I would be prepared to defend the position that beneath the perfect bite, as beneath all perfection and beauty, there is the formal exactitude of mathematics. Beauty in a human face—as is true, certainly, of all bodily beauty—is a matter of the proportions of parts to parts. Any compound, whatever it be, that does not by some means or other exhibit measure and proportion is the ruin both of its ingredients and first and foremost of itself. What you are bound to get in such cases is no real mixture into a whole, but only immiscible parts to parts (*Philebus* 64e).

And you know, Marcus, it only gets worse with time, I added. That overbite is going to stick out more and more, parts to parts.

So an expert, such as Dr. Kolodny, Plato said, taking up the reins once again, could look at a bite that is presently giving a person no problem, functionally speaking, and, projecting into the future, be able to predict problems which do not yet exist, so to speak. He can see not only the present bite but also the entailed future bite.

Exactly, I agreed. Though in the case of some people it wouldn't take too much expertise to predict problems down the road. They're kind of staring you right in the face.

But perhaps not in the face of the person whose bite it is, Plato said, which for a person with manners like his was hitting pretty hard. Bad alignment must be one of his major obsessions.

Sometimes that is the very person who is least able to grasp the problem, Plato continued.

Well, it stands to reason, I said. They don't have to look at themselves.

So in the case of a perfect bite, Plato continued, which even I was beginning to think might be belaboring the point, the proper person to consult is not the person whose bite it is, but rather the expert in bites, Dr. Kolodny.

Yes, he's an excellent orthodontist. I can highly recommend him, I said, thinking that this would end the matter once and for all. But I was wrong.

So we have now established, Plato continued, that there is a difference between a good bite and a bad, and the person who can best judge the bite is not the person whose bite it is, no matter how right the bite might feel, but rather the expert in bites, Dr. Kolodny. Now tell me, have you ever wondered whether, just as there is a difference between good bites and bad, there is likewise a difference between good actions and bad, or even, more broadly speaking, between a life well-lived and lives which are not well-lived?

Yes, I said. Of course, I've considered that. It would be hard not to consider that, being in the profession that I'm in. You'd be surprised by how many of my celebrity authors, who people imagine are living these fabulously perfect existences, what with infinite amounts of money and fame and people groveling at their feet night and day, are screwing up at almost a Charlie Sheen level of ass-dumbness. I mean, I'm not mentioning any names here, since I consider it kind of like doctor-patient privilege, except, of course, you, Rhonda, she said, since I know that whatever I tell you isn't going any farther. Cheryl has said this to me dozens of times, and I'm not sure whether she means it as a compliment to my personal integrity or as a comment on my meager social life.

And have you not further considered that the person who can best judge this difference, when it comes to actions and to lives, Cheryl went on, assuming her Plato style, is not the person whose actions and life these are, no matter how right they might feel to him or her, but rather the expert in such matters, the so-to-speak Dr. Kolodny of actions and of lives?

Well, I don't know that I'd go that far, I answered Plato. I don't know that the two sorts of knowledge are all that comparable. Dr. Kolodny has special technical knowledge. He went to dentistry school for something like four years, and we're talking after college, and then he had to go and study orthodontics on top of that for, I don't know, maybe two or three additional years. So we're probably talking something like ten or eleven years of higher education.

And the reason Dr. Kolodny's extensive training was required, Plato continued, must be that the subject of teeth alignment—of what constitutes the perfect alignment and the many ways that that perfection can fail and what must be done to correct that failure—is complicated. This is why this knowledge requires such an expert as Dr. Kolodny, who must have mastered many years of technical study.

Well, yeah, I agreed.

But isn't the knowledge of what distinguishes between good actions and bad, not to speak of the knowledge of lives that are supremely worth the living and those that are not, at least as complicated as knowledge of the proper alignment of teeth? Plato asked me in his soft but pressing way. It's sort of like getting attacked by a pillow, Rhonda. If the knowledge of the difference between good and bad bites requires such extensive expertise, demanding many years of study, why doesn't the knowledge of which I speak also require many years of study? You asked me why I harp on philosophers, and this is my answer to you. Philosophers are those who spend at least as much time mastering a complicated subject as Dr. Kolodny did, only their subject is not what makes for a good bite but rather it is what makes for a good life.

By this time, I was beginning to get an inkling of what Plato might be driving at, and I was getting a sort of creepy feeling about it, too, despite his downy-pillow tactics. Plato, it sounded like to me, was an elitist of the most extreme sort. It even began to dawn on me that his implying that back where he comes from people don't think of the word "tyrant" as necessarily negative might be saying more about him than about Greece. Maybe Plato, for all his soft voice and good manners, was some sort of would-be tyrant. And for some reason, instead of dancing around him, the way I usually do with my authors, letting them go to town with their VOOMs, I decided to just call it as I saw it.

Who are these philosophers to tell people how to live? I asked him in no uncertain terms. Who are they to tell *me* how to live? It's my life, and I think I know what makes it supremely worth the living, to use your phrase, better than any expert in philosophy does.

But now, said Plato, raising his eyebrows a little, you sound like our friend Marcus, who tells us that he knows perfectly well that his bite is good without Dr. Kolodny telling him any differently. Just as you yourself pointed out that the bad biter is the last person to know his bite is bad, could it not be true that the person who is living his or her life badly is the last person to know that it is bad? Not, he rushed on to assure me, that I am saying that you are in any way such a person. I am simply making the point that how a life seems to someone who is living that life may not be accurate. Seeing a life from a vantage that is, so to speak, within that life may not provide the proper perspective for judging whether that life is indeed worth living. So an expert who, by definition, has assumed a different perspective by reason of his or her knowledge is in a better position to judge than the person whose life it actually is.

Well, I really have to beg to differ with you here, I told him, again in no uncertain terms, because the two situations, a bad bite and a bad life, aren't at all alike. How to get the most out of your life isn't something you study in a textbook, like Dr. Kolodny studied how to fix alignments in a textbook, taking all those years to memorize diagrams of teeth and bones and nerves and who knows what all else. There isn't any textbook filled with that kind of specific information that would tell you what the difference is between a good life and a bad. The two situations couldn't be more different from each other.

And yet it is your belief, he said in that soft-spoken way of his, which frankly was beginning to remind me less of a pillow and more of a tiger or something creeping up on you right before it makes its leap for your throat, that some people know how to live and some people do not. Am I right?

Yes, I said. Like I said before, I know some people whose names will go unmentioned who don't know what to do with all the incredible advantages they've been given, all their fame and money and getting anything they want. They may know how to get the world to do their bidding, but as far as I'm concerned they don't know the first thing about how to live their lives.

Is it your belief, he asked me, that those people who do know how to live well, such as you yourself, are born with that knowledge, so that it just comes naturally to them, like the fear of falling or the desire for warmth and human contact, or is their knowledge of how they ought best to live knowledge that had to be acquired during their lives?

No, I don't think anyone is just born knowing how to live, I answered him. If they were, it would be a lot easier to be a parent. But I don't think it's philosophy that teaches us how to live. It's common sense and common decency.* I mean, look at me. I didn't take any philosophy in col-

---

* In Plato's dialogues, Cheryl's opinion is voiced by many of Socrates' interlocutors, including, interestingly, a certain Anytus who makes a late entrance in the *Meno*, a dialogue that is concerned with the question of whether or not virtue can be taught. What is so interesting is that Anytus, who enters briefly into the discussion, being Meno's host in Athens, is identified by most scholars as the very Anytus who will be one of Socrates' three accusers. See chapter ζ. The *Meno*'s Anytus certainly behaves in a way consistent with such an action, showing great impatience with Socrates' form of questioning. The eponymous character of the *Meno*

lege. I shopped it a few times but it never grabbed me. I remember one professor spent her whole first lecture talking about valid and invalid arguments, and I don't think I've ever been more bored in my life. Are you going to tell me that I needed to know the difference between valid and invalid arguments in order to teach Jason and Valerie what they should and shouldn't do? If you went to a philosophy department, I bet you wouldn't find that everybody there was some kind of saint, rushing out after their classes to perform acts of mercy and charity. You'd probably find the same percentage of jerks there as you'd find anywhere else. For all I know, even more.

But perhaps they are not true philosophers, then? Plato asked softly.

Now you're just playing with words, Plato, I said. They're drawing paychecks from their universities to teach philosophy, aren't they?

And to you that signifies that they are philosophers?

Well, what else, then? I said. They're not orthodontists! They're no Dr. Kolodny who knows how to straighten teeth! They're philosophers who know how to make philosophical arguments and be able to tell people which are valid and which aren't, in case anybody was interested in asking them, which I don't see people lining up to do. I don't see how that particular skill has to do with anything that people actually care about, or something that people need to know in making any decisions, much less knowing how to live lives supremely worth the living.[*] I don't expect orthodontists to necessarily have the most beautiful smiles in the world, and I don't expect philosophers to be better behaved than other people either. It's just one more subject to take in college—or not. I mean I don't want to be rude, or to undermine your confidence right at the beginning of your book tour, but really what you're saying is pretty out there, and, more importantly, I don't think it's going to go down well with most of your audiences. Maybe that's

---

is also quite interesting. A handsome young man, quite wealthy—the dialogue shows him with his several slaves (82a), one of whom Socrates is able to prod to derive a geometrical solution (82b–85b)—Meno opens the dialogue by asking Socrates if *aretē*—that is, virtue— can be taught. Socrates casually calls Meno a rascal (81e), and elsewhere (*Crito* 53d) Meno is described as lawless. Debra Nails writes: "It is Xenophon who depicts Meno as so thoroughly scurrilous as to deserve his end; whereas other generals were beheaded, Meno was tortured alive for a year before being tortured to death" (*The People of Plato*, p. 204). Xenophon represents Meno as being prepared to take all shortcuts to get ahead, including betraying friends (*Anabasis* 2.6). As always, Plato chooses his characters carefully.

[*] Cf. "The fact that you might give one set of answers rather than another to standard philosophical questions will say nothing about how you will behave when something other than a point of philosophy is in dispute." Stanley Fish, "Does Philosophy Matter?"

good, since controversy sells books and all, but you also don't want to offend the people who are potential buyers. I mean, I don't blame you ivory-tower types for overestimating the importance of being smart. After all, that's what you've got going for you, that you're smart, at least a certain kind of smart. So you naturally think that that's what everything comes down to, and that anybody who isn't your kind of smart is really out of luck since not only is he or she not going to go to a fancy eastern elitist school that opens all sorts of exclusive opportunities, but even the door to living a good and decent life is slammed shut in their faces, too. That's what you're implying, isn't it? That if a guy or gal doesn't have the sort of smarts that you have, then they don't have a hope in hell of living a life worth living? That is not going to go down well with your audiences, Plato. Hasn't your publisher warned you about that? You know, I've escorted runway models who have written their memoirs, and from their perspective they can wonder how people who aren't drop-dead gorgeous can live lives that are worth living since that's what made *their* lives so supremely worth living. Ditto football players who can't imagine for the life of them how people who aren't huge masses of flesh and muscle who can ram into people and knock them unconscious can possibly live lives worth living. I don't see how being your kind of smart is any different. I mean, in your line of work—and I gather from what Marcus said that you're a mathematician, too, and not just a philosopher, which I'm sure is very impressive, I'm not saying it isn't—but in your line of work you obviously have to be super-smart to succeed, and maybe that's why you're being misled into thinking that a person has to be super-smart in order to know how to live her own life. You have to be super-smart to live your life as a philosopher well and a supermodel has to be super-gorgeous to live her life as a supermodel well, but you don't have to be super-gorgeous and she doesn't have to be super-smart. Think how unfair it would be otherwise. It's like you're saying that someone who doesn't have a higher-than-average IQ can't live a worthwhile life. Well, that sucks big-time for them, doesn't it! Or maybe you think that the philosopher-kings should just dictate to them how to live? Is that the idea here? That everybody in the big swell of the IQ curve should just hand their lives over to the philosopher-kings the same way Marcus here should hand his teeth over to Dr. Kolodny?

Frankly, I don't know what got into me, Rhonda, Cheryl said, at the same time summoning the waiter and ordering another Long Island Iced Tea. I never argue with my authors. I usually never even notice what they're saying. It's all in one ear and out the other and onto the next VOOM. But I guess Plato had struck a nerve. I mean, here he was suggesting that, because I'd never taken a philosophy course, I didn't know how to live my own life. He didn't come right out and say it, but that was the implication. I didn't need any philosopher droning on and on about valid and invalid arguments to tell me that that was the implication. And I wondered whether his friend Socrates used to go around spouting similar ideas, and if it ticked people off so much that they ended up convicting him on trumped-up charges, especially since, from the little Plato had told me about his friend's trial, it sounded like he went out of his way to keep implying that only someone superior like him knew what life was all about right up to and even after his conviction. I mean, I'm not excusing anybody here for putting a man to death for being such a royal pain in the ass. I'm just saying that the story might be a little more complicated than Plato presented it.

Anyway, I noticed that when I'd finished my little diatribe, Marcus was grinning at me, and not at all snidely for once. Somehow I seemed to have gained a lot of points with him by lashing out against the whole idea of there being experts who could tell you how to live your life. I wasn't altogether sure, though, how kindly my author would take my little outburst, but he seemed okay with it, in fact strangely impressed, because he said to me, well, if all that you say is correct, and you can prove to me that anybody with common sense and common decency knows the answers to these questions, then I think that it is you who ought to be giving the talk at Authors@Google and not me. We should switch places, and I will assume the role of your media escort.

You know, it's funny that you should say that, I told him, even though I knew he was teasing, as if the whole idea of our switching places was simply ridiculous, but screw him, I thought. I often have the same thought when I'm listening to a lot of my authors, I told him. The only thing is, I haven't written any books, so I'm not, technically speaking, an author.

As you say, a mere technicality, he said, raising his eyebrows, which probably meant he was messing with me. I myself have always written with the greatest of misgivings, he said, far preferring the give-and-take of conversations such as this one in which we are now engaged, in which real progress of understanding can be made. In an ideal world, in which we are freed from the bias against authors who have authored no books, and it is, therefore, you who are to speak at the Googleplex, tell me what you would say on this question of how a person ought to live his life.

How a person ought to live his *or her* life, I corrected Plato, since he'd specifically asked me to help him avoid saying anything that came off sounding sexist.

Thank you, he said with a smile. You agree that there are ways that a person ought to live a life, and ways that a person ought not to live?

Of course, I said.

And it is not just a matter of different personal preferences concerning, for example, whether one prefers to live within the city walls or rather out in the countryside, or whether one prefers a life which exposes one to the thrill of many risks or a life of relative safety.

You mean like different strokes for different folks?

That is nicely put, said Plato.

I can't take credit for it, I told him. But you know, about some things it *is* just a matter of different strokes for different folks. I mean, I don't make any judgments about your wearing a toga, for instance. As far as I'm concerned, that's your business. And I take it you're not married either?

No, he said.*

And never have been? I asked.

He shook his head no.

Well, I don't make any judgments about that sort of thing either, I told him. We're in the San Francisco Bay Area, after all. I'm not asking you about what went on between you and your friend Socrates or between you and your friend Dion, staying up all night and watching the sunrise together. That's none of my business, though I notice it's only men you talk about, never any women. But I'm not making any moral judgments here. If that's your orientation or stroke or whatever you want to call it, then who am I to say differently?

But about some things you would want to say differently, or so I suspect, he said softly. For example, if I were to decide that you were so fine

---

*Historians can't be completely certain that Plato never married since Athenian women rarely show up in the record. And then Plato does stipulate twice in the *Laws* that male citizens must marry.

a media escort that I needed to make you my personal attendant at all times, and so decided to kidnap you, you would probably not respond to my decision by saying "different strokes for different folks."

Are you talking about making me your *slave*? I asked him.

My slave, yes, he answered. My unremunerated personal escort for all times. I'm assuming that you would voice some objection.

Well, yeah, I said.

And what would your objection be?

My objection would be that we don't do slavery.

So I've noticed. But what if I could get away with it? What if it were the custom in my land, and nobody gave any thought at all to enslaving others, most especially if they were barbarians,* and so, even though you don't do slavery in the San Francisco Bay Area, I saw nothing wrong with it and had the means to get away with it? Would you have something to tell me about why it was wrong nevertheless?

What, are you kidding me? I asked him.

Not at all. Would you have something to say to try to appeal to my sense of right and wrong?

I think I'd have a thing or two to say. First off, I'd say just who are you to call anyone a barbarian? One person's barbarian is another person's brother. And second, I'd say what gives you or anybody the right to take anybody else as a slave, even if you somehow think that he or she *is* a barbarian? Barbarians have just as much a right to live as you

---

* "Barbarian" was the onomatopoeic word that the ancient Greeks used to refer to all foreigners, because to Greek ears non-Greek languages sounded like so much *bar bar bar*. Plato, in the *Republic* (469b–c), made what was then the rather radical argument that Greeks should not take other Greeks as slaves, even if they were defeated in war. A war between Greek *poleis* should be regarded as a civil war, and so the conventional rules of war—wholesale rape and pillage and enslavement—should not be permitted. He also argued in *Laws* that slaves should not be treated with violence, since it's in our behavior toward those over whom we have control that our moral character is most deeply revealed—which doesn't mean, he hastened to add, that we should spoil our slaves or treat them as if they were free (777c–d). As Peter Singer points out, Plato's argument advanced the morality of his time, even if, from our vantage point, it was not very far. Singer further points out that the Bible likewise, in discussing the laws of slavery, distinguishes between those who are within one's own tribe and all others. "When your brother is reduced to poverty and sells himself to you, you shall not use him to work for you as a slave. . . . Such slaves as you have, male or female, shall come from the nations round about you; from them you may buy slaves. You may also buy the children of those who have settled and lodge with you and such of their family as are born in the land. These may become your property, and you may leave them to your sons after you; you may use them as slaves permanently. But your fellow-Israelites you shall not drive with ruthless severity. Here is a code that could be disinterestedly recommended to Israelites, but hardly to Canaanites." Peter Singer, *The Expanding Circle* (Princeton, NJ: Princeton University Press, 2011), p. 112.

do. What possible difference does barbarian or not even make? What difference does anything you can say about a person make in terms of whether you're allowed to make them your slave? A person is a person. Everybody's life is just as important as anybody else's, and if you don't know this simple truth for yourself, then just go ask them.

Brava, Plato said softly, and I couldn't for the life of me tell if he was teasing me or not. I mean, he's impressed because I know that slavery is wrong? So I asked him again, What, are you kidding me?

Not in the least, he said. That was magnificent, he said, as if he really meant it. And then he repeated what I'd just said, word for word, as if this were some kind of revelation. A person is a person, everybody's life is just as important as anybody else's.

Well, listen, Rhonda, I like to be showered with compliments as much as the next person, but frankly this was ridiculous, which is what I told him. Who doesn't know a person is a person? I said to him.

You'd be surprised, he answered me. There is so much you take for granted now, far more than is stored, I begin to suspect, in the information-clouds of Google. There are treasures of hard-earned knowledge stored right there in your view of the world.

And you're going to argue, Marcus piped up at this point, that all this amazing stuff that's now stocked in Cheryl's mind, in the aisle labeled "morality," like in a modern convenience store, was imported there by you philosophers.

But I just finished telling you, I explained patiently, that I never took a philosophy course. Anything that's amazing in my mind—and frankly I can't see what's so amazing about anything I've just said— didn't get there because of the fifty minutes I heard some philosopher drone on about valid and invalid arguments.

You go, girl, Marcus said to me. I couldn't tell if he was being sarcastic or not, but, given the general tendency of the guy, I figured he was.

I have as much right to my opinion here as anyone else, I said. In fact I have more right, since what we're discussing is how what's in my mind got there.

I couldn't agree with you more, Marcus said, and this time I could tell he was sincere. I think you're right on target with your skepticism. You've got the right intuitions there. Do you mind if I run with them a little?

Be my guest, I said, because I'm going to tell you something,

Rhonda. Having to answer Plato's questions is exhausting. The guy is like the Energizer Bunny when it comes to asking questions. Oh, and I ought to mention that right around now there was a little commotion because two Googlers had shown up to take my author on his tour, and I could see that he was really torn. He was involved in the kind of conversation that, you could see, he just lives for. I mean you could see it was his life's blood. But he was also eager to get an insider's look at the Googleplex. That whole cloud idea had caught his fancy for some reason. Before I had wanted him to skip the tour, since I thought it would be exhausting for him, but then the kind of conversation he was intent on having was every bit as exhausting, at least for me. He asked me if it would be possible to take the tour after he gave his Authors@Google talk. No, I told him, that's out of the question, since that's when you're signing the books. He was all for skipping the book signing, but obviously that wasn't going to happen on my watch. When it comes to that sort of thing then I'm the one who has to be the tyrant, since I answer to the publishers, and believe me, they keep track of how their authors do with the various media escorts, comparing our sale records with one another. Marcus tipped the balance by telling Plato again that not only did he want to challenge his view but he could also explain the way that Google works, the way the search engine is able to deliver the right information to people with such accuracy.

You're not going to get that on a tour of the Googleplex, he told Plato, since the answers you're looking for lie in the sphere of abstraction, which, of course, is right up your alley.

With that, Plato was so happy that he actually allowed himself a real smile, not just the hint of one, and I sent the tour guides, who seemed genuinely disappointed, on their way. Apparently, they were big fans of Plato's, too. Go figure.

Plato got right down to business, asking Marcus what were the points he wanted to challenge.

Okay, let me just reiterate the major points that have emerged so far, at least as I see them, Marcus said. You assume that if there's knowledge, especially if it's non-trivial knowledge that's difficult to come by, then those who have the knowledge are the few, the experts in that knowledge. Is that right?

Yes, Plato said. That this is so seems to me almost a tautology.

We shall see, said Marcus. Okay, then you also assume that there is such a thing as knowledge about how we ought to live our lives, that it's not just a matter of personal preference or cultural norms, say, and that this knowledge is non-trivial and hard to come by. Am I right?

Yes, Plato said. It must certainly be non-trivial, since so many people get it wrong, living lives that are almost at a Charlie Sheen level of ass-dumbness.

He didn't! I said to Cheryl.

He did. Repeated me word for word. Even Marcus had to all-out laugh, instead of snicker. Okay then, Marcus went on. So then the implication of these two assumptions is that if anybody has this knowledge—and I suppose it's consistent with your two assumptions that nobody has it—but if, in fact, anyone has this knowledge, it can only belong to those who have managed to think their way to it. Am I right?

Yes, said Plato.

So, according to you, this knowledge is something like mathematical knowledge. It's just as objective, its truth not determined by personal preferences or societal norms, and it's just as non-trivial and difficult to access, being a matter of reasoning.

Yes, said Plato. I agree to all of this. In fact, I would go further and claim that it is not only *like* mathematical reasoning but that mathematics itself goes into the knowledge.

Okay, Marcus said, but let's just stick with the weaker proposition, for the purpose of my refuting you.

Yes, agreed Plato. That makes dialectical sense.

Okay, Marcus said, and then took several moments to breathe. He was, as my mother used to put, overexcited. Okay, he said again. What all of this implies, as Cheryl was quick to point out, is that only those who are gifted in reasoning can discover how we ought to live. There's no other way to access these truths, meaning that non-philosophical people who can't follow philosophical arguments have to accept the conclusions from the philosophers.

Yes, said Plato. Hence the enormous obligations to others that philosophers have.

Including obligations to tell others how they should be living, to legislate morals for them.

Not in every respect, no, since as was also pointed out by Cheryl, there are many matters that lie entirely in the sphere of personal or cultural preferences. And many such decisions, which are therefore entirely a matter for an individual to decide for himself or herself or to let society decide for him or her, will help to make for a life worth living. So that, for example, if I am of such a nature that I need to test my manhood—I hope it is not sexist to mention manhood in this context, he asked, looking at me.

Why don't you just say test your courage? I said to him. Women test their courage, too.

Yes, you're right. If I am of such a nature that I need to test my courage by taking repeated risks, then I am free to do so insofar as the risks I take do not involve any moral transgressions. A great deal of the substance of a life worth living is made up of decisions of this sort.

But not all of them, Marcus prodded.

No, not all of them, Plato agreed.

And these are the contributions to a life worth living that the philosophically unintelligent can't decide for themselves, but must instead turn the decision over to the philosophically intelligent.

Not quite, said Plato. *Nobody,* the philosophically intelligent any more than the philosophically non-intelligent, can decide these matters for himself—or herself—since there are objective facts of the matter. Those who can access such knowledge are no more able to change the facts of how we are to live than anybody else. Like everybody else, they must abide by them. The one difference is that they are able to discover, through the special talents and training that are theirs, what the facts are. So they are not imposing their personal will on others, any more than mathematicians are imposing their wills on others by informing non-mathematicians what the mathematical truths are. They are simply sharing their knowledge with others, knowledge that others cannot access for themselves, lacking the requisite cognitive skills, a matter both of talent and training. This seems to me no more unfair than that the mathematically intelligent share their knowledge of mathematics with the mathematically unintelligent. So, for example, you, as a software engineer, possess considerable mathematical intelligence, I would assume.

Yes, Marcus said, I suppose you could say that.

And by working here at the Googleplex, Plato said, you are able to provide the benefits of your mathematical intelligence to others who are lacking this kind of intelligence, who make use of your powerful search engine while simply regarding it as techno-magic.

The difference, Marcus said, is that not everybody believes himself to be good in math or even cares to be. In fact, only those who are, in fact, good in math give a hoot about whether they're good at it or not. But everybody cares about living his life well and has strong views about how best to do it. And if you tell him that he's just lacking the cognitive ability to figure it out for himself, he's going to have some harsh words for you.

That is exactly right, Plato said. This is a way in which philosophical skill is entirely different from all others. When someone thinks that he or she has great talent at, say, the flute, when in fact there is none, people either laugh at the person or are annoyed, and family members try to restrain the person as if he or she were crazy (*Protagoras* 323a). But all people have a stake in believing themselves masters of much of the domain of philosophy, most especially the questions of how life should be lived. To think oneself to be anything less than a master seems to diminish one's very humanity. This is true to such an extent that a person who lays no claim to such knowledge seems not human at all (ibid. 323b). This might well be called the predicament of philosophy. For the fact that each and every person is committed to the belief that they are masters of this domain no more shows that they are indeed such masters than that . . .

Yes, yes, I know, Marcus said, smiling. Than that I'm able to decide for myself whether I need the expert attentions of Dr. Kolodny.

Precisely, answered Plato, with his suggestion of a smile. So now, which of the steps in my argument do you want to challenge?

To tell you the truth, just about all of them, Marcus said. But since our time is limited—how long do we have? he asked me.

Twelve more minutes, I said firmly.

Right. So since our time is *extremely* limited, Marcus said, I'm going to restrict myself to only one kind of objection, and mainly because of the promise I made that I would explain something to you about the secret behind Google's success. It just so happens that that secret provides grounds for rejecting your claim that non-trivial knowledge that can't be gotten to by regular individuals can only be accessed by experts.

I am intrigued, said Plato.

Okay, well, I'll start with Google. You know about the World Wide Web, I take it, Marcus said, which seemed a bit patronizing to me. I mean, my grandmother knows about the World Wide Web.

I know that it contains vastnesses of information, Plato said, and that each person has a screen on which can be projected portions of that information.

Vastnesses upon vastnesses of information, Marcus responded. The last time anybody counted, which was back in 2008, there were more than a trillion Web pages, and who can even estimate what it is by now, since we've been adding information by orders of magnitude. But that doesn't even begin to calculate the amount of information that Google has in its cloud storage. The aim is to store all the world's information, which means copying the contents of the thirty-three million books in the Library of Congress, or, even more ambitiously, if you count every pamphlet and piece of ephemera and miscellanea ever printed and in all the world's languages, eventually approximately 129,864,880 books. It means every restaurant menu, every telephone book, the archives of newspapers and magazines, the merchandise for sale in every store. And I'm not even counting all the videos on YouTube, which Google bought in 2006, and all the pictures, including a photograph of every street corner and road on the planet, which is the plan of Google Street View, photographed in high resolution and kept as up-to-date as possible. When Google says it wants to make available all the world's information, you should understand us quite literally.

But getting all this information is the easy part, Marcus continued. The hard part is, given the vastness of the storage, how to scan through it all and get the user the exact bits of information that he needs—and don't correct my sexist pronouns, Cheryl. That's bullshit. Okay. Wait a minute. Have you actually used Google yet?

No, Plato said. Is it hard to learn?

There's nothing to learn, Marcus and I both said simultaneously. Marcus pushed his Mac over in front of Plato. Just type in some word or some words. Anything you're curious about. And, Rhonda, what do you think Plato typed in?

I have no idea, I said to Cheryl. I'm still trying to process the information that he had never used Google.

Yeah, she said, I know. Anyway, it was just the one word: *Socrates.* That was his search.

Wow, I said.

Yeah, Cheryl said, taking another long sip of her drink. I think even Marcus was overwhelmed by that. We were all silent for a moment as Plato just stared at the page, transfixed. I couldn't see the page because Plato was across the table from me, but I imagine that the results included a row of images of Socrates to be found on the Web, and maybe that's what put that expression on his face.

Okay, Marcus said after a few moments, and I noticed that he had brought his voice down a few decibels. Let me explain a bit to you what's going on here. The first thing to see is that there are 4,700,000 results for your query. These are all the places in Google's storage cloud in which the word "Socrates" is mentioned. Enormous, huh? Your friend Socrates is a popular guy, even though, as you can see, some of these results are not really about him. Like here, the fourth result down is some sort of online business where you can download forms for a do-it-yourself divorce or rental agreement, and which for some reason or other named itself "Socrates." Are you okay? Marcus asked suddenly, noticing Plato's face, which frankly, Rhonda, I just don't know how to describe. I mean I had used the word "stricken" before, but I just don't know what word to use for the expression he had on his face as he was staring at the results he got by searching for Socrates.

Plato slowly turned to look at Marcus. What do I do now? he asked quietly.

Marcus gave me a quick glance and I discreetly shook my head no. I didn't think it was a good idea to click on any of the results because who knew what effect it would have on Plato, who, by the way, had to speak in less than fifteen minutes.

Wait a minute, I said to Cheryl. Are you sure he wasn't asking something else when he asked, what do I do now?

No, I don't think so, Cheryl said. I mean just like he said, all that traumatic stuff with Socrates was ancient history. I don't think he was about to be having a nervous breakdown just because he Googled the guy.

Okay, I said. Go on.

So Marcus said, Let me just explain a bit to you what's going on here. See the numbers here: what they tell you is that the search engine scanned through its trillion-plus Web pages and found those 4,470,000 results in 0.10 seconds. But that's only the beginning, okay, since it's not going to be any use to you for the search engine to just dump all those 4,470,000 results on you, right? It's got to sort them for you, put them in a usable order, hopefully progressing from what's likely to be the most useful to what's least likely to be useful. That's what the search engine has got to figure out, and Google's search engine figures it out better than any other search engine. That was the secret of its early success. Okay. But how does it know how to do this? How can it get inside your head and try to get the information to you that you most need? Did Google hire a panel of experts, a bunch of talented scientists or mathematicians or philosophers or literary scholars to read each of the trillion Web pages out there and write a review of it, which Google used to decide whether ordinary people are going to want to see that page when they search for something? No! This is where the original genius of Google came in, the genius of discounting experts. Google has an algorithm that automatically assigns a number to every single page on the Web, a number that corresponds more or less to its usefulness, and which depends on how many *other* pages link to that particular page. The more links to the page, then the more useful it is, all things being equal. But, of course, all things are *not* equal—they rarely are, are they?—since not all of those linking pages are going to be equally important. How do we sort out which ones are? Well, by in turn ranking *them* in terms of how many other pages link to them. So if a page that has many pages linked to it is linked to another page, its linkage is going to count more, be more heavily weighted, than a page that doesn't have that many links. Google's algorithm—the simple one, the one with which it all began a decade ago, which is ancient history in this world—assigned a value to every Web page ranking its usefulness, based on the number and usefulness of the pages that linked to it. Then when you type in a word or words, Google can deliver you the results ranked in their order. Okay, you following? Sorry, of course, you are. Anyway, that was the original idea for the algorithm, but it's gotten a lot more complicated now. There are actually a few hundred signals that the search engine is using now. And one of the most important signals is how each individual user responds to the results he gets. So if you were to click on the third result, rather than the first or second, that's a kind of vote you're making with your click, telling us that for you the ordering was wrong. All of the votes, in the form of hundreds of millions of users' responses to the results they get in the order they come in, is information that goes into the algorithm that Google uses to get the information to you. There's no expert in the system somewhere who knows anything or decides anything. Somebody once said,

I can't remember who, that when you read a really great book you feel as if it's reading you while you're reading it. That seems like bullshit to me—what does it even mean?—but when it comes to Google it's really true. Google is using you while you're using Google, using you to perfect itself.

So it's just as I thought, I said. There's something creepy going on here.

Bullshit, Marcus said. Google is gathering knowledge and, as Plato here will tell you, knowledge is, in itself, a good thing.

Google is gathering information, Plato said very softly. It's not clear it's gathering knowledge.

And also, I said, whatever it's gathering, whether you call it information or knowledge, it's not necessarily a good thing, at least not according to what Siva Vaidhyanathan wrote in his book. According to him, Google is not only using us, but selling us.

Yeah, well, frankly Siva doesn't know what he's talking about. Google doesn't sell its information to its advertisers.

Because that would be evil? I asked sarcastically, because I'd been forced to hear on all those countless Google tours that their corporate motto is "Don't be evil."

Maybe that, too, but mainly because if the users caught on they'd probably start a protest which would take the form of using another search engine. As we like to remind ourselves here, choosing another engine is only a click away. Anyway, we're getting off-track. The point I'm trying to make to Plato is that the knowledge about Web pages that Google is using here is crowd-sourced. That's the important concept I'm trying to get to here. There's knowledge without a knower. The idea is that sometimes the crowd, each individual registering its own response, can come up with an answer that's better than any single individual in the crowd, no matter how smart or expert, could come up with. Sometimes the aggregate is superior to any single member and the only way to get at the right answer is by letting each member have a vote. Think of it as a rolling plebiscite. That's the general idea behind democracy, after all. Who should be the ruler? Whoever the voters end up voting for is the answer, and you can't ask them whether they got it right or wrong. Given the rules of democracy, the answer is the one delivered by the crowd, end of story.

Ah, yes, I had a student who liked to make a similar point, Plato said.

That one you liked so much? I asked him. Dion?

No, another, also quite talented in his own way.

Aristotle? Marcus asked him.

Yes, Plato said, Aristotle. He said that it is possible that the many, no one of whom taken singly is a sound man, may yet, taken all together, be better than the few, not individually but collectively, in the same way that a feast to which all contribute is better than one supplied at one man's expense (*Politics*, Book 3.11, 1281a39–b17).*

Yeah, well, I'd say Aristotle put his finger right on it. Crowd-sourcing is the way to get a lot of the answers you guys are looking for.

Guys and *gals*, I said, earning a nod of approval from Plato and a glare from Marcus. Don't distract me with trivia, he said. I'm taking on *Plato* here.

Then who are you to decree that Cheryl's objection is trivial? Plato asked. Shouldn't you crowd-source that opinion?

Okay, Marcus laughed, I see that you've already jumped ahead to the conclusion I was about to draw.

Draw it anyway, said Plato. It will make it dialectically cleaner.

Okay, said Marcus, though with a little of the air let out of his tires. Where I was heading was that the kind of knowledge you think belongs in the domain of philosophical experts belongs in the hands of the crowd. There are facts of the matter, okay, about what's right and wrong, but no individual can get at them in isolation just because everybody is so invested in his own life and is biased to see things from his own point of view. That's what makes ethical knowledge so tricky: Everybody's view is warped by their commitment to their own life. There's just no way to abstract sufficiently from your own circumstances, you can't help but skew the vantage point because of who you are and

---

* "The principle that the multitude ought to be in power rather than the few best . . . For the many *(hoi polloi)*, of whom each individual is not a good *(spoudaios)* man, when they meet together may be better than the few good, if regarded not individually but collectively, just as a feast to which many contribute is better than a dinner provided out of a single purse. For each individual among the many has a share of excellence *(aretē)* and practical wisdom *(phronēsis)*, and when they meet together, just as they become in a manner one man, who has many feet, and hands, and senses, so too with regard to their character and thought. Hence the many are better judges than a single man of music and poetry; for some understand one part, and some another, and among them they understand the whole. There is a similar combination of qualities in good *(spoudaioi)* men, who differ from any individual of the many, as the beautiful are said to differ from those who are not beautiful, and works of art from realities, because in them the scattered elements are combined, although, if taken separately, the eye of one person or some other feature in another person would be fairer than in the picture. Whether this principle can apply to every democracy, and to all bodies of men, is not clear."

where you're standing. In some sense, Cheryl touched on this when she said that you valorize intelligence because that's your most valued personal characteristic, just like the model feels about her beauty and the football player about his talent for using his body weight to mow down other human bodies. You kind of seemed to be getting at that in the Myth of the Cave. I mean, the people are all chained to their own points of view so they can't share their knowledge, but then you just make an entirely wrong turn, at least in my humble opinion, because you have it all depend on the one guy, the one who singly manages to make it out of the cave.

Well, said Plato, that's not entirely the case. He's pulled along at the first stage of his trip.

Yeah, but the suggestion is that it's by some other smart guy—or gal, Cheryl. The point you're making is that it's only superior reason that can get a person out of the cave. But what you don't consider is that the only way to get out of the cave is to crowd-source, which is the only way of canceling out the peculiarities of the individual members, the way they're skewed toward their own vantage points, including the smart guy who thinks his smarts are all that matters. There's some ideal algorithm for working it out, for assigning weights to different opinions. Maybe we should give more weight to people who have lived lives that they find gratifying and that others find admirable. And, of course, for this to work the crowd has to be huge; it has to contain all these disparate vantage points, everybody who's staring from their own chained-up position in the cave. It has to contain, in principle, everybody. I mean, if you're including just men, or just landowners, or just people above a certain IQ, then the results aren't going to be robust. Including only philosophers, a very peculiar group of people, is definitely going to skew the results. The crowd knows what no individual—not even Plato himself—can know.

Marcus was talking at the rate of about a hundred words a minute, and by the time he had finished, he was dripping with sweat, and Plato, in contrast, looked, I don't know how to describe it, he looked more than ever like he was carved in marble, sitting so still and staring so intently at Marcus that I was surprised that Marcus wasn't at all nonplussed. I was, frankly, worried about Plato. I'm supposed to make sure that my author is at the top of his game when he talks to audiences, and let's face it, I had failed. I mean I'm not blaming myself, because what was I supposed to do? I was sabotaged every step of the way. But still the guy looked barely alive. This was the very beginning of his book tour, his publisher had obviously put some money into him, which, believe me, publishing houses are doing less and less these days, and he looked like he had been, I don't know, stung by a stingray and was simply paralyzed (*Meno* 80b).* I figured it was overload. First, he sees his long-dead friend Socrates up there on the screen, and then he has Marcus coming at him like a steam engine, or maybe like Google's search engine. Overload.

Plato, I said gently, I'm afraid we've got to get going. He immediately stood up, but then he just stood there, stock-still, and I swear to God, standing there in that toga, he really did look like a statue, one which was about to be toppled over by the barbarian hordes, which Marcus, who was really buzzed, looked like he could play the part of to perfection. Plato, I said again. It's time to start moving over to the auditorium. And that's when I got the surprise of my life. Plato hadn't been numbed by Marcus's harangue. He had been invigorated. He wanted to keep going.

Couldn't we three simply continue our dialogue in front of the audience? he asked me. It would be like the old days, Socrates in the agora, in discussion with one or another of the citizens, while a crowd gathered to listen. It would be much truer to the spirit of philosophy than my standing before an audience and lecturing. Nobody can learn anything of importance by assuming such a passive stance. But if we were to continue our dialogue, then the lived conversation of our actively thinking would have a better chance of drawing others in.

Your publisher isn't sending you on tour to promote the true spirit of philosophy, I told him firmly. I noticed that a little contingent of Googlers were approaching us, no doubt wondering why Plato wasn't already standing at the podium.

---

* "Socrates, even before I met you they told me that in plain truth you are a perplexed man yourself and reduce others to perplexity. . . . If I may be flippant, I think that not only in outward appearance but in other respects as well you are exactly like the flat sting ray that one meets in the sea. Whenever anyone comes into contact with it, it numbs him, and that is the sort of thing that you seem to be doing to me now. My mind and my lips are literally numb, and I have nothing to reply to you. Yet I have spoken about virtue hundreds of times, held forth often on the subject in front of large audiences, and very well, too, or so I thought. Now I can't even say what it is. In my opinion you are well advised not to leave Athens and live abroad. If you behaved like this as a foreigner in another country, you would most likely be arrested as a wizard."

You are performing your difficult job, with a very difficult subject, very well, he said to me, and this time there was no denying the twinkle in his eyes. But allow me, please, to perform *my* job, which is also with a difficult subject, as well as I can, at least for just a few moments more.

But there's a conflict here, I said to him. Maybe it's even an ethical conflict.

Yes? said Plato, and now his eyes were twinkling for all they were worth.

You and I can't *both* do our jobs responsibly. You have a responsibility to do yours, just like I have a responsibility to do mine. It's a dilemma! I have to confess, Rhonda, that I felt kind of psyched. Whatever this philosophy game is, I was playing it with a pro.

It *is* a dilemma, said Plato. And since you are bound to feel your responsibilities just as keenly as I feel mine, perhaps we ought to crowd-source here. We know how you will vote, and we know how I will vote. So it is up to Marcus to decide what is the right thing to do.

Let's keep talking, said Marcus.

Do you have a principle on the basis of which you made your decision?

I need a principle? Marcus asked.

Oh, yes, Plato said. It doesn't count as an ethical decision unless there's a principle behind it. Otherwise it is arbitrary.

You want a *logos*, Marcus said, grinning. I don't buy that either. Not if morality is a matter of crowd-sourcing, but okay, you want a principle, I'll give you a principle. The philosophical worth of our continuing the discussion outweighs the inconvenience of keeping people waiting a few extra minutes to hear you address them. They've waited a long time, they can wait a little longer.

Plato turned to me with a smile and asked, Does Marcus's principle apply some balm to your smarting moral scruples?

Okay by me, I said, although these people who have come to collect you for your talk—and there were about six or seven of them who were standing there waiting for us to come with them—may disagree. But you know what, Rhonda? They stood there, insanely smiling, happy to eavesdrop, as if even they could appreciate that this isn't something you get to see every day, not even at the Googleplex. The three of us sat down again, with the little cloud of Googlers hovering over us.

So, Plato said to Marcus, you have explained the concept of crowd-sourcing to me very well, so well that I think I grasp it. Your idea is that the answers to ethical questions—including the question of what it is to be living a life worth living, what are the kinds of actions that such a life demands, and what are the kinds of actions that such a life prohibits—are to be answered not by any ethical experts but rather by something like the Google search engine. Only this will be an ethical search engine. Have I understood you reasonably well?

Yes, Marcus said, grinning. Reasonably well.

And just as the Google search engine has an algorithm that accords different weights to different votes, so, too, will your ethical search engine. So, for example, even if the majority of people think that a life of sensual indulgence is supremely worth the living, and so commended all acts which led to such indulgence, if these people ranked low on either satisfaction with their own lives or low in the esteem with which they were regarded by others, their votes would be given a proportionately lower weight. Is that right? he asked Marcus.

Yes, basically, Marcus agreed. Of course, I'd have to work out the math.

Of course, agreed Plato. But mathematics would be the only expertise required. The mathematics, applied to the data of crowd-sourcing, would settle all ethical questions, eliminating any need for experts. Because to both of you the entire idea of ethical expertise is suspect.

Exactly, Marcus said.

Exactly, I agreed.

In fact, let me amend that. I'd say it's even worse than suspect, Marcus said. I'd say—and I'd back this up by crowd-sourcing—that it's flat-out ethically wrong. You saw the moral outrage with which Cheryl reacted to your proposing moral experts? I'd say that's the wholesale reaction you'd get.

Even though, I said to Marcus, ruling out moral experts would prematurely end your reign as philosopher-king?

I didn't know I was in the running, he said.

Yeah, I said. Plato fingered you early on as a possible candidate for the slot.

Well, I can't say I'm not honored to be considered, but frankly I don't think I'm quite right for the job.

I'm with you there, I agreed.

Mainly for the reason that I don't think anybody is. My ethical search engine can do a better job than any one person in arriving at ethical answers. There's no one person I would trust more—not even myself—than I would my Ethical Answers Search Engine, or EASE.

Cheryl spelled it out for me, and asked me if I got it, and I gave her a nod to show her that I wasn't a total idiot.

I had to hand it to Marcus, she said. EASE is pretty clever, and Marcus couldn't have looked more pleased with himself. EASE sounded to me like the kind of VOOM that might get Marcus a nice fat book advance. He was just beaming, and you couldn't really blame him, since Plato didn't seem to have a reply ready for him. He continued to sit there, staring down at his folded hands. I took this as a sign that the conversation was finally over, and it was time to reassert myself, so I stood up, and Marcus did, too, but Plato just continued to sit there.

Plato, I said, gently, but in a voice that would brook no dissent, just the way I do with my kids, it's time to get going. You're not here to tête-à-tête with me and Marcus. You're here to sell books.

Indeed, he said, though he just continued to sit there. I started gathering up all my stuff in a very no-nonsense way, then he said, in that soft way of his, I have just one more question I would like to ask about EASE.

Okay, I said, but we simply have to start walking. You can ask your question while you walk, can't you?

Of course, he said. In fact, I highly recommend walking while thinking. The grounds of the Academy are lined with pathways, and I always encourage ambulatory cerebration.

Okay, I said. Then let's get this ambulatory cerebration on the road.

So, Plato said the moment we got outside, the contingent of Googlers trailing after us, I have a problem with getting all ethical answers with EASE.

I thought you would, said Marcus, grinning.

You have equipped EASE, just like Google's search engine, with a preferential ordering system, have you not?

Yes, Marcus said, just like with Google's search engine.

Well, there are substantive philosophical presuppositions that are slipped into that ordering system. For example, EASE is presupposing that the lives most worth the living are lives that both bestow a sense of worthiness to those who are living them and also, simultaneously, command admiration from others.

No, said Marcus. I didn't say that. I just said that we'd give heavy weights to the opinions of those who rate their own lives the most satisfactory and also were judged by others as living well. I didn't say that such people were necessarily living the most worthwhile lives.

Well, it seems to me hard to avoid that conclusion. How can we preferentially weight their opinions about the worthwhile life without presupposing some such substantive claim? What you're presupposing is that if enough other people think you're living a worthwhile life, then you *are* living a worthwhile life, with the opinions of those other people counting all the more if still *other* people think that *those* people are living a worthwhile life. That claim is implicit in your weighting system. You're building in the claim that worthwhile lives are like desirable search results. Perhaps that is so. Perhaps it is not so. Either way, it is an ethical claim implicit in the preferential ordering you've built into EASE.

So you're saying that in programming EASE in this way *I'm* acting as the moral expert? Marcus asked, still grinning. You're saying that I am, after all, taking on the job of philosopher-king?

Precisely, said Plato.

But wait a minute, I said. Why do you have to order the various opinions like that, weighting some opinions more heavily than others? That doesn't even seem fair to me. Why don't you just make it completely democratic, since everybody has just as much right to their opinion of what makes life worthwhile as everybody else? So then it would just be a matter of counting up votes, nobody's vote counting more. I think that's how you should do it with EASE.

No, that's not going to work, Marcus said. Plato's just going to come right back at you and say that what you've just said, that everybody has just as much right to their opinion of what makes life worthwhile as anybody else, is itself an ethical statement. You've even put it in ethical terms, talking about fairness.

I turned to Plato and asked him if Marcus was right, if that's what he'd say. And he smiled and asked me, What do *you* think? Do you think that is what I would say? And I had to admit that yeah, that's what he'd say because yeah, saying that everybody should be counted equally was the kind of thing that EASE would first have to tell us so that we could program EASE to use it in order for EASE to tell us anything.

So it's a chicken-and-egg sort of thing, I said to Cheryl.

Right, she said. Chicken-and-egg is exactly what it is. And I guess since I'd already gone on record that I think a whole lot of people don't know how to live their lives, I don't know if I'd really want EASE to count everyone the same. I'd just said that because it seemed it might be a way of getting around Plato's objection to Marcus's way of programming the thing.

Then Marcus said, While you're at it, Cheryl, you can ask Plato whether he's going to say that any algorithm you use for extracting an ethical answer from crowd-sourcing is going to have ethical presuppositions built into it. So I turned to Plato and he raised his eyebrows, and he didn't even have to ask me his question out loud about what did *I* think he'd say, and I just said, yeah, that's what you'd say.

So nothing was settled, I said to her.

That's an understatement, she said. Everything was *un*settled, most of all me. On the way over to Plato's event, instead of worrying about hustling my author I just kept thinking about how everything had been left open in a way that just really galls me. It's like when I open the refrigerator door, and I see that all the tops on the jars haven't been screwed on, the mayo and the mustard and the pickles and the milk with their lids just carelessly perched on their tops, ready to slide off, which is this really annoying habit that Michael has. The toothpaste, too. He's just incapable of screwing the tops on things. I wanted to shout at Plato: Will you screw the frigging tops back on!

Cheryl said this loudly enough so that our waiter came over at a pace that counted in that establishment as an Olympic sprint. Cheryl waved him away.

I mean, I liked Marcus's idea of crowd-sourcing, she said, since the alternative seems so ridiculous to the point of revolting.

You mean about there being moral experts, I said.

Right, she said, that there are moral experts who, knowing stuff that regular people don't know, can straighten out people's twisted lives the way Dr. Kolodny can straighten twisted teeth. That's totally ridiculous.

I agree, I said.

And to add insult to injury, she said, there's the ridiculous idea that somehow these moral experts magically transmitted their knowledge into my mind, without my knowing how it got there, so I can just reach for it, like in the aisle labeled "right and wrong." Right, tell me another. Or maybe what I really mean is, I want another, she said with a laugh, her bangles doing their song-and-dance as she raised her arm to summon back our waiter, while I sat there pondering.

It wasn't the nature of moral truth I was pondering, but whether my control freak of a friend was drunk. Yes, she was definitely drunk. Her next words confirmed it.

I'm sorry for being late, Rhonda, she said. See, what happened is that after I left Plato I felt like I just needed to sit in my car and, you know, think. I guess I kind of lost track of the time.

Cheryl stared at me, and I stared back at her. Cheryl never loses track of time and Cheryl never apologizes.

There's another alternative, I finally said to her.

To what? she said.

To this dilemma that Plato forced you into. It's not just a choice between whether the crowd knows or the experts know. It's also possible that nobody knows, and maybe that's because there's nothing *to* know. Everybody just makes it up as they go along, mostly in the way that will make them feel better about themselves.

I was thinking about that, too, while I was sitting there in the car, Cheryl said. I was thinking about that a lot.

So maybe there's your answer, I said.

Do you think so? she said. I don't, and I'll tell you why. I'll tell you where all of my thinking in the car led to. It led me to an author I had back in January. This author really got to me, maybe because she was one of my authors without a VOOM. She just had her life.

The waiter brought Cheryl her drink and she took a long sip before she went on.

She was about our age, Rhonda, maybe a little younger. It was hard to tell because of the life she'd had. She'd written a memoir, but part of that memoir was written on her face. I actually read her book after I'd spent the day with her. It was pretty harrowing. Her stepfather had abused her and her younger sister starting when they were little kids. The guy, who was very well off and well educated, had gone after the two of them starting when they were six and four. As bad as it was for her, she found it that much worse, even when she was a kid, that it was happening to her little sister.

Was there a mother around? I asked her.

There was. The way she described it, the mother sort of knew, but kept herself from really knowing, meaning she didn't want to know.

She'd lived kind of hardscrabble, and was just tired out and wanted the security this guy brought, and, long story short, it was in her self-interest not to know, though on some level she had to.

She was in a state of denial, I said.

Call it that, but I don't know if it gets her off the hook for being his accomplice in crime or not. And the guy was some sort of sicko, which may mean he's less responsible for his actions than otherwise, so maybe that makes her even more evil than he is. Or maybe it doesn't. I don't know how we're supposed to think about these sickos. Are they just sick, leave it at that, or are they evil, or are they sick *and* evil?

Ask Marcus's EASE, I said.

What I really wanted to do was go back and ask Plato, Cheryl said.[*]

So what happened to the kids? I asked her.

What do you think happened? she said. They grew up to have miserable lives. Talk about lives not worth living. The woman who wrote the memoir spent most of her life battling with various addictions, including alcohol, cocaine, crystal meth, you name it. She was unable to hold down a job, even though she's really, really bright. She even lived on the streets for a while, which she writes about. Then she got it together, which is a story in itself. You should really read the book, Rhonda. I thought Oprah would choose it for sure, but that hasn't happened, at least not yet. Anyway, compared to her sister she's a raving success story. The sister's the one who really got screwed up. She's a poet and a musician, who writes her own music, but she's just so entirely messed up that she has multiple personalities and tried to kill herself a few times. If she ever gets it together enough to write a book, it would be a real best seller. No way Oprah would stay away from that one.

---

[*] Plato gave Cheryl an answer in the *Timaeus*. He not only ascribes an organic cause to much mental illness, locating them in disorders of "the marrow" in the head (the brain) which links to marrow encased in the bones throughout the body (his way of explaining, without knowing about nerves, how communication takes place between the brain and the rest of the body); but he goes further and says that when mental disorders have genuine organic etiologies, such that the person's will is rendered inoperative, then the person cannot be regarded as evil. "And if the seed of a man's marrow grows to be overflowing abundance like a tree that bears an inordinately plentiful quantity of fruit, he is in for a long series of bursts of pain, or of pleasures, in the area of his desires and their fruition. These severe pleasures and pains drive him mad for the greater part of his life, and though his body has made his soul diseased and witless, people will think of him not as sick but as willfully evil" (86c–d). This is an extraordinary passage, pointing the way toward neuroscience. Plato gets the physiology wrong, of course, but he does get that it is neurophysiology that is determining the aberrant behavior, and he draws humane conclusions with which it would take psychiatry millennia to catch up. What a lot of tormenting of the mentally ill could have been avoided if only people had paid as much attention to this passage of the *Timaeus* as Galileo paid to the passages that inspired him to the new mathematical sciences.

Tragic, I said.

I think Oprah is great for books, Cheryl said.

No, I meant about those two little abused sisters.

Tragic is an understatement, Cheryl said. And that's the point. That sicko of a father completely ruined the lives of two innocent little girls just for his own selfish pleasure. I mean, even granted his brain is wired wrong, he could have resisted.[*] And if he couldn't, then he should have just cut off his own dick. *That* would be moral.

Maybe, I said. Hard to say.

Not for me, she said. I don't find it the least bit hard to say. And then there's the mother, who's maybe even more immoral, since her brain *isn't* wired wrong. How could a mother with a normal-working brain not have done everything to protect those little kids? Well, obviously she was capable of not protecting her kids, since that's what she did, namely not protect them. But she shouldn't have been capable of not protecting them. It almost seems like you could prove it with numbers that she shouldn't have been able to, if you add up the total misery of the situation. That's how she should have been thinking. She should have thought, okay, my life is easier if I pretend that what's happening isn't happening, but there are two other people, who just happen to be my kids, whose lives are going to be forever wrecked. That's how I figure it.

But that's not how people think about what to do, I said. Their own misfortunes loom way larger, since that's the misery they're going to have to experience in reality instead of just by imagining it in their heads.

Well, maybe they *should* think that way, Cheryl said. Maybe when they don't, then their brains aren't working right either, and maybe a guy like Plato could prove that to them.

I don't know, I said. Even if you're right and Plato could prove something or other, I don't know what difference it would make.

What are you *talking* about? Cheryl said. Of course, it would make a difference! That would mean we're not just making it up as we go along. That would prove that it's my author's stepfather and her mother

---

[*] For Plato, in the relevant passage of the *Timaeus* (86b–87b), this is the essential question. He suggests there that truly evil behavior is sufficient grounds for judging a person mentally unsound for organic reasons, though "bad education" plays a subsidiary role as well. "But it is not right to reproach people for them, for no one is willfully evil. A man becomes evil as a result of one or another corrupt condition of his body and an uneducated upbringing. No one who incurs these pernicious conditions would will to have them" (86d–e).

who are making it up as they go along. Making it up as they go along is what the scum of the earth do.

Making it up as we go along is what all of us do, to some extent or other, I said.

Yeah, well, I've got news for you. Not *everything* is made up. Nobody's making up the fact that those two kids suffered, and they're still suffering and probably always will. I mean, they're not even going to get the opportunity to figure out what's the best life for them to live, they're just so wrecked and through no fault of their own. So whose fault is that? You can't tell me those parents are blameless.

But look, I said, just using words like "fault" and "blame" and "blameless" you're already coming at the situation fully loaded with all sorts of assumptions. Where'd you get them? Not from Marcus's EASE.

What's wrong with my coming at them all loaded up with assumptions? she asked me. Wasn't that Plato's point, that Marcus wasn't going to be able to get everything he needed out of EASE? But just because we can't get them out of EASE doesn't mean we don't have them. Plato's point was that we can't just develop some technology that would . . . I don't know quite how to put it, she said.

Relieve us of our capacity for individual judgment? I said.

Right, exactly, she said. Like you just said. There's no app coming out of the Googleplex that's going to relieve us of our capacity for individual judgment.

Too bad, I said, considering how hard individual judgment is.

Not for me, she said. I don't have any trouble at all judging.

There's a difference between judging other people and judging yourself, I said. So far, you've only demonstrated how easy it is for us to judge other people.

Just what are you trying to insinuate? she asked. Are you saying I don't apply the same standards to myself that I apply to everybody else?

Nobody does, I said. Everybody makes excuses for themselves they wouldn't be prepared to make for other people. The extenuating circumstances are just so obvious in our own cases.

I think that what you're doing, Rhonda, is accusing the whole world of being hypocritical. Which frankly, Rhonda, I'm just a little bit surprised to hear you say and which might be revealing more about yourself than you intend to.

Cheryl folded her arms over her chest and narrowed her eyes in an appraising stare. Was she trying to determine whether that tête of mine was unworthy of any more tête-à-tête outings?

All I'm saying, Cheryl, is that it's a whole lot easier to be objective when it comes to other people's behavior. It's easy to sit here and wonder how that mother of your author isn't able to see her kids and herself as clearly as we do. But it's entirely different when it's your own life. I'm just not as confident as you are that I see myself with the same objectivity that I can focus on others.

Well, thank goodness I am, Cheryl said. And if that mother were to say to me that it's better for her to look the other way because otherwise she's going to have to divorce that perv and go out on her own and work to support herself and her two little kids instead of having him to cushion her existence, then I'd say to her, What the hell are you talking about? Do you have any idea what the effects of child abuse are? You're telling me that the financially pinched life of one person outweighs the suicidal misery of another? Will you just stop and think about that for a moment, lady? Get out of your fog and think!

Okay, I said, let's say you could even prove that, yes, she really ought to protect her kids no matter that it makes her life harder, and you go to the mother with your proof. Do you actually think that would get her to act differently? Do you actually think that you could whip her into shape with some flimsy proof?

I don't know, she said. Maybe. There used to be things that everybody thought were okay, and then just about everybody changed their minds about them, and could see that they were flat-out wrong. Maybe it's because there was some kind of proof that someone discovered.

Like slavery? I asked her.

Right, she said. Slavery. That's a perfect example. Even the Bible thinks slavery is okay, no qualms at all, so long as it's the right people that you enslave, but now we know it's not. I mean, we *know*. *I* know, who never really gave it any thought. Plato might have said "brava" to me as if I'd discovered the cure for cancer, but I didn't do anything special to deserve any praise for knowing slavery is wrong. It's not open to discussion anymore. So how did that happen? How did a person like me get so smart?

Cheryl paused as if she was maybe waiting for me to answer her.

I think we, uh, fought a war about it, I said. And the guys who thought slavery was wrong won the war?

No, that's not right at all, Rhonda. The guys who thought slavery was wrong were right, and the guys who thought it was okay were wrong. It wasn't winning the Civil War that made the difference of who was right and who was wrong. You know it's wrong, I know it's wrong, just about everybody knows it's wrong. Probably the same people who don't know slavery is wrong are the same people who don't know the earth is round, or people whose brains are just all wrong, like that stepfather's brain. So how did we all get this knowledge? I'm thinking it was some kind of proof. I mean, who wouldn't want to own a slave? I sure would. Even just one would make all the difference in the world. Wouldn't you want a slave?

So you think that Plato knows that slavery is wrong? I asked her.

Of course he knows! she said, as if I'd questioned whether the pope was Catholic or Donald Trump was rich. I mean, if you and I know it's wrong, then someone like Plato certainly knows it.*

I guess, I said, though don't forget how surprised he was to see men and women treated so equally. You said yourself that he's really old-school.

Look, Rhonda, she said, there's old-school, and then there's old-school. A guy would have to have been living in a cave not to know that slavery is wrong.

Still, they used to not know, I said, as you keep stressing. So it's not obvious.

Yeah, right, she said, but that was exactly Plato's point. If it's not obvious, and if there are so many self-serving reasons for not seeing it, then it's going to take a really good argument to break through all the resistance. You need these super-arguers, which believe me Plato

---

*Cheryl is giving the historical Plato too much credit, though, of course, the Plato whom she encounters on his book tour, so eager to catch up on all the progress, scientific as well as ethical, that has been made in the last 2,400 years, would presumably soon understand that slavery is wrong. John Locke is sometimes credited with being the first philosopher to consistently argue against slavery, but neither is he altogether consistent nor was he the first. Orlando Patterson, in a private communication to me, writes: "The philosopher who traditionally gets pride of place for voicing the first strongly anti-slavery views is Montesquieu in Book 15 of *The Spirit of the Laws*. He did clearly state that it was evil, but his subsequent discussion explaining why slavery persisted rather complicates the matter. It is not entirely clear whether Montesquieu was merely summarizing the views traditionally given to justify slavery or was advocating a kind of pragmatic defense of the institution. If the latter, then pride of place should go to Bodin who, writing over 180 years earlier (*Six Books of the Commonwealth*, 1576), condemned slavery in terms much harsher than Montesquieu, or anyone else for that matter, before the Quaker abolitionists of the mid 18th century. What's more, Bodin argued forcefully that the presence of slavery weakened the authority of the King. He was certainly far more consistent than Locke and I'm personally prepared to grant him pride of place among early modern philosophers."

is. You need people who are thinking of these arguments all the time because that's what they do for a living.

Yeah, well, super-arguers can argue for really immoral things, too.

True, she said, but if you've got all these super-arguers going after each other's arguments in this professional way, then they're going to be able to eventually find the mistakes. That's what these people are trained to do. You just have to let them loose at each other.

So then why did it take so long for them to discover slavery is wrong? If it was even they who discovered it. Because, frankly, I think the slaves themselves may have had some good arguments to make about why slavery is wrong. It's wrong to cut the slaves out of the picture here. It wasn't just the arguers arguing among themselves.

Yeah, but the slaves couldn't have done it on their own. You needed the arguers to argue that what the slaves had to say was even worthy of being heard. That was all part of the argument.

But you didn't answer my question. Why did it take so long, if those arguments were so good?

Maybe the arguments get better over time. Maybe at first it's all very tentative, and people attack the arguments and other people attack the attackers, and in the process the arguments improve until they just break through finally and nobody can deny them. Maybe the arguers have to argue themselves into believing their own arguments.

Arguments don't change a thing, I said. Nothing changes until feelings change.

But feelings don't change until something strong happens to make them change. It's like that mother of my author. It was working for her not to see what was happening to her kids, and her feelings were all working toward her not seeing.

So how do you know we aren't still like that mother? I asked her.

What are you talking about? Cheryl said.

Well, if we can look back and say that about other people in the past, that they were like that self-deceiving mother, how do we know we aren't any different about all sorts of things we feel perfectly okay about right now because it's in our interest to feel perfectly okay about them? Why should we be any different from people in the past?

Wait a minute, Cheryl said. You know what, Rhonda? You just asked a good question. I don't know what to say.

She sat there drumming those fingertips so they were clicking loudly in that silent interior. I noticed our harried waiter looking over at us, wondering what was up now.

Well, Rhonda, Cheryl finally said, you just got me straight to the point where I didn't want to go.

Where's that? I asked her.

What you just said, that other people are going to look back at us and ask how we could do the things they wouldn't think of doing. That's simply awful.

Why awful? I asked.

Why awful! You don't think it's awful that people who aren't even born yet are going to look back and wonder how we just didn't *see*?

They're going to know all sorts of stuff we don't know. They're going to be using all sorts of technology we can't begin to dream of, and seeing all sorts of stuff we can't begin to imagine.

No, Rhonda, you're comparing apples and oranges. They're not going to condemn us for not having the technology they do. The kind of things I'm talking about are things they're going to condemn us for not seeing. They're going to say we were too self-serving to see them, just like that mother of the abused kids. Doesn't that drive you to distraction?

No, not really. It's kind of what you'd expect if anything that you've been saying is true.

No, it isn't what you'd expect at all! What I expect is that I've been able to teach Valerie and Jason how they should live their lives, that I've taken care of all of that just the way I've taken care of, well, straightening their teeth.

By handing their teeth over to Dr. Kolodny.

Right, the best orthodontist in the San Francisco Bay Area.

That's been established, I said.

And that's about the only thing that has been established, she said. She was still drumming those clicking fingernails on the tabletop. They're deep purple and they match her lipstick.

What is so awful to contemplate, she went on after a pause, is that someday Valerie and Jason, or their kids or grandkids or great-grandkids, are going to look back at us and wonder how we just couldn't see how wrong it was.

It? What it? I asked her.

Well, how the hell do I know? she all but exploded. We're the ones who aren't seeing it yet! Some future guy running around in some crazy getup is going to make some argument that's going to sound completely bizarre and ivory-tower-ish, until *it* starts making a little bit of sense to a couple of others, and then to more others, until it seems so obvious that people won't need any argument at all. They're going to feel it in their marrow.

So I guess the crowd-sourcing thing really isn't going to work, if we're all waiting for some guy in a crazy getup to show us the error of our ways.

Yeah, she said, Marcus's EASE is total toast.

Too bad for Marcus, I said.

Oh him. Cheryl flicked her wrist in dismissal, sending her bangles a-jangling. He didn't seem too upset about any of this. He took it all in his stride, just like he did when he found out he wasn't going to be a philosopher-king. I think at the end of the day I was a lot more upset. I mean picture it, Rhonda. We're on this path, walking to Plato's Google event, and I stop walking, just stand there stock-still, and everybody stopped with me, the whole contingent, Plato and Marcus and all the Googlers who are trailing after us. It was just bizarre. I'm standing there on a walkway in the middle of the Googleplex, running late for an event with an author in a toga, talking about some crazy ethical search engine that doesn't even exist but is just the pipe dream of a software engineer in dreadlocks who's just turned down the nonexistent position of philosopher-king, and I'm feeling really upset because EASE can't give us the answers that it's supposed to. And it's all because of Plato, going on about how bad teeth are like a bad life and how we have to find the Dr. Kolodny who can straighten us out. I mean, I just can't explain it, Rhonda, the whole effect it had on me. I'm just standing there, as if we have all the time in the world, which we certainly don't. And, I mean, I'm the media escort!

What did Plato do? I asked her.

Oh, he was perfectly happy to just stand there next to me, biding his time. He said I walk like a free person and not a slave.*

---

* "Well, look at the man who has been knocking about in law courts and such places ever since he was a boy; and compare him with the man brought up in philosophy, in the life of a student. It is surely like comparing the upbringing of a slave with that of a free man. Because the one man always has what you mentioned just now—plenty of time. When he talks, he talks in peace and quiet, and his time is his own. It is so with us now: here we are beginning on our third new discussion; and he can do the same, if he is like us, and prefers the newcomer to the question in hand. It does not matter to such men whether they talk for a day or a year, if only they may hit upon that which is. But the other—the man of the law courts—is always in a hurry when he is talking; he has to speak with one eye on the clock" (*Theaetetus* 172d–e).

Again with the slaves, I said. It's a regular obsession with him. That and the crooked teeth. Do you have any idea what he was talking about?

Well, as a matter of fact I do, because I asked him. He said, a slave doesn't own his or her own time, and so he or she can always be known on the street by his or her rushing, but a free person can walk and talk at his or her own leisure, stopping when he or she will.

You've got to hand it to him, I said, for remembering to keep saying "his and her" and "he and she" to the point of its being downright awkward.

And the freest of all is the philosopher, Cheryl continued on, quoting Plato and not paying attention to me, who thinks so little of the ceaseless flow of time as to step out of it. This is why the philosopher often appears ridiculous in the practical affairs of life, because he or she has stepped out of the rush of time (*Theaetetus* 172c–173b). And then he said to me in just the sweetest way: Cheryl, I think that this is what has happened to you. Isn't that just the nicest thing for him to have said to me? And he used my name to address me directly. That was the first time he'd done that.

I didn't see why Cheryl was so bowled over by this, though maybe you had to have been there.

This author certainly did have an effect on you, I said.

You have no idea, she said.

Are you sure you're not a little sweet on him? I asked her.

Don't be ridiculous, Rhonda, she said, he's old enough to be my . . . well, I don't even know what he's old enough to be to me.

They ought to bottle that trick of stepping out of the flow of time and sell it at Bloomie's cosmetics counter, I teased her, but she was too distracted to react.

Anyway, when he mentioned the word "time," she continued, even though he seemed to be telling me that being out of time is a good thing, the word itself was like an alarm clock going off in my brain, and I got us moving again and finally delivered him to the auditorium, which, by the way, was hugely—and I mean *hugely*—crowded. They'd put him in their biggest venue, and still there was standing room only, which was gratifying, at least for me. I don't think Plato gave it a thought. And the Googlers seemed almost as hyper as when the fantasy writer George R. R. Martin came to speak to them. They'd made up T-shirts they were all wearing with Greek letters and showing two guys in togas, one with his finger pointing up and the other with his finger pointing down. I don't know whether one of them was supposed to be Plato or what. Neither of them looked the least bit like him. Anyway, they were all wearing these T-shirts, grinning like a bunch of idiots so that I hoped that Plato was finally getting the message that this wasn't really the place to go shopping around for a philosopher-king. Marcus took off his oversized Grateful Dead T-shirt to reveal that he'd been wearing the Greek T-shirt the whole time, with a grin that revealed the full extent of the job Dr. Kolodny has ahead of him. Right before they all surrounded my author and swooped him away from me, I asked him how we're supposed to get our answers if EASE can't give it to us.

So did he answer you? I finally had to ask her after a considerable pause, accompanied by the tapping of her fingernails.

Not really, she said. I'm not sure giving answers is in his bag of tricks. He seems to be more about messing with your mind so that you can't stop thinking about his questions. And if he thinks I can afford to keep stepping out of time like this, with my schedule, then, well, he's just way off.

Maybe that's why he needs slaves, I said. A slave would help if you're stepping out of time all the time. Kidding, Cheryl, I hastened to add, given the glare she directed at me.

Here's what he said to me, she finally said. He said, I do not say your questions cannot be answered, Cheryl. I only say they cannot be answered with ease.

Her voice had gone so quiet when she spoke the last line that the long silence just seemed a natural extension of it.

Was that "EASE," all capitals, or just lowercase "ease"? I finally asked her.

I'm not sure, she said. When all is said and done, I'm just not sure.

There was an odd look plastered on my friend's face. I couldn't read it at all. Maybe it was her timeless look, or maybe she was just plain plastered.

γ

IN THE SHADOW OF THE ACROPOLIS

Hippolocus begat me. I claim to be his son, and he sent me to Troy with strict instructions: Ever to excel (αἰὲν ἀριστεύειν), to do better than others, and to bring glory to your forebears, who indeed were very great. . . . This is my ancestry; this is the blood I am proud to inherit.

—*Iliad* 6.208

Socrates was part of the urban scene. *Yelp Ancient Athens* would have highlighted him under Street Theater. He performed daily without ever passing the hat (*Apology* 33a–b), a regular at the agora, which was the center of the city, the site of its commercial, political, and cultural life, crowded with the mix of the city's inhabitants—its many slaves, its non-citizen foreign residents or metics, and its male citizens, ranging from aristocrats to *thētes*, who were the common laborers,* and its free-women, though these last were mostly kept in their households and out of sight, especially if highborn. The agora was spread out beneath the sheer rock outcropping of the Acropolis on which the monuments to Athens' new-gained imperialist glory were displayed, including the pièce de résistance, the Parthenon, erected from the white Pentelic marble that radiates a golden light when struck by late afternoon sun.

The architectural splendors had arisen, phoenix-like, out of the ruins to which the older shrines of the Acropolis had been reduced in 480 B.C.E. by the invading Persians. During the second incursion of the Persians, the one that had brought mighty Xerxes himself into Greece

---

*A keystone of Athenian democracy was the abolition of any property qualifications for citizenship. There were some property classifications in Solon's original constitution, but they were allowed to lapse into disuse so far as the rights of citizenship went. See pp. 151–152 below for what substituted for property in determining Athenian citizenship.

to finish the job left undone by his father, Darius, who had died while making his follow-up war preparations, the Athenians strategically abandoned their city for the nearby island of Salamis.* In the battle of Salamis, their naval forces, into which the Athenians, persuaded by the statesman and general Themistocles† had poured their capital, defeated the massive forces of the invaders. Xerxes, perched on the gold-encrusted spectator's throne he had set up on shore to view his anticipated victory, watched the bulk of his navy drown in the narrow straits, as the Greeks, through skill and cunning, outflanked his heavier boats. Xerxes beat it back to Persia, leaving a substantial force behind to get on with conquering the Greeks. The final routing of the Persians came soon after, at Plataea in 479.‡

Athens hadn't, of course, vanquished the Persians all on its own. Other Greek states had joined in the Hellenic effort, with Sparta, whose military prowess on land was unsurpassed, sharing with Athens in the glory of driving the barbarians out of Hellas. But Sparta, at the close of the Persian Wars, had no wish to assume any ambitions that kept its citizen-soldiers away from Sparta. Its way of life, exclusively devoted to military values, was supported by an extensive helot population who did all the agricultural and other life-sustaining tasks and were always in danger of rebelling, making the Spartans disposed to stay put in

---

* It had been the Ionian states on the coast of Asia Minor, absorbed into the massive Persian empire, which had precipitated the war. When the Ionian states rebelled, Athens came to their rescue, and Darius resolved to teach all the Greeks a lesson. His first attack had come in 490 B.C.E. and he was defeated. Xerxes, whose empire by then reached from India in the east down to northern Africa in the south, had mounted a stupendously large force, intending to use his defeat of Greece to move into Europe. Herodotus, who wrote his account of the Persian Wars some fifty years later, estimated the Persian soldiers to have numbered more than a million, and the entire number of "barbarians," including those who fed and ministered to the troops, to have been well over five million. Modern scholars consider this a gross overestimate, estimating the fighting force to have been somewhere between 100,000 and 300,000, but the invading force was, without a doubt, daunting; and Herodotus' overestimation is a measure of just how extraordinary the Greeks had come to think of themselves for their defeat of the Persians.

† Themistocles was later subjected to ostracism, one of the safeguards of Athenian democracy by which they tried to prevent any one person from growing too influential, swaying the crowd until he assumes the power of a tyrant. Ostracism was put to the vote—the first vote was on whether to have an ostracism in any given year, and if the result was affirmative, the next vote was whom to ostracize. The ballots were broken pieces of pottery, *ostraka*. After ten years the ostracized individual could come back and resume his life as he'd left it. Themistocles, however, never returned and, ironically, ended his days in Persia.

‡ The Greeks also vanquished the Persians at the sea battle of Mycale, off the Ionian coast— traditionally on the very same day as the land battle of Plataea, though modern scholars are skeptical of this last piece of lore. In any case, the victory of Mycale probably quickly followed the news of the victory at Plataea.

Sparta.* The very single-mindedness of their militarism led to their isolationism.†

Athens was anything but isolationist. It had transformed its Delian League of allies, formed to keep the Ionian states from falling again under Persian tyranny, into tax-paying tributaries. More wealth flowed in from the rich silver mines discovered nearby in Laurium. These mines were worked by a multitude of slaves—according to scholars, upwards of twenty thousand, primarily captives of war. "The average life of a slave in the Athenian silver mines was less than a year. You entered the mine and never stood up again; you lived the rest of your life crouching, and soon you died," as the philosopher Alexander Nehamas described this underside of Athenian grandeur in an interview, summing up: "We don't admire the Greeks for their morality."‡

But you were not meant to dwell on hidden squalor and pitiless exploitation as you gazed up at the perfectly proportioned Parthenon, so massive and yet seeming to float in an idealized form of materiality. What you were meant to think of were the most glorious possibilities for human achievement. You were meant to think of *aretē*, the most daunting reaches of human excellence, of a sort to provoke other men's astonishment and put your name on many lips. You were meant to feel that here, in this most exceptional of all the city-states, among the most exceptional of all the peoples of the world, the Greek-speakers among the barbarians, the Ethos of the Extraordinary had been realized as nowhere else on earth. The ethos that undergirded the extraordinary strivings and achievements can be most briefly—and harshly—summarized as the view that the unexceptional life is not worth living. Ordinary lives don't matter as extraordinary lives do. And if that is true, it follows with the certainty of a geometrical proof that no people could be living lives that mattered more than the citizens who lived in the shadow of the Acropolis.

---

* The helots were ritualistically abused. For example, every autumn, during the coming-of-age ritual for Spartan young men known as the Crypteia (a word meaning secret, from which we get our word "cryptic"), the helots could be hunted down and slaughtered with impunity. They were conquered Laconians, traditionally said to have been the residents of nearby Messenia, and thus fellow Hellenes, though, as with everything having to do with the ethnicity of the ancient Greeks, there is a world of discordant scholarship.

† Sparta was, in its own way, as intent on achieving the extraordinary as Athens, but its conception of the extraordinary was rigorously collective rather than individualistic. This collective aspect of Sparta's notion of virtue invited Plato's respect. There was, I'd argue, a conflict within Plato between Athenian individualism and Spartan collectivism. This conflict will surface in chapter δ, under the prodding of co-panelists Sophie Zee and Mitzi Munitz.

‡ *Bomb Magazine* 65 (Fall 1998): 36–41.

The Greek word *ethos* means "habit" or "custom," and there is a kind of paradox in demanding that the extraordinary become habitual, customary. And yet this desideratum was an active force in the Athens that carried out its life beneath the Acropolis. It was a distinct aspect of the normative culture—the values-culture—not only of Athens but of the Greek-speaking peoples of the other *poleis* as well, being implicit in the Homeric works that were integral to Panhellenic culture (though notions of collective versus individual extraordinariness vied, the one associated more with Sparta, the other with Athens). The desideratum provided the context for the famous statement that Plato has his Socrates make in his *Apology,* that for a human being, the unexamined life is not worth living. This statement, which Plato has Socrates make after the Athenians had already voted him guilty of defying the norms of their city, represents a radical revision of the Ethos of the Extraordinary. "Unexamined" is substituted for "unexceptional." Only one kind of extraordinary will do. The revision could not have been better designed to unsettle and enrage his fellow citizens. You'll remember the trial didn't end well for Socrates. But the revision is not so radical as to depart from the ethos altogether. That's what made it so intolerably unsettling for his contemporaries, especially in the circumstances in which they found themselves in 399. This will be further explored in chapter ζ. For now, I just want to assert that Plato's Socrates was being very Greek when he issued his harsh judgment regarding what lives aren't worth living. Some intellectuals are apt to be pious about the statement, but when one examines it, it turns out to be, like so many piously regarded propositions, quite heinous.

The Ethos of the Extraordinary was in the making long before Socrates came on the scene to stir things up. Its dim beginnings can be found in the Homeric age. The *Iliad*'s poet gives voice to it most especially in the character of the petulant, self-absorbed boy who spends the bulk of the epic's action sulking in his tent. This, of course, is Achilles, son of a mortal father and an immortal mother.

Achilles is regarded as Greece's greatest mytho-historical hero well into the classical age in which Plato wrote. Plato has Socrates explain the choice that he makes when he is on trial for his life by comparing it to the choice of Achilles (*Apology* 28c–d). And what is this choice? Homer presents it as the choice between a short but extraordinary life or a long and ordinary one. In Book 9 of the *Iliad,* Achilles explains, "Mother tells me, the immortal goddess Thetis with her glistening feet, that two fates bear me on to the day of death. If I hold out here and I lay siege to Troy, my journey home is gone, but my glory (*kleos*) never dies. If I voyage back to the fatherland I love, my pride, my glory (*kleos*) dies."*

Here, in the choice of Achilles, the pre-Socratic version of the Ethos of the Extraordinary is laid out in the starkest of terms. Achilles is the greatest of the heroes, the "best of the Achaeans," not only by virtue of his physical speed and his prowess in war and his godlike beauty, and his metaphysically mixed parentage, even though these characteristics all mark him out as extraordinary. These traits, though necessary for any figure the Greeks would celebrate as the greatest of all the legendary heroes, are not sufficient. The item on his curriculum vitae that puts him over the top is the choice he made: the short but extraordinary life. And the proof of his most extraordinary life is the very work devoted to him, the *Iliad,* also called *The Song (Kleos) of Achilles.*

Kleos means both "glory" or "fame" and also "the song that ensures that glory or fame." The noun is cognate with the Homeric verb *kluō,* meaning "I hear." *Kleos* is sometimes translated as "acoustic renown"— the spreading renown you get from people talking about your exploits. It's a bit like having a large Twitter following. In the Homeric version of the Ethos of the Extraordinary—the pre-Socratic/pre-Platonic version of the Ethos of the Extraordinary—to live a life worth living was to live a *kleos*-worthy life, a song-worthy life. Being sung, having one's life spoken about, your story vivid in others' heads, is what gives your life an added substance. It's almost as if, in being vividly apprehended by others, you're living simultaneously in their representations of you, acquiring additional lives to add to your meager one.

After all, here's the predicament, no less true for us than for those who gathered round to hear the ancient bards. Your human life is a confounding smallness, bounded at either end by the everlastingness of time that has been emptied of yourself.† All of that time in which

---

* Unlike Homer, Plato's Socrates doesn't measure the difference between the ordinary and the extraordinary in terms of *kleos,* that is, the acclaim of others. Rather the difference, for him, between the ordinary and extraordinary is measured in moral terms (*Apology* 28b–c). This change in the measure of the extraordinary is of the essence of Socrates' and Plato's departure from Athenian norms. As has been remarked, Myles Burnyeat, among others, has argued that Socrates was guilty of the crime with which the Athenians charged him—namely his rejecting their values in favor of new ones. And we can see here, in Socrates' both embracing Achilles' choice of the extraordinary and rejecting the Homeric interpretation of what counts as extraordinary, exactly where the departure from Athenian norms resides.

† The Greek conception of the afterlife, such as it was, wasn't designed to offer solace. One of the most searing moments in the *Odyssey* occurs in Book 11, when Odysseus, traveling to

you are not seems a kind of vanquishing of your person, which is, in comparison, such a nothing, even as it is, at least to you, the everything.

This situation can be taken quite to heart. I'm reminded of a neighbor I once had, a pediatrician by training. One day he stood on my front porch and told me how anguished he used to feel when contemplating all of the time in which he won't exist, until the moment that he realized—and his tone was one of revelation—that, after such a time as he no longer is, there won't *be* any more time. My emphatic addition of the words "for you"—as in, "there won't be any more time *for you*"— were just as repeatedly and emphatically resisted. No, he kept rejoining, there simply won't be any more time, full stop. Since he was a doctor whose patients were young children, and also a devoted parent, I wondered whether he had thought through the consequences of the vast annihilation he insisted would succeed him. I asked him whether the cosmic implications of his not existing went backward in time as well: was there a history that preceded him? No, I was told. Time exists just so long as he does. I've often thought back to that conversation we had on my front porch and wondered what was going on in my neighbor's mind. What was he really thinking when he kept repeating those words, "no more time"? Did he mean them as literally as he kept insisting he did? It was at least thirty years ago, and I still find myself wondering what exactly he was thinking. But I do know that his anguish at conceiving time-minus-him must have been quite intense to make him think so hard to reach a conclusion so far-fetched and against which he would entertain no possible objections.

My neighbor might have been unusual in the conclusions he drew to paliate his angst; but he wasn't unusual in his angst. It's out of such existential thrashings that the great normative ferment of the ancient world was born, not only in Greece but across wide stretches of the globe. In *The Way to Wisdom*, the philosopher Karl Jaspers describes the extraordinary conceptualizing that appears to have gripped the ancient world from about 800 to 200 B.C.E. This time period spawned alternative normative viewpoints, both spiritual and secular, offering possible solutions to the kind of question that tried my neighbor's soul. Many

---

the underworld, meets various shades, among them the most heroic of them all, Achilles. Odysseus greets him with reverence, "blessed in life, blessed in death," but Achilles quickly disabuses Odysseus of this mortal illusion, telling the living man that he would rather be enslaved to the worst of masters than be king over all the dead. The moral to be drawn is that whatever solace there is for human limitations must be sought in our earthly lives.

of these visions persist intact into our day, standing ready to contain and shape our confoundment so that most people, unlike my neighbor, don't need to try to think it all out from scratch for themselves.

The Greek-speakers were part of this normative ferment. Their contributions are almost always represented by their great thinkers—by Pythagoras and Plato, by Aeschylus and Aristotle—but this is not doing justice to the pre-philosophical ethos out of which the great names emerged and which oriented the secular strain of Greek normative reflection. It wasn't only the philosophers and tragedians whose names we now can recall that confronted the question of what can be done to amplify the thin sliver of the one small life—your own!—amidst all the time that will not know who you were, how you loved and hated and succeeded and failed and feared and longed and won and lost. The concern with doing something to rescue yourself from being blotted out by all that vastness of unknowing and uncaring time was felt long before Socrates and Plato arrived to challenge the ethos that had formed itself around it. What can people do to withstand time's drowning out the fact that they once had been? The Ethos of the Extraordinary answered that all that a person can do is to enlarge that life by the only means we have, striving to make of it a thing worth the telling, a thing that will have an impact on other minds, so that, being replicated there, it will take on a moreness. *Kleos.* Live so that others will hear of you. Paltry as it is, it's the only way we have to beat back uncaring time.*

Our own culture of Facebook's Likes and Twitter followings should put us in a good position to sympathize with an insistence on the social aspect of life-worthiness. Perhaps it's a natural direction toward which a culture will drift, once the religious answers lose their grip.

---

* Plato has Socrates, himself quoting the priestess Diotima, explain the obsessive love for *kleos* in pretty much these same existential terms, connecting it with a desire to defeat death: "In every way, this zeal, this love, is in pursuit of immortality. . . . When, if you will, you look at people's fondness for fame, you may be surprised by their irrationality, unless you keep in mind what I have said and consider how terribly inflamed they are by a love of becoming a famous name and 'laying down immortal glory for eternal time,' and how they're ready to face any danger for this—even more than for the sake of their children, to squander their money, to endure any pain, and even to die. So you think . . . that Alcestis would have given her life for Admetus, or that Achilles would have sought out death after Patroclus died, or you own Codrus would have sought to be the first to die for the sake of his sons' kingdom, unless they thought there would be the immortal memory of their *arête* that we now have? That's a long way from being the case. On the contrary, I believe all these people engage in these famous deeds in order to gain immortal virtue and a glorious reputation, and the better people they are, the more they do so, because they love immortality" (*Symposium* 208c–e). Plato isn't himself endorsing, as Diotima does, this obsessive pursuit of *kleos*. In fact, he introduces the above quote by writing, "Like a perfect sophist, she said . . ."

The ancient Greeks lived before the monotheistic solution took hold of Western culture, and we—or a great many of us—live after. A major difference between our two cultures is that, for the ancient Greeks, who lacked our social media, the only way to achieve such mass duplication of the details of one's life in the apprehension of others was to do something wondrously worth the telling. Our wondrous technologies might just save us all the personal bother. *Kleos* is a tweet away.

The Greek-speakers were not unique in their existential pondering. At roughly the same time that they were developing their own approach to the problem of human mattering, there were other peoples with other distinctive approaches to the same existential concerns. In particular, across the Mediterranean from ancient Athens there was a constellation of tribes also working out a vision that has continued to flourish in our day, in all the multiple iterations and variations offered by the Abrahamic religions. They called themselves the *Ivrim,* from the Hebrew word for "over," indicating their location on the other side of the Jordan River. Their sense of separateness was central to their sense of themselves, and it resulted in their keeping so low a profile that not even Herodotus—who, an early ethnographer, was fascinated by the many belief systems flourishing in his day—gave any indication of knowing of their existence. But there they were, slowly evolving a worldview that eventually transformed one of their regional gods into the singularly awesome Yahweh, a transcendence providing metaphysical grounding for both physical and moral reality, his unquestionable will establishing the rules by which we were to live. This god is sometimes described by the Hebrews as a jealous god, but unlike the Greek gods, he never deigns to be jealous of his devotees, but only of "other gods." He is too remote for comparisons between himself and humans even to be entertained. He inhabits a sphere of holiness incomprehensible to the human point of view, an inhuman purity so alien that it's fraught with mortal dangers. He is so unlike us that the very notion of a graven image is an affront to his singularity. Even his name holds fearful possibilities for humans; laws will be enacted as to who can utter the one true Name and under what elaborate circumstances arranged with intricate precautions.[*] In fact, the common way in Orthodox Juda-

---

[*] Known as the Tetragrammaton, because of the four Hebrew letters composing it, the name was permitted, according to the Mishnah (*Berakhot* 9:5), for everyday greetings until at least 586 B.C.E., when the First Temple was destroyed. In time, its pronunciation was permitted only to the priestly caste of Kohanim, traditionally regarded as descendants of Aaron, brother

ism, still, for referring to him is Ha-Shem, the Name. And yet, from his position of remotest transcendence, he is engaged in human concerns and has intentions directed at us, his creations, who embody nothing less than his reasons for going to the trouble of creating the world ex nihilo. He takes us (almost) as seriously as we take us, thus settling, for believers, the question of our mattering

So there were the (pre-philosophical) Greeks, pondering ways of endowing their limp slivers of mortality with a stiffening coating of mattering, and there were the Hebrews, considering a similar project, though they ended up with quite a different approach. The Hebrews offered a transcendent answer in terms of a god; the Greeks offered a secular answer in terms of the possibilities for enlarging a life in strictly human terms; this Greek pre-philosophical answer was incorporated into, and refined by, secular philosophy; and Western culture has been wildly oscillating between these approaches—the Hebrew and the Greek—ever since.

And these Mediterranean peoples weren't the only ones provoked by an existential confoundment not altogether dissimilar to my former neighbor's. This was the period not only of the philosophy-creating Greeks, and of the major and minor prophets settled on the other side of the Jordan, but also of Confucius and Lao Tzu in China, of the Buddha and Jainism, the Upanishads and the Bhagavad-Gita in India, of Zoroastrianism in Persia. Pythagoras, Confucius, and the Buddha were contemporaries of one another. Jaspers calls this period the Axial Age *(Aschenzeit)* because, he affirms, all moral and religious thought ever since has revolved around it. "The new element in this age is that man everywhere became aware of being as a whole, of himself and his limits. He experienced the horror of the world and his own helplessness. He raised radical questions, approached the abyss in his drive for lib-

---

of Moses and first high priest, who would pronounce it in their public blessing of the people. After the death of the high priest Shimon the Righteous around 300 B.C.E. (Babylonian Talmud, Tractate *Yoma* 39b), the name was pronounced only by the high priest in the Holy of Holies on Yom Kippur (*Mishnah Sotah* 7:6; *Mishnah Tamid* 7:2). The sages then passed on the correct pronunciation of the name to their disciples only once (some say twice) every seven years (Babylonian Talmud, Tractate *Kiddushin* 71a). Finally, upon the destruction of the Second Temple in 70 C.E., the name was no longer pronounced at all, though on Yom Kippur, when the liturgy of the Temple is recalled, the congregation and its leader prostrate themselves in simply remembering, but not enunciating, its pronunciation. Since the vowels are omitted from the Tetragrammaton, Jews traditionally believe that the correct pronunciation is no longer known, that "Jehovah" is, most likely, not the correct pronunciation. Still, in the home in which I was raised, we were never allowed to say that there was a Jehovah's Witness at the door, but rather referred to the visitor as a Witness, just in case, so awesome was the power the word was felt to hold within itself.

eration and redemption. And in consciously apprehending his limits he set himself the highest aims. He experienced the absolute in the depth of selfhood and in the clarity of transcendence."[*]

I would put Jaspers's point this way: What arose as a major preoccupation during the Axial Age, provoking the emergence of powerful normative responses, is the question of what it is that makes a human life matter—if, in fact, it does. The possibility that it doesn't matter provides the psychic energy that went into these wide-ranging responses.

Here's another psychically charged question: If there is mattering, is it differentially distributed? Do some of us matter, while others of us don't? That's a galling proposition. I can better tolerate my life's not mattering if I'm sure that everyone else's doesn't ultimately matter either. But if there is an unequal distribution of mattering, are those who matter born into mattering or is mattering a state that must be achieved? And if it is to be achieved, then how are we to do it?

Such preoccupations are still our preoccupations, so it's not so surprising that the normative viewpoints that arose as responses to such preoccupations still resonate with us. Mattering is a state devoutly to be wished for. We no sooner know *that* we are, than we want that which we are to matter. In one of my earlier books I dubbed this the "will to matter" and also proposed something I dubbed "the mattering map" to explain how the will to matter functions within us.[†] I've been pleased to see the idea of the mattering map adapted to various explanatory purposes—even, to my delight, made use of in behavioral economics to explain the inadequacy of the rational-actor model.[‡] But nowadays, I'm most interested in the will to matter as it illuminates both the persistence of religion and the emergence in ancient Greece of secular philosophy. The will to matter is at least as important as the will to believe, if we're to understand the continuing force of the normative systems that emerged during the Axial Age.

Why did preoccupations with mattering emerge at just this period in history over large swaths of the world—from China and India and

[*] Karl Jaspers, *The Way to Wisdom* (New Haven: Yale University Press, 1954), p. 100.
[†] *The Mind-Body Problem* (1983; repr., New York: Penguin, 1993). Reissued in e-book form by Plymptom.com.
[‡] See, for example, G. Loewenstein and K. Moene, "On Mattering Maps," in *Understanding Choice, Explaining Behavior: Essays in Honour of Ole-Jørgen Skog,* ed. Jon Elster, Olav Gjelsvik, Aanund Hylland, and Karl Moene (Oslo, Norway: Oslo Academic Press, 2006). Reprinted as "How Mattering Maps Affect Behavior," *Harvard Business Review* (September 2009).

Persia, all around the Mediterranean, and into Europe? This is a question that lies far beyond the scope of this book. All that I would like to maintain here is that the pre-philosophical Greek ethos, out of which arose the genius of Greek philosophy and Greek tragedy, is of a piece with the wider existential activism of the Axial Age.

Nevertheless, I'll toss out some recent ideas that social scientists have proposed as possible hypotheses for the Axial Age's normative activity. The first thing that one notices is that the affected regions all saw the emergence of large societal formations, organized around urban centers. These polities all introduced a level of anonymity and impersonality into human life, so different from tribal village life, where all relations were determinately personal—oftentimes borderline incestuous. With such a thickness in human relations, existential ponderings may have been stifled.* Could the move toward larger polities have been a nudge in the direction of the kind of existential confoundment that I was fortunate enough to witness that day on my front porch?

But as social scientists have also pointed out, the emergence of larger polities can't offer a sufficient explanation in itself, since there are regions that saw large-scale societies without these same developments, for example Egypt. Some social scientists, most notably David Graeber, have pointed out that the core period of Jaspers's Axial Age corresponds almost exactly to the period and places in which coinage materialized, with minting overseen by governments which then used the wealth on military ventures that often resulted in large captures of people who were converted into slave labor—often sent to mine the ore that would be turned into coins. He calls this the "military-coinage-slavery complex." Here, too, the changes are in the direction of impersonalization: "To understand what had changed we have to look, again, at the particular kind of markets that were emerging at the beginning of the Axial Age: impersonal markets, born of war, in which it was possible to treat even neighbors as if they were strangers.† Perhaps

---

* Archaeologists say that belief in the supernatural—animistic spirits of nature lodged in animals, wind, trees, rivers, sun, moon—extend back at least 30,000 years to Cro Magnon man, whose cave paintings, requiring their gaining access to torturously inaccessible places, are interpreted as expressions of supernatural beliefs. Why else go to all that inconvenience? But I am here concerned with the quite different kind of preoccupations that motivated the normative responses of the Axial Age. I discuss the relation of the will to matter to religion at greater length in "Feminism, Religion, and Mattering," *Free Inquiry* 34, no. 1 (December 2013–January 2014).

† David Graeber, *Debt: The First 5,000 Years* (Brooklyn, NY: Melville House, 2011), p. 238.

(although this is not exactly Graeber's conclusion) the introduction of markets and of money—providing an impersonal measure of worth—intensified the impersonalization inherent in the emergence of the large polities, again coaxing forth poignantly existential questions.

Another line of approach derives from data revealing that all the regions affected by the Axial Age's normative ferment were unusually well fed:

> Studies indicate a sharp increase in energy capture (how much energy people extract from the environment) that occurred at the same time in three distinct regions of Eurasia, the Yellow–Yangzi rivers, the Ganga valley, and the eastern part of the Mediterranean. At the end of the first millennium B.C.E. these regions reached a production level (25,000 kcal per capita per day) that largely surpassed that of previous societies, which ranged from 4,000 kcal for hunter-gatherer societies to 15,000 kcal for states such as Egypt and Uruk.... This suggests a tentative scenario in which the spread of moral religions followed a sharp increase in the standard of living in some Eurasian populations.
>
> What would be the connection between these two developments? Empirical studies on the impact of economic development on individual preferences, in a variety of different cultural contexts, suggest that material prosperity allows people to detach themselves from material needs (food, protection, affiliation).[*]

In other words, once the basic material supports for life have been established you're free to start wondering what it all means. But of course this elevated "energy capture" would presumably occur just exactly in areas in which urbanization and the military-coinage-slavery condition were satisfied, so it's difficult to determine which is the causally significant factor.

Fortunately, sorting all this out isn't my problem to solve. The point I'd like to make is simply that what happened in the Greek city-states was part of something larger, a confrontation with existential dilemmas that involves a certain abstraction from the daily grind of life, an ability to remove yourself sufficiently from the midst of your own life in order to ask whether your brief sojourn here amounts to anything. Such musings don't necessarily start out on a plane of moral and spiri-

---

[*] Nicolas Baumard and Pascal Boyer, "Explaining Moral Religions," *Trends in Cognitive Sciences* 17, no. 6 (2013): 172–180.

tual sophistication and transcendence. It would be surprising if they did. They more than likely begin with questions put emphatically in the first person and accompanied with agitated emotion: Do *I* matter? So many have come before me, with no record of their having existed at all. Why shouldn't I assume that the exact same thing will happen to me? But that makes death—already a terrifying proposition—infinitely worse. The annihilation of death is so complete that it seems to spread its cold terror into life even as it's lived. These are powerfully unsettling thoughts and powerful responses emerged. Zoroastrianism, Confucianism, Taoism, Buddhism, Jainism, Hebraic monotheism: all had (and still have) widespread appeal because this starkly personal question was being asked, and is still being asked, by vast numbers of people, perhaps everyone with a filled belly and the relative safety and leisure to ponder and agonize.

Greek philosophy is always included among these great normative paradigms of the Axial Age. But also included should be the Greek Homeric ethos that was the precursor to Greek philosophy. Like the approaches we now call religious or spiritual, the Ethos of the Extraordinary emerged out of a confrontation with the brevity and impermanence of human life. It does nothing to deny or mitigate these conditions. The only solace it offers is itself brief and impermanent. *Kleos.* What is most startling about it is the clarity of its *non*-transcendence. It accepts the general indifference of the cosmos, and what solace it has to offer is given in strictly human terms: do something outstanding so as to attract the praise of others, whose existence is as brief and impermanent as your own. That's the best we can do to enlarge our lives:

And two things only
tend life's sweetest moment: when in the flower of wealth
a man enjoys both triumph and good fame.
Seek not to become Zeus.
       All is yours
If the allotment of these two gifts
Has fallen to you.
     Mortal thoughts
Befit a mortal man.[*]

---

[*] Pindar, "Isthmian V," for Phylakidas of Aigina, pankration, 478? B.C.E., lines 7–12, from *Pindar's Victory Songs*, trans. Frank J. Nisetich (Baltimore: Johns Hopkins University Press, 1980), p. 311.

This is from one of the epinician odes of Pindar, the greatest of Greece's lyrical poets. Pindar was born in the sixth century B.C.E., and his poems are seen as shedding light on the values that attended the transition from the archaic into the classical age. The epicinian odes were composed to honor the victors in the Panhellenic games—the Isthmian, Nemean, Pythian, and of course Olympian—and they are chock-full of expressions of the Ethos of the Extraordinary. (The word "epinician" derives from *epi,* upon, and *nikê,* victory. I have a Greek-American friend who named her daughter "Nike" and is often asked why she chose to name her offspring after a sneaker.) The greatest lyric poet of his day didn't disdain casting his immortal poetry at the feet of a grunting discus thrower or mesomorphic wrestler. Why should he, since to win glory at these games was to demonstrate your standing as outstanding? It doesn't matter that the achievement of the *kleos*-worthy life exposes that life to more dangers than an ordinary life:

> Great danger
> does not come upon
> the spineless man, and yet, if we must die,
> why squat in the shadows, coddling a bland
> old age, with no nobility, for nothing?*

The odes are fascinating for the dark thoughts and intimations of death that constantly wreathe themselves together with the laurels:

> But he who has achieved a new success
>   basks in the light,
>   soaring from hope to hope.
>   His deeds of prowess
> Let him pace the air,
>   while he conceives
> plans sweeter to him than wealth.
> But the delight of mortal men
> flowers,
> then flutters to the ground,
> shaken by a mere
>   shift of thought.

* Pindar, "Olympian I," for Hieron of Syracuse, race for single horse, 476 B.C.E., lines 81–85, in *Pindar's Victory Songs,* trans. Frank J. Nisetich, p. 84.

Creatures of a day!
What is someone?
   What is no one?
Man: a shadow's dream.
   But when god-given glory comes
A bright light shines upon us and our life is sweet.[*]

Though the gods are incessantly mentioned, this ethos presents a life worth living in terms that are drawn far more from the world of men. What is desired is *not* the attention of the immortals, but rather the attention of one's fellow mortals. The gods come prominently into the picture because they either promote or prevent this good—that is, the achievement that brings fame—from being attained, but the good itself isn't defined in terms of the gods. The good belongs to the world of mortals; it's their attention and acclaim one is after.

In fact, as most of the tales of the gods attest—going back to the *Iliad*—it's far better *not* to attract the immortals' attention, which more often than not results in disaster. Aeschylus' *Prometheus Bound* is a long meditation on just how undesirable it is to attract the attention of the gods. The play opens with a demonstration of how little humans count to Zeus. He despises us so much that he horribly punishes the Titan Prometheus for the love he showed the "creatures of a day" by having given them the gifts of fire and of hope. Without those two gifts the race of men would have perished, which was Zeus' genocidal wish. In retribution, the parvenu tyrant of a god—"tyrant" is used repeatedly to describe Zeus—has arranged that Prometheus be chained to formidably inaccessible cliffs, an eagle swooping down daily to treat his liver as so much foie gras. And then the tormented young Io comes wandering onto the stage, in the midst of her own torment. She can find no rest and is not only being driven across the globe but into raving madness, tormented by a swarm of biting insects. The poor child had done nothing to deserve her agony other than being so lovely as to invite Zeus' rape, and now his jealous wife, Hera, is wreaking her revenge on the girl.

The Chorus of Oceanides, who have come to pay a condolence call on the Titan, has this to say:

---

[*] Pindar, "Pythian 8," for Aristomenes of Aigina, wrestling, 446 B.C.E., lines 88–96, in *Pindar's Victory Songs*, trans. Frank J. Nisetich, p. 205.

For me, when love is level, fear is far
May none of all the gods that greater are
Eye me with his unshunnable regard;
For in that warfare victory is hard
And of that plenty cometh emptiness.

In other words, the last thing any mortal needs is the attention of a god.

I wouldn't for a moment suggest that Greek religion wasn't an over-whelmingly important presence throughout all the city-states. Religion came in many forms, both public and secret, both Panhellenic and specific to individual *poleis*. And like everything else having to do with the Greeks, the matter of their religious practices is both complicated and controversial; it's still keeping scholars busy. Again, none of these complexities and controversies are mine to sort out. All I wish to affirm is that, whatever psychological, intellectual, social, and political purposes Greek religion served—and it served all of these—it did not address in any profound way the existential concerns of the Axial Age. This is why what survives for us from the ancient Greeks isn't their polytheistic religion. There are no multitudes of people who revere Zeus or Apollo or Athena, as there are those who revere Jehovah, or seek their life's purpose in the teachings of the Buddha or Confucius or Lao Tzu. What survives for us from the Greeks is what their thinkers made of the secular approach to the existential dilemmas, a secular approach that was pre-philosophically implicit in the Ethos of the Extraordinary. Greek religion, per se, didn't minister to the will to matter, except insofar as helping to strengthen group identity within the *polis,* observing both its public religion and its mystery rites.

There is a big jump to get from the *kleos*-centered conception of a life worth living to the sort of reasoned life of Plato's vision, though both kinds do require enormous exertion. (The extraordinary doesn't come easily, unless it's physical beauty, which, being a form of the extraordinary, counted for a great deal to the Greeks.)[*] For Plato, the

---

[*]The only two characters called godlike in the *Iliad* are Achilles and Helen, both of them remarkable for their beauty, though Achilles was remarkable in other ways as well. Helen earns her divine status merely by virtue of her beauty, since it was sufficient to make her name live on forever. There's a poignant scene in the *Iliad,* when Helen remarks to Hector, the greatest of the Trojan heroes who will soon lose his life because of Helen's tragedy-inspiring beauty, that they all seemed to be living a story for generations still unborn, "so even for generations we will live in song" (*Iliad* 6:358). She says it mournfully, since there is misery to be experienced and that's never pleasant, but with a sense, too, of achievement. She will not pass into nothingness so long as she lives on in song and memory. *Kleos*-measured *aretē* is never to be equated with happiness.

kind of exertion that is required to attain the best of lives is philosophy itself. It's in the light of philosophy that one is to make oneself over. It's in the light of the truth, arduously attained, that one's inner being is transformed and true *aretē* attained, with the *kleos* conferred by the multitudes firmly filed away as beside the point.

Aretē is an important word in the history of philosophy, since, with the modifications to it that Socrates and Plato introduced, it comes close to what we mean by our word "virtue," which is the word that Benjamin Jowett substituted for *aretē* throughout his famous transla-tions of Plato's dialogues. The philosopher Alexander Nehamas, him-self a superb translator, mildly admonishes Jowett for his treatment of the term, pointing out that "virtue" doesn't quite capture all of the nuances of *aretē*. After all, the word is applied not only to people and their actions, but to things like horses and knives. Can one speak of the virtue of pinking shears? Nehamas also stresses the social aspect to the term *aretē:* the response of one's social circle is part of the very meaning of the term: "We could do no better, I suggest, than to think of it as that quality or set of qualities that makes something an *out-standing* member of the group to which it belongs. *Aretē* is the feature that accounts for something being *justifiably notable.* Both suggestions, which come to the same thing, involve three elements: the inner struc-ture and quality of things, their reputation, and the audience that is to appreciate them. And this is as it should be. From earliest times, the idea of *aretē* was intrinsically social, sometimes almost the equivalent of fame (*kleos*). That dimension of the term is clear in the Homeric epics, but it survives in the classical period as well."[*]

Our term "distinguished" comes a bit closer to this notion of *aretē*. When we call, say, a philosopher distinguished, we are not just saying that she is worthy and not just saying that she is recognized, but we are saying that she occupies the intersection of both—that she's recog-nized *and* worthy; even that she's recognized *because* she's worthy. In the case of *aretē*, the direction of the "because" can seem a little vaguer, so that it can sometimes seem almost as if someone is regarded as worthy *because* they are recognized. The recognition isn't just a mea-sure of the worthiness but the worthiness itself. (The contemporary analogue is found in those who see celebrity as an end in itself.) But in

[*]Alexander Nehamas, *The Art of Living: Socratic Reflections from Plato to Foucault* (Berkeley: University of California Press, 2000), p. 78.

any case, the social aspect, as Nehamas put it, is an intrinsic element of the notion.

And it is this social aspect of *aretē* that both Socrates (almost certainly) and Plato (certainly) would see gone from the notion. For Plato's Socrates, there is no contradiction in saying that someone has *aretē* even though it goes unrecognized and unappreciated. The acclaim of others is irrelevant. So, for example, Socrates says in the *Apology* that the city can do him no harm, even if their disapproval of him is so great that they sentence him to death, which is exactly what they did. They can't deprive him of what it is that makes a person truly worthy, the sort of outstandingness that he has earned from the kind of life that he has lived, even though his fellow citizens condemn that life as death-worthy. Plato has Socrates repeating such statements throughout the dialogues, assertions of the *kleos*-independence of *aretē*. Here is how the point is put in the *Gorgias*: "And yet I think it better, my good friend, that my lyre should be discordant and out of tune, and any chorus I might train, and that the majority of mankind should disagree with and oppose me, rather than that I, who am but one man, should be out of tune with and contradict myself" (482c). *Aretē*, in other words, has nothing to do with *kleos;* the kind of true distinction to be achieved has nothing to do with being regarded as distinguished.

There were some philosophical precedents for Socrates and Plato's philosophical deviation from the more general conceptions of outstandingness, though these didn't amount to the rejection of *kleos* as its measure, but rather a protest that philosophers ought to be accorded more *kleos*, as the sixth-century Xenophanes complained:

> If a man were to win a victory by speed of foot
>   or by performing the pentathlon, there where Zeus's
>   precinct is
> beside Pisa's stream at Olympia, or by wrestling
>   or by sustaining boxing's painful bouts
> or that terrible contest that they call the pankration,
>   he would, in that case, be more glorious for his
>   townsmen to gaze upon,
> and he would win the right to sit in the front row in full
>   view at assemblies,
>   and he would be given meals at public expense
> by the city, and a gift that would be for him as an
>   heirloom—

even if he won in the chariot race, all these things
  would fall to his lot,
though he would not be my equal in worth; for superior to
  the strength
  of men and of horses is the expertise that *I* lay claim to.
But thought on this point is very haphazard, and it is not
  right
  to give preference to strength over serviceable
  expertise.
For neither if the people should have a good boxer among
  them,
  nor a man good at the pentathlon or at wrestling,
nor yet again in speed of foot, which is most honored
  of all the deeds of strength which men perform in
  contests,
not for that reason would the city be any better
  governed.
  Small is the joy that a city would get from such a man
if he should be victorious in the games beside the banks of
  Pisa,
for not in this way are the city's storehouses fattened.[*]

But the Socratic/Platonic deviation from the more common use of
the word *aretē* is far more radical than Xenophanes would have it. It
entails a major revision of the dominant normative ethos—so radical
that Socrates could quite rightly be regarded as a heretic in regard to his
society's values. His use of the word *aretē* moves it far closer into the
range that we would call ethical. A number of writers before Socrates
had used the word and its associated vocabulary in ethical contexts, but
Socrates was probably the first to identify *aretē* with what is—in terms
of a person's moral makeup or moral character—analogous to the
health in that person's body. A person doesn't have to be recognized as
healthy in order to be healthy, and so it is with Plato's Socrates' *aretē*. In
the myth of the Ring of Gyges, presented in the *Republic*, Plato argues
that even if a person could get away with all manner of wrongdoing
while maintaining a good reputation because of a magic ring that renders him invisible, still he should not do any of these awful things,
since by destroying his *aretē* the man will destroy himself. *Aretē* then

---

[*]Xenophanes, Fragment 2. This translation is taken from Andrew M. Miller, *Greek Lyric: An
Anthology in Translation* (Indianapolis, IN: Hackett, 1996), pp. 108–109. The Greek original
is in elegiac couplets.

is entirely independent of social regard. And in the *Gorgias*, Socrates is presented as asserting something so radical that his hearers think it has to be a joke. He would, he says, rather be treated unjustly than treat others unjustly (469c). But if *aretē* is conceived of as analogous to the health of the body, then Socrates' statement is hardly absurd. The injustice that we do involves us far more intimately than the injustice that we suffer. I don't only act out of my character; my character reacts to my actions. Each time I lie, for example, even if I'm not caught, I become a little bit more of this ugly thing: a liar. Character is always in the making, with each morally valenced action, whether right or wrong, affecting our characters, the people who we are. You become the person who could commit such an act, and how you are known in the world is irrelevant to this state of being. (*The Picture of Dorian Gray* is a very Platonic book—except that Dorian's great love of beauty ought to have induced in him such a revulsion at his characterological ugliness as to forestall his immoral actions. That is Plato's great hope: that love of beauty can, when rightly cultivated and educated, battle immorality.)

These Socratic/Platonic departures from the *kleos*-centered notions of *aretē*, which carry with them a sophisticated moral theory, signal a limit to the control that others—that a whole *polis* of others—can exert over your sense of your life and what it's ultimately about. No wonder that 501 Athenian jurors could be convinced that Socrates was a serious threat to their values. He as much as told him so by saying that nothing they could do to him would harm him, thus firmly setting aside the social aspect of their notion of human excellence.

But before moving on further to the way in which Socrates and Plato diverged from their society's *kleos*-centered version of the Ethos of the Extraordinary, let's consider that unregenerate version a bit more, if only to better appreciate how Socrates and Plato departed from it.

The first question—leading to a whole series of questions—is why did the Greeks develop such a demanding ethos in the first place, one which foisted on them the quasi-paradoxical requirement of becoming habitually extraordinary? Aren't there easier ways to assure oneself that one can achieve mattering? Couldn't they have devised a religion—or adapted the religion that they already had—so that it could better minister to the kind of existential exigencies that erupted so forcefully in the Axial Age? And yet their religion, as important as it was in other regards (most particularly in supplying a sense of identity, both Panhellenically as Greeks and as citizens of the individual *poleis*), remained, as far as the existential questions are concerned, inert. Perhaps (speculating wildly here) their Ethos of the Extraordinary itself stymied the growth of a more existentially sophisticated religion. Their existential answers derived from a different aspect of their culture.

There is a historical contingency that seems relevant. The Greeks of the Iron Age, which was when the Homeric tales were composed, lived among the ruins of a dauntingly superior society, the Mycenaean palace societies of the Bronze Age. These Bronze Age Greeks are the heroic protagonists of the Homeric tale. It's they, the superior predecessors, who travel to Troy to retrieve the Mycenaean queen, Helen, stolen by the prince of Troy, Paris. (And this because of some typically irresponsible behavior on the part of the immortal Olympians. Those gods!) The Bronze Age civilization—fabulously wealthy and cultured and literate (they wrote in Linear B)* had been mysteriously destroyed, and the Homeric tales were composed during the regressed and chaotic period that followed. The Homeric tales that form such a substantive part of *to hellēnikon,* or hellenicity, were composed at a time that represented a giant step back from what had been; and it was a time that saw itself in those inferior terms, which is why its attention is fixed on the heroic past. The calamity—or, more likely, series of calamities—that destroyed the palace economies had plunged the Greek world into a state of darkness, shattering stability along with the knowledge of writing, which would have yielded us some evidence for what had really happened to the powerful Mycenaean civilization, which had traded with people across the Mediterranean.† Their architecture was so formidable that it came to be called by the peoples who walked among the ruins "Cyclopean": How could mere humans have built such edifices without the participation of the one-eyed giants? The engineering that went into the extraordinary beehive-shaped shaft graves of the

---

* Most of the clay tablets inscribed with Linear B were found in Knossos, Thebes, Mycenae, Pylos, and Cydonia. The extant writing is almost exclusively concerned with administrative matters connected with the palaces. It's been hypothesized that the writing is all the work of a small guild of professional scribes, who were employed by the palaces, and that when those palaces were destroyed, literacy too was destroyed. Linear A is an even earlier script used in the Minoan civilization. Although Linear A and Linear B are clearly related to each other— with Linear B, a syllobary, being more advanced because it uses fewer symbols—Linear A has still not been deciphered.
† Their pottery and vials of perfumed olive oil were found in Egypt, Mesopotamia, and as far west as Sicily.

royals, originally packed with magnificently wrought golden objects, is still capable of leaving us gaping in wonder today, an attitude to which I, who recently gaped in wonder, can attest. The lintel stone over the entrance to the gravesite in Mycenae known as the Treasury of Atreus, also sometimes called the Tomb of Agamemnon (misidentified as such by Heinrich Schliemann), weighs 120 tons. Is it any wonder that the relatively primitive people of the succeeding period would stare at these remains, often inscribed with writing that they couldn't read, and form stories of heroes who exceeded anything that they could ever have known, stories that provoked wonder at the extraordinary forms that human life could achieve—at least at one time, though not in theirs. They lived among the daunting physical evidence of such superior possibilities.

There is a kind of parallel discussed by Stephen Greenblatt in *The Swerve: How the World Became Modern*. The humanists who seeded the European Renaissance—first and foremost Petrarch (1304–1374), joined by his contemporaries Giovanni Boccaccio (1313–1374) and Coluccio Salutati (1331–1406)—were conscious of living in an age overshadowed by a glorious past, among whose ruins they lived, whose ways of thinking and living they were eager to recapture for themselves, which led to their obsessive seeking of ancient lost writings. "The urgency to the enterprise reflects their underlying recognition that there was nothing obvious or inevitable about the attempt to recover or imitate the language, material objects, and cultural achievements of the very distant past. It was a strange thing to do, far stranger than continuing to live the ordinary, familiar life that men and women had lived for centuries, making themselves more or less comfortable in the midst of the crumbling, mute remains of antiquity. Those remains were everywhere visible in Italy and throughout Europe: bridges and roads still in use after more than a millennium, the broken walls and arches of ruined baths and markets, temple columns incorporated into churches, old inscribed stones used as building materials in new constructions, fractured statues and broken vases. But the great civilization that left these traces had been destroyed." Coupled with the dazzled admiration for the past was the acute sense of an unworthy present: "For his own present, where he was forced to live, Petrarch professed limitless contempt. He lived in a sordid time, he complained, a time of coarseness, ignorance, and triviality that would quickly vanish from human memory." Zeal for the achievements of ancestors was converted into equally zealous ambition. "To prove its worth, Petrarch and Salutati both insisted, the whole enterprise of humanism had not merely to generate imitations of the classical style but to serve a larger ethical end. And to do so it needed to live fully and vibrantly in the present."* Salutati, who would channel his aspirations into ambitions for his beloved city-state of Florence, wrote, "I have always believed that I must imitate antiquity not simply to reproduce it, but in order to produce something new." The ideal of the "Renaissance man," with isotropic achievements shooting out in every direction like the rays of the sun, captures the kind of ambition that can be unleashed when a sense of former greatness sets the bar.

The Homeric normative vision, its ideal represented by a young man who would choose the short but song-worthy life, was an age that judged its own relative insignificance by the measure of a glorious past. Its heroic specimens of humanity must have lived, in that long-ago and legendary time, on the most intimate of terms with the denizens of Mount Olympus, sometimes so intimate that they mated with the gods or were the offspring of such couplings. It wasn't the distance to the gods that was elongated in the imagination, but rather the distance to the mortals who had disappeared from earth. The Mycenaeans who had built the brilliantly engineered bridges and roads, the massive palaces and treasure-house tombs, who left behind stones inscribed with uninterpretable writing, represented realizations of the human possibilities that made them seem closer to the gods than to the sort of men who populated a time of coarseness and ignorance, destined to vanish from human memory. These epic poems created during the prehistoric period—until recently called the Greek Dark Age—were sung by bards and perfected over the course of the illiterate centuries, with certain phrases polished into formulaic idiom chunks—what the linguists call "collocations"—to aid both with memorization and with the constraints of the dactylic hexameter. These bards were called "rhapsodes," meaning those who stitched together songs, and the finished works we now call the *Iliad* and the *Odyssey* were perhaps stitched together in their final form by one and the same person, whom we call Homer, who, if he truly existed—everything is questionable about this prehistorical

---

* Stephen Greenblatt, *The Swerve: How the World Became Modern* (New York: W. W. Norton, 2011). Quotes from pp. 117–119 and 124.

time—lived sometime around 750–700 B.C.E. The *Iliad* and *Odyssey* can be thought of as a kind of wiki-epic, with many mutually anonymous authors collaborating. (It is a matter of fierce debate, going back at least to the first century C.E., whether Homer, assuming he existed, was himself illiterate.)*

In a mere few centuries, the Greeks went from anomie and illiteracy—lacking even an alphabet—to that explosive outpouring of creativity which was to set the bar for the Romans and thus for the humanists of whom Greenblatt writes. The Homeric tales were an integral part of *to hellēnikon* that united all the *poleis*. But whereas before they had sung of these heroes of a remote and superior age that had vanished forever, they now could imagine becoming avatars of the heroic themselves. As would happen once again with the humanists of the early Renaissance, admiration for the past was converted into ambitions for the present, the Ethos of the Extraordinary reconfiguring from backward-looking ancestor-awe into norms of action. This conversion was perhaps partly tied to the enormous shift that the reemerging stability brought about, over the course of the eighth and seventh centuries B.C.E., as the city-states emerged out of the anomie of the previous few centuries and literacy returned.† Centuries of singing about heroes had prepared the way for an ethos that celebrated the extraordinary possibilities that human life can attain. The Greek-speaking peoples emerged into the historical period prepared to evolve into a society of unprecedented ambition and restlessness. Not only kings could aspire to inspiring tales of wondrous feats; that aspiration—if not its

---

*The Jewish historian of the Judeo-Roman Wars known as Josephus argued, on the basis of the premise of Homer's illiteracy, that the Greek people were not so very ancient, having only quite recently acquired writing. "However, there is not any writing which the Greeks agree to be genuine among them ancienter than Homer's Poems, who must plainly be confessed later than the siege of Troy; nay, the report goes, that even he did not leave his poems in writing, but that their memory was preserved in song, and they were put together afterward, and that this is the reason of such a number of variations as are found in them" (*Against Apion*, 1.2.12), http://www.gutenberg.org/files/2849/2849-h/2849-h.htm. The debate continues unabated today, with much ink spilled over one ambiguous line in Homer's *Iliad,* which may or may not refer to writing. (*Iliad* 6.168–169). But, of course, Homer's knowing *of* writing doesn't entail that he, or any of the anonymous bards of the Iron Age knew *how* to write. For a display of contemporary passions stirred up by this subject, see "Homer's Literacy," Joseph Russo in reply to Hugh Lloyd-Jones, *New York Review of Books*, March 5, 1992.

† Literacy was regained, with the import of a Semitic alphabet by way of the Phoenicians, which the Greeks adapted to their own language, using the signs for sounds that they didn't have to indicate vowels, for which the Phoenicians lacked symbols. In adapting this imported alphabet to their own needs, the Greeks produced a system of writing for which there was a one-to-one correspondence between sound and symbol—and they were the first to do so. Literacy, however, remained the possession of the aristocracy.

realization—was the birthright of all Homer-quoters. The heroes of the former age no longer functioned to cut a Greek down to size but to inspire him to assume heroic proportions himself. (And I hope Plato will forgive me the male pronoun here. One of the mysteries, still, of the Greek Ethos of the Extraordinary was the large gap between the female paragons, whether mortal or immortal, presented in their epics and dramas, on the one hand, and the female possibilities for achieving extraordinary lives, on the other. Plato was all for closing that gap. See his discusssion in *Republic* 451c–457b/c, which ends with the words "male and female guardians must share their entire way of life and . . . our argument is consistent when it states that this is both possible and beneficial.")

But still, how did such a change overtake the Greeks of the historical age? How did they go from a mythology that stood in passive awe before the human possibilities of greatness to actively undertaking the realization of those possibilities for themselves? It must have been, for the most part, gradual, as all such processes are, but if one is to point to one historical event as relevant, then the choice is obvious. No collective experience so transformed the Greeks' perception of themselves as their unlikely defeat of the Persians. In vanquishing the vastly superior forces of this world empire, the Greeks had given their poets something contemporary to sing about. Herodotus initiates his *Histories,* which is to say initiates the practice of history itself, with these words: "These are the researches of Herodotus of Halicarnassus, which he published, in the hope of thereby preserving from decay the remembrance of what men have done, and of preventing the great and wonderful actions of the Greeks and the Barbarians from losing their due meed of glory." The Greco-Persian Wars helped to transform the Ethos of the Extraordinary from a mythologized memory into a working normative framework. Aristotle, writing his *Politics* at least a century after the wars, observes the effect that the triumph over the amassed forces of the Persians had on the self-confidence of the Greeks, spilling over into the life of the mind that was of particular concern to him. "Proud of their achievements, men pushed farther afield after the Persian wars; they took all knowledge for their province, and sought ever wider studies" (*Politics* I.341).

And nowhere were this pride and this pushing more assertively on display than in the Athens of the fifth century, living out its days beneath the splendor of its Acropolis. The sense of Athenian exceptionalism added a political dimension to the Ethos of the Extraordinary. Athenian exeptionalism allowed for the extraordinary to be spread around, distributed among all Athenian citizens, solving the central paradox that the ethos presented in its desideratum that everybody must achieve an exceptional life—reminiscent of the fictional Lake Wobegon as a place "where all the childen are above average." All Athenian citizens, merely by virtue of being Athenian citizens, could rest assured that they were above average. Athenian exceptionalism granted its citizens a kind of participatory extraordinariness. The unique form of government, democracy, which it evolved by fits and starts, gave each Athenian citizen an extraordinary degree of participation in shaping policy. Many important decisions—for example, whether or not to go to war and whom to send as generals—were voted on by all the citizens. Other decisions were made by the Council of 500, but the council, too, was composed of average citizens who were chosen by lot and served for a year, each of the ten artificially constructed tribes contributing fifty members.* So if Athens was singularly great, it was a singularity to which all its citizens could singularly lay claim. In his *Histories,* there is an extraordinary passage in which Herodotus of Halicarnassus takes much the same view of the Athenians as they take of themselves, breaking out into a paean of praise for the special freedom that the Athenians had achieved with their democracy, and crediting it with having made them the first among the Greeks.†

And if the bulk of the (otherwise ordinary) citizens didn't intuit this participatory extraordinariness for themselves, there was their extraordinary leader, Pericles, whose very name means "surrounded by glory," to articulate it for them. "In sum, I say that our city as a whole is a lesson for Greece, and that each of us presents himself as a self-sufficient individual, disposed to the widest possible diversity of actions, with every

---

* Instead of having the unwieldy 500 sitting day after day for the year of their office, each tribe sat on the administrative and executive council for one-tenth of the year.

† "Thus did the Athenians increase in strength. And it is plain enough, not from this instance only, but from many everywhere, that freedom is an excellent thing, since even the Athenians, who while they continued under the rule of tyrants, were not a whit more valiant than any of their neighbors, no sooner shook off the yoke than they became decidedly the first of all. These things show that, while undergoing oppression, they let themselves be beaten, since they worked for a master; but so soon as they got their freedom, each man was eager to do the best he could for himself. So fared it now with the Athenians." *The History of Herodotus,* V, 78, translated by George Rawlinson, http://classics.mit.edu/Herodotus/history.5.v.html.

grace and great versatility. This is not merely a boast of words for the occasion, but the truth in fact, as the power of this city, which we have obtained by having this character, makes evident. For Athens is the only power now that is greater than her fame when it comes to the test. . . . We do not need Homer, or anyone else, to praise our power with words that bring delight for a moment, when the truth will refute his assumptions about what was done. For we have compelled all seas and all lands to be open to us by our daring; and we have set up eternal monuments on all sides, of our setbacks as well as of our accomplishments."* So declared Pericles in his famous Funeral Oration, burying the dead of one of the early battles of the Peloponnesian War—a war which can itself be seen as a result of the Ethos of the Extraordinary and the over-reaching to which it led, both individually and collectively.†

It was under Pericles that the *kleos*-worthy urban renewal of the Acropolis had taken place.‡ All that an Athenian citizen—an otherwise average Timon, Dicaeus, or Heiron—had to do in order to feel lifted above the ordinary was cast his eyes upward to the thirty-foot colossus of Athena, sculpted by the genius of Phidias from the bronze of the

---

* Paul Woodruff, trans., *Thucydides: On Justice, Power, and Human Nature; Selections from "The History of the Peloponnesian War"* (Indianapolis, IN: Hackett, 1993), ii, 41 (p. 43).

† See Appendix B.

‡ Plutarch describes how Pericles pulled the common people, whose champion he always represented himself as being, into his great schemes for using the riches of the empire to build Athens into a city of splendor. "And it was true that his military expeditions supplied those who were in the full vigor of manhood with abundant resources from the common funds, and in his desire that the unwarlike throng of common laborers should neither have no share at all in the public receipts, nor yet get fees for laziness and idleness, he boldly suggested to the people projects for great constructions, and designs for works which would call many arts into play and involve long periods of time, in order that the stay-at-homes, no whit less than the sailors and sentinels and soldiers, might have a pretext for getting a beneficial share of the public wealth. The materials to be used were stone, bronze, ivory, gold, ebony, and cypress-wood; the arts which should elaborate and work up these materials were those of carpenter, moulder, bronze-smith, stone-cutter, dyer, worker in gold and ivory, painter, embroiderer, embosser, to say nothing of the forwarders and furnishers of the material, such as factors, sailors and pilots by sea, and, by land, wagon-makers, trainers of yoked beasts, and drivers. There were also rope-makers, weavers, leather-workers, road-builders, and miners. And since each particular art, like a general with the army under his separate command, kept its own throng of unskilled and untrained labourers in compact array, to be as instrument unto player and as body unto soul in subordinate service, it came to pass that for every age, almost, and every capacity the city's great abundance was distributed and scattered abroad by such demands. So then the works arose, no less towering in their grandeur than inimitable in the grace of their outlines, since the workmen eagerly strove to surpass themselves in the beauty of their handicraft. And yet the most wonderful thing about them was the speed with which they rose. Each one of them, men thought, would require many successive generations to complete it, but all of them were fully completed in the heyday of a single administration." *Parallel Lives: The Lives of Plutarch*, 12–13, vol. 3, Loeb Classical Library edition, 1916, pp. 39–40, trans. Bernadotte Perrin, http://www.perseus.tufts.edu/hopper/text?doc=Perseus: abo:tlg,0007,012:12.

spoils of the Persians defeated at the battle of Marathon. There she stood, the symbol of their collective ascendancy, holding her ground between the Parthenon and the Propylaia, her spear in her right hand held aloft, its tip flashing in the sun-bronzed sky, bronze to bronze. Athena's crested helmet and spear could be seen for miles out to sea.

If, as Aristotle attests, a great deal of Greece exploded into ambitions raying out in every direction, the epicenter of this explosion was Athens. It became, in the wake of the Persian defeat, an imperial power, forcing tribute from its allies, who, when they visited Athens and looked up at the magnificence displayed atop the Acropolis, might very well have felt something quite different from the the Athenian citizen's pride of exceptionalism. Every year, when crowds from all over Greece arrived for the drama festival, bringing "tribute," they might have reflected on how Athens had removed the Delian treasury to Athens, and the building campaign it had permitted, and thought of Athenian exceptionalism more in terms of theft and greed.*

While Sparta receded into its insularity—even recommending that the Ionian *poleis* be allowed to revert back to Persian dominion so as to provoke no further foreign wars—Athens, in the decades after the wars, exploded outward. The very notion of a shared Greek culture didn't really emerge until Athens, after the Persian Wars, reshaped Greek culture in its own image.† A speaker in Plato's *Protagoras* describes Athens as "the prytaneum of Hellas" (337d), its central hearth and shrine.

---

* Plato gives vent to a similar assessment: "And they say that *they* have made the city great!" he has Socrates reflecting on looking back on the political leaders of the last fifth of the fifth century, including Pericles. "But that the city is swollen and festering, thanks to those early leaders, that they don't notice. For they filled the city with harbors and dockyards, walls, and tribute payments and such trash as that, but did so without justice and self-control" (*Gorgias* 518e–519a).

† "It is a culturally-based Athenoconcentric notion of Hellenicity that is central to the doctrine of 'Panhellenism'—a term coined by modern scholars to describe the various appeals made by late fifth- and early fourth-century intellectuals to foster Hellenic unity and to submerge interstate differences in a common crusade against the 'eternal enemy,' Persia." Jonathan M. Hall, *Hellenicity: Between Ethnicity and Culture* (Chicago: University of Chicago Press, 2002). Herodotus, in a famous passage of *The Histories* (Book 8, 144), speaks of *to hellēnikon*, or Hellenicity, and specifies as its elements shared blood, shared language, shared culture, and shared religion. He offers the term and its analysis in the context of relating the response the Athenians gave to their Spartan allies during the second of the Persian Wars, when the Spartans were concerned that the Athenians might be on the verge of making a separate peace with the Persians. Don't worry, the Athenians say. We'd never do such a thing, not even if it were to our distinct advantage, because of *to hellēnikon*. But of course Herodotus is writing in the post–Persian Wars years, when the notion of *to hellēnikon* has emerged in tandem with Athenian hegemony.

Pericles speaks of Athens as "the school of Hellas."* The epitaph for Euripides declared Athens "the Hellas of Hellas."†

Athens was not just a city, in our sense of the term, but a *polis,* or city-state. All of the *poleis* of ancient Greece—of which there were, according to the best modern tally, 1,035, though not all were coexistent‡—were independent states, with their own (very active) armies and their own forms of government. Each *polis* had a city center, its *astu,* usually walled and containing an acropolis, or "city on an extremity," high up and thus defensible, probably the reason that the original settlement had grown around it. The *astu* was surrounded by extensive territories, the *khora,* which included farmlands, olive groves, vineyards. The more established cities—Athens, Sparta, Thebes, Corinth, Argos—also had colonial settlements, other cities that they either established or seized and that would pay taxes to the "metropolis," the mother *polis,* and were required to be allies in war. Most *poleis* weren't large. The average number of citizens was between 133 and 800.§ Athens had a citizen population of about 30,000, while its total number of inhabitants was in the vicinity of 100,000, which means that only one out of three residents had the rights of citizenship. Though there were no property qualifications, as there were in the Greek oligarchies, citizenship in the Athenian democracy was hard to come by. Women, children, and slaves were excluded from citizenship, as in all *poleis.* So too were foreign residents, the metics, who were often among the richest of those living in Athens. Athenians prided themselves on the myth that they, alone of all Hellenes, were autochthonous, literally "sprung from the earth," by which they meant that they had always occupied the same soil. Being born of an Athenian father had long been a requirement for citizenship, but in 451 B.C.E. Pericles tightened this law, the pride in autochthony having strengthened following the Persian Wars and Athenian imperial hegemony. Now citizenship required both father and mother to be

---

* Thucydides, *History of the Peloponnesian War,* Book II, 41.1., trans. Paul Woodruff.
† "All Hellas is the funeral monument of Euripides; yet the Macedonian ground holds his bones, since there he met with the end of his life. But his fatherland was the Hellas of Hellas, Athens." This is cited in "The Classical Tragedians from Athenian Idols to Wandering Poets," by Johanna Hanink, in *Beyond the Fifth Century: Interactions with Greek Tragedy from the Fourth Century* BCE *to the Middle Ages,* edited by Ingo Gildenhard and Martin Riverman (Berlin: De Gruyter, 2010), p. 54.
‡ See Mogens Herman Hansen and Thomas Heine Nielsen, eds., *An Inventory of Archaic and Classical Poleis* (Oxford: Oxford University Press, 2004).
§ See E. Ruschenbusch, "Die Bevölkerungszahl Griechenland in 5 und 4 Jh's," in *Zeitschriften für Papyrologie und Epigraphik* 56 (1984): 55–57.

Athenian-born, making citizenship even more exclusive and desirable, just as Athens was asserting itself throughout Hellas as the standard for what made all Hellenes great.

Political experimentation was rife in ancient Greece, and one of the most radical experiments was the one that Athens pursued, off and on, but mostly on, dating roughly from, say, the reforms of Cleisthenes in the sixth century B.C.E.* until the eventual triumph of Alexander of Macedonia in the third century B.C.E. We get our word "democracy"—rule by the people—from the form of government, the *demokratia*, that Athens pursued, often to the dismay and derision of the other *poleis*.

So potent had the sense of Athenian exceptionalism grown that participation in the life of the city might almost seem, to Athenian citizens, to provide the very definition of *aretē*—again a point that Pericles explicitly makes in his Funeral Oration, cataloguing all of the Athenian achievements, from the exceptionalism of its democracy to its superiority-bred magnanimity: "So that we alone do good to others not after calculating the profit, but fearlessly and in the confidence of our freedom" (ii.40). Its vanquished enemies should almost take pride, he suggests, in being vanquished by such specimens of humanity.

The suggestion is that all of humankind ought to aspire, if it were but possible, to what Athens was achieving, and most certainly all Hellenes, though they could not hope to become voting Athenian citizens, ought to look to Athens as providing the model of *aretē*.† And it was true that artists and thinkers flocked to Athens, eager to be in the center of the world, even without the benefits of citizenship. Aristotle, who came to study at Plato's Academy and founded his own Lyceum, had been born,

---

* Cleisthenes broke up the ancestral Attic tribes in 510 B.C.E. and established local residence as the qualification for voting. This proved to be the most essential reform in putting Athens on the road to democracy, and even today the parts of our world still dominated by tribal arrangements are resistant to democracy. The successive reforms, as important as they were—the abolishing of property qualifications for suffrage, the universal eligibility among citizens for holding office, the popular courts having judgment over the conduct of magistrates, and the payment for office holding (so that the poorer citizens, the *thētes*, could afford to miss work in order to cast their votes), as important as they were, only complete the process begun by Cleisthenes. Some date the start of democracy a bit earlier, to the time of Solon, earlier in the sixth century B.C.E., and his reforms, which eased the debt burdens of farmers by abolishing the practice of borrowing against one's own freedom. Democracy developed by innovative accretions in Athens, so it is impossible to put a precise date on when it began. Many of Solon's reforms managed to survive the tyrannies of Pisistratus and his two sons, who were succeeded by Cleisthenes.

† See Jonathan Hall, *Hellenicity,* for an interesting discussion of whether the Athenians thought their superiority a matter of nature or nurture and how the discussion shifted over time.

as has been mentioned, far north in Stageira, in Chalcidice, his father the physician to the king of Macedonia. Athenian superiority drew to itself more superiority, intensifying its sense of its own exceptionalism, which typically happens with imperial powers. The ever-higher achieving that this allows is the most handy way for imperialism to defend itself, should it feel the need, and this is true even now.*

And it was in the shadow of the Acropolis that Socrates went about his daily work, which was to try to sow doubt in his fellow citizens that they had any notion of what their lives were about. You're quite right that the unexceptional life is not worth the living, but what is the exceptionality that matters? *Aretē* can't be relegated to Athenian politics. It's not enough, he's constantly haranguing the Athenians, to be a citizen of Athens. You haven't crossed the finish with the voice of the crowd chanting the sweet sound of your name, assuring you that you've achieved what most matters. You haven't even gotten to the starting line.

Plato had probably known about Socrates since he was a young boy. His older brother, Glaucon, was a Socratic enthusiast before Plato was of an age to appreciate the point of Socratic antics.† The soldier and historian Xenophon, who was also devoted to Socrates and left us his own anecdotes and impressions in several works, treats Plato's older brother as something of a doofus. Xenophon recounts that Socrates interceded when Glaucon was in danger of making an ass of himself before the Assembly, or *ekklêsia*, the central institution of Athenian democracy, held under the open skies, on the hill called the Pnyx, with a wide open view of the Acropolis. Glaucon had wanted to set himself up as a man with whom to be politically reckoned, "and not one of his relatives or other friends could prevent him from getting himself dragged down from the tribunal and making himself ridiculous," writes Xenophon. But Socrates—for the sake of Plato, Xenophon tells us, and it is the only time that Xenophon ever mentions Plato—intervened to inform Glaucon that he hadn't a clue regarding the affairs of state.‡ Though Xenophon scoffs in his telling, still it is a point in Glaucon's favor that,

---

* See, for example, the many writings of Niall Ferguson, such as *Civilization: The West and the Rest* (New York: Penguin, 2012); or *Empire: The Rise and Demise of the British World Order and the Lessons for Global Power* (New York: Basic Books, 2004).
† What the age gap was between Plato and his older brother we don't exactly know. See Debra Nails, *The People of Plato*, pp. 154–156.
‡ *Memorabilia* III, 6.

though he fancied himself a would-be politician, he could be convinced that he didn't know what he was talking about. In any case, this is only Xenophon's version of Glaucon.

Plato makes Glaucon one of Socrates' major interlocutors in the *Republic*, together with another brother of Plato's, Adeimonstos, who is not quite as prominent in the dialogue. Plato's Glaucon is not in the least a doofus. He's shown as enjoying an excellent memory, a decent knowledge of mathematics, above average musical ability, and a great deal of political idealism. In fact, both brothers, according to one scholar, exemplify exactly the right combination of thoughtful resistance and just as thoughtful receptiveness that allows Plato's Socrates to progress from negative elenchus, or refutation, to positive creativity.*

In the first book of the *Republic* an irascible sophist named Thrasymachus becomes so apoplectic at the high-minded moral argument he's hearing from the lips of Socrates that he charges into the discussion like an analytic philosopher on amphetamines, insisting on what a contemporary philosopher would describe as the cognitive meaninglessness of the normative propositions—those involving the word "ought"—that Socrates is spouting. Allow me to disabuse you of your delusions, Thrasymachus thunders, while Socrates pretends to recoil in terror (336d). There are no objective facts of the matter concerning how people *ought* to live their lives. There are simply the facts of how they want to live their lives, people pursuing their own self-interest as best they can, with the strong succeeding as anyone would if only they could. So live as you wish, live in any way that you can get away with, shouts the sophist, his own bullying style of argument dramatizing his views. Just so long as you're able to get away with it, you're in no danger of getting it wrong, there being nothing *to* get wrong.

Blustering Thrasymachus soon runs out of steam, and Plato hands the discussion over to his brother Glaucon to develop the amoral line in quieter, subtler tones, evincing a sort of wistful wish to be talked away from the point of view he's developing, whether in self-advocacy or as the devil's advocate. "Perhaps the most striking feature of Plato's double portrait here is his brothers' shared detachment from the case they make with such vehemence."† Plato has Glaucon devise a thought-

---

* See Ruby Blondell, *The Play of Character in Plato's Dialogues* (Cambridge: Cambridge University Press, 2002), pp. 190–226.
† Blondell, *Play of Character in Plato's Dialogues*, p. 190.

experiment involving the Ring of Gyges, which renders its wearer invisible and so able to get away with anything.

> Now, no one, it seems, would be so incorruptible that he would stay on the path of justice or stay away from other people's property, when he could take whatever he wanted from the marketplace with impunity, go into people's houses and have sex with anyone he wished, kill or release from prison anyone he wished, and do all the other things that would make him like a god among humans. Rather his actions would be in no way different from those of an unjust person and both would follow the same path. This, some would say, is a great proof that one is never just willingly but only when compelled to be. No one believes justice to be a good when it is kept private, since, wherever either person thinks he can do injustice with impunity, he does it. Indeed, every man believes that injustice is far more profitable to himself than justice. And any exponent of this argument will say he's right, for someone who didn't want to do injustice, given this sort of opportunity, and who didn't touch other people's property would be thought wretched and stupid by everyone aware of the situation, though, of course, they'd praise him in public, deceiving each other for fear of suffering injustices. (360b–d)

Glaucon is not arguing the nihilism of bilious Thrasymachus. The sophist had argued that there were no objective moral truths, whereas Glaucon seems to acknowledge, albeit vaguely, that there might be such truths, but if there are they're completely beside the point in changing our behavior. Even if you've convinced me that something is wrong, why should I change my behavior, especially if I'm strong enough—or sneaky enough—to get away with it? It's only the fear of the damage we can do to ourselves and our reputations—our acoustic renown— that has any effect on our wills, argues Glaucon, almost as if against his will. You can all but hear him pleading: Please, please convince me otherwise.

Thrasymachus and Glaucon by no means put forth the same views. Thrasymachus speaks for an unregenerate Ethos of the Extraordinary that licenses unmitigated individualism. He's an Athenian Ayn Rand. The extraordinary person is the one who can act as he will. He embodies *aretē*, the sort of human attributes that make for "acoustic renown," and there is no external standard by which to judge him. Everything about Thrasymachus, his opinions as well as his high-decibel manner of putting them forth, highlights the antisocial nature of his version of the Ethos of the Extraordinary.

An ethos encouraging *kleos*-measured extraordinariness can easily lead to antinomian individualism. This was not just a theoretical but a practical problem, most especially in Athens, whose sense of collective extraordinariness was very much a matter of its extraordinary individuals. (Sparta is to be contrasted with Athens in this regard. Sparta's exceptionalism eschewed individualism.) The Athenian predicament— the potentially destabilizing effect of celebrating individual extraordinariness for the sake of its extraordinariness—might be dubbed its "Alcibiadean Problem," after the famous Alcibiades. This undeniably extraordinary individual left behind a trail of grand-scale treachery and turmoil. He betrayed Athens to its archenemy Sparta, then betrayed Sparta to Athens, then betrayed both Athens and Sparta to Persia. Alcibiades, beautiful as a god, brilliant in both oratory and military prowess, rich and charismatic, always insisting on his own way and more often than not getting it, carrying on as if he, too, just like Achilles, came from metaphysically mixed parentage: how could the Athenians have failed to love him, no matter how many times he used and abused them, misled and mistreated them?[*] Alcibiades was not only Athens' Achilles. He was Athens' Achilles' heel.

The *kleos*-worthy man was several decades dead when Plato wrote the *Republic,* having lived out the short but extraordinary life that could have been predicted for him, killed by an assassin (or so goes one of the tales of his ending). Thrasymachus' view could not muster any ground for finding fault with Alcibiades, but Thrasymachus' view was not the majority opinion of Athens' citizenry on the issue of such men as Alcibiades, whose extraordinariness allow them to get away with everything.

And this is where Glaucon joins the conversation of the *Republic,* representing an Athens that realizes the dangers of antinomian individualism. Glaucon's Athens recognizes its Alcibiadean Problem—that there are those who, by virtue of their personal advantages, wear a kind of Ring of Gyges—and he offers a political solution. For the good of the *polis,* certain compromises must be made. The good of the *polis* trumps the free-range extraordinariness of individual men, and so men have

---

[*] See chapter ε for more details of Alcibiades' *kleos*-saturated life.

freely ceded some of their freedom. They have made a social contract of sorts, and it is from this that all the notions of justice and injustice are derived.

Plato's Glaucon, in developing his politicized view of justice, anticipates Hobbes. In the state of nature, so to speak, there is neither justice nor injustice. The state of nature is a Thrasymachean world, in which individuals all seek to impose their own wills, causing suffering as they need to in order to get their way. To impose one's will is good, while having another impose his will on you is evil. But men discovered that the evil was greater than the good, Glaucon says (359a), by which he seems to mean both that, qualitatively, the pain of having others impose their will on you is more experientially intense than is the pleasure of your imposing your will on others; and also, quantitatively, there are just more people who get imposed upon than who get to impose.* And for this reason, men

> decide that it is profitable to come to an agreement with each other neither to do injustice nor to suffer it. As a result, they begin to make laws and covenants, and what the law commands they call lawful and just. This, they say, is the origin and essence of justice. It is intermediate between the best and the worst. The best is to do injustice without paying the penalty; the worst is to suffer it without being able to take revenge. Justice is a mean between these two extremes. People value it not as a good but because they are too weak to do injustice with impunity. Someone who has the power to do this, however, and is a true man wouldn't make such an agreement with anyone not to do injustice in order not to suffer it. For him that would be madness. This is the nature of justice, according to the argument, Socrates, and these are its natural origins. (359a–b)

The view that Plato has Glaucon develop is startlingly modern. It not only anticipates the Hobbesian theory of the social contract, but puts the reasoning behind the social contract in game-theoretic terms, defining what we today would call a prisoner's dilemma. The "rational players" in Glaucon's argument make their decisions based on trying to avoid the worst possible payoff, of behaving morally while others exploit

---

* We could also invoke the second law of thermodynamics. There are fewer states of the world that are beneficial than those that are harmful; it is easier to damage and cause harm than to create and cause good.

them, and the second worst, of living in a world where they both exploit and are exploited, even if those same decisions exclude their chances of obtaining the best possible payoff, in which they exploit other innocents who don't exploit them back. In order to avoid getting maximally screwed a rational person will forsake the chance of maximally screwing.* This seems, both in Plato's day and in ours, as good a description of the nature of politics as any, if politics consists of a viable, stable social order. But does it describe the nature of justice? Plato thinks not, and he spends the bulk of the *Republic* refuting the Glauconian—the Athenian—view that the demands of the political dictate the nature of morality. Quite the contrary, he argues, the demands of morality dictate—or at least *ought* to dictate—the nature of the political.

The view that morality was essentially political was far more common in Athens than Thrasymachus' brand of nihilism. Thrasymachus' nihilism was much more in line with the teaching of various sophists, and most Athenians mistrusted and despised the sophists. (One of the charges Socrates feels he must defend himself against during his trial is that he is a sophist. This is why he's so insistent on never having accepted any fee for his services.)

Glaucon's view, in which *aretē* has been socialized and spread around, is much more in line with the view of most Athenians, though far more sophisticated—and more detached from the specifics of Athenian exceptionalism than anything you'd have been likely to hear in the agora that Socrates wandered, incessantly pressing his questions. It's still a requirement at the philosophy seminar table that, in order to refute a conclusion, you have to put forth the best possible argument for it. That's what Plato is trying to do with Glaucon's attempt to enfold morality into the political. Plato erects theoretical sophistication around what was, for most of Athens' citizens, no more than the presumption that all of this business of what it is to live a life worth living

---

*The prisoner's dilemma, which Plato has Glaucon anticipating, involves the following scenario. You and your partner in crime have been arrested, and you're each put in solitary and cannot communicate with each other. The prosecutor offers you a deal. The police don't have enough evidence to convict you, so they're counting on one of you to rat out the other and will reward you with your freedom if you do, while your unfortunate partner will be sent away for ten years. Or vice-versa: If you stay loyal to him while he rats you out, then you go away for ten years while he walks. If each of you implicates the other, then you'll both get six years. But if you both remain loyal, the police can only convict you of some lesser charge, and the two of you will be out in a year. You're best off if you desert a loyal partner, worst off if you're the one deserted, and still badly off if you both desert; cooperating is the least bad outcome for the two of you combined.

had been taken care of, that they didn't have to think about any of these questions since they had the great good fortune of being Athenians, an exceptional *polis* which defined *aretē* for all the world. (There are Americans who feel the same way, as well as citizens of other nation-states.) Socrates may have had the patience to try to talk his fellow citizens out of their complacency, but Plato wants the views he's refuting to be worthy of his refutation.

In the *Republic,* Glaucon recognizes that his game-theoretic notion of justice hasn't the force to restrain someone who wears the Ring of Gyges and so can get away with anything. It can't even restrain an Alcibiades. Nor can Glaucon even argue that it *ought* to. What rational reason would a person who can always get his way with screwing anyone he wants have for abiding by the decisions of those who are just as likely to get screwed as to screw? What's rational for them won't be rational for him. Glaucon wants Socrates to tell him whether there are notions of goodness and justice that apply even to those who, by reason of their special advantages, escape his game-theoretic construction of morality; and if there are, he wants Socrates to tell him whether these notions have any muscle such that they can compel a rational person to act differently in accordance with them, even if, if he chooses not to, no ill will come to him since he can get away with anything. If moral truths are socially fabricated ("as a result, they begin to make laws and covenants, and what the law commands they call lawful and just"), then a person who is assured of no social opprobrium will do what he wants. Isn't that rational, Socrates?

In the course of the *Republic* Plato will argue that not only do moral truths, when revealed as they really are rather than contextualized into a social construct, have the force to compel a rational person's actions; but moral truths have the force to compel the rational arrangement for the *polis,* which will, in turn, help promote the good behavior of its citizens. The good *polis* is made by the good person, his moral character intact, and the good *polis,* in turn, helps turn out good persons, their moral characters intact. Plato goes a very far distance in the *Republic* beyond both the nihilism of Thrasymachus and the social constructivism of Glaucon.

And Plato has his older brother stay around for the long haul of the *Republic,* becoming Socrates' yes-man during the course of the famous Myth of the Cave (514a–520a) and receiving Socrates'—and so

Plato's—final benediction in the closing lines of the *Republic:* "And so, Glaucon, the story wasn't lost but preserved, and it would save us, if we were persuaded by it, for we would then make a good crossing of the River of Forgetfulness* and our souls wouldn't be defiled" (621b–c). Once again the story is brought round to death in relation to the question of what it is to have lived a life worth living. And one wonders how it felt for Plato to address such words to his older brother, who might well have been dead when Plato wrote them. Was the preserved "story" too late to save the brother whom Plato allows to gaze at the utopian vision he has Socrates construct of the *kallipolis,* the beautiful city, which is most decidedly *not* Athens? Did Glaucon live out his life in keeping with what Plato had first had him argue, the view that only an idiot will let moral truth compel his actions and shape his life rather than the opinions that his fellow citizens have of him, his name on their lips, a person of substance in the agora and the Assembly, this older brother from whom Plato first heard tell of the man who devoted his life to trying to convince the fine proud citizens of his city that they all had no idea of what they were doing with their lives?

You will have noticed that I am putting Socrates' questions—and thus the birth of philosophy as overseen by Plato—into a historical context, created by both the wider Axial Age's preoccupation with existential dilemmas, and the Athenian response which grew out of its Homerically shaped ethos. I was trained as a philosopher never to put philosophers and their ideas into historical contexts, since historical context has nothing to do with the validity of the philosopher's positions. I agree that assessing validity and contextualizing historically are two entirely distinct matters and not to be confused with one another. And yet that firm distinction doesn't lead me to endorse the usual way in which history of philosophy is presented. This consists of an idealized conversation between philosophers—Socrates and Plato and Aristotle and Augustine and Aquinas, and Descartes and Spinoza and Locke and Leibniz and Berkeley and Hume and Kant and . . . The philosophers talk across the centuries, exclusively to one another, hermetically sealed away from any influences derived from non-philosophical discourse. The story is far more interesting than that.

---

*This is the river Lethe—which means "forgetfulness" or "oblivion," one of the five rivers that flow through the underworld of Hades.

A distinction that the philosopher of science Hans Reichenbach introduced has always seemed a very useful one to me in keeping the two sets of questions—historical influences and assessments of validity—unconfused. Reichenbach distinguished between, on the one hand, "the context of discovery" and, on the other, "the context of justification." When you ask why did some particular question occur to a scientist or philosopher for the first time, or why did this particular approach seem natural, then your questions concern the context of discovery. When you ask whether the argument the philosopher puts forth to answer that question is sound, or whether the evidence justifies the scientific theory proposed, then you've entered the context of justification. Considerations of history, sociology, anthropology, and psychology are relevant to the context of discovery, but not to justification. You have to keep them straight, though sometimes relations between them can be established that are helpful. Sometimes, for example, examining the context of discovery can help flush out unstated premises in an argument, presuppositions that were regarded as so intuitively obvious in the context of the thinker's mind-set, whether for cultural reasons, personal reasons, or the interactions between the cultural and the personal, that they didn't bear being stated. But still the assessment of those intuitions in terms of the argument's soundness isn't accomplished by work done in the context of discovery. And conversely, one doesn't diminish a philosopher's achievement, and doesn't undermine its soundness, by showing how the particular set of questions on which he focused, the orientation he brought to bear in his focus, has some causal connections to the circumstances of his life. In a previous book, I tried to show how the circumstances of Spinoza's life—in particular, his being a member, until the age of twenty-three, of the Jewish-Portuguese community of Amsterdam, all of whom were refugees from the Spanish-Portuguese Inquisition, which was raging at the time—influenced his radically rationalist approach in philosophy, and in particular his preoccupation with the issues of personal identity. Some readers so confused the context of discovery and the context of justification, thinking I was perhaps arguing that Spinoza's philosophical positions were groundless because his personal history helped determine which problems he set out to solve—that I thought it might be useful to articulate Reichenbach's distinction here. There's a golden mean to be struck between the extremism of historicism, on the one hand, and the extremism of philosophical insularity, on the other, and Reichenbach's distinction helps to dissolve the false dichotomy.

And so, bearing in mind that I am here smack in the context of discovery and not the context of justification, let me summarize:

Socrates radically departed from the Homeric Ethos of the Extraordinary, most especially as it had been intensified and politicized by imperialist Athens. The Greek adjective *atopos*, meaning out of place, strange, and the noun *atopia*, the quality of being out of place, strangeness, are used by Plato repeatedly to describe Socrates.[*] Socrates was strange in a way that only those normatively at odds with their society are strange. Athens put him on trial for violating their values, and they were right that such a violation was exactly what he had in mind. How more transgressive can you be than to go after so central a notion as *aretē*, the linchpin of the Ethos of the Extraordinary? One of the charges against him may have been put in religious terms—he introduces new gods—but, as Plato takes pains to show us in his setting up of the *Republic*,[†] the Athenians were tolerant of new gods. Socrates was a heretic at another level, far more deeply disturbing to his fellow Athenians, especially at that moment in their history, as we'll see in chapter ζ.

Just how dangerous was it to debate what it means to live a life worth living in the shadow of the Acropolis? The answer to that particular question lies in the bottom of a vial of hemlock.

---

[*] Socrates' *atopia* is referred to by Alcibiades at *Symposium* 215a, and 221d. Socrates is described as *atopos* at *Gorgias* 494d, and *atopotatos* (superlative adjective) at *Theaetetus* 149a and *Phaedrus* 230c.
[†] See p. 379 below.

δ

PLATO AT THE 92ND STREET Y

Bringing together the country's foremost thought leaders, educators and experts to exchange ideas and inspire action in the global community, the 92nd Street YM/YWHA is a hub for innovative discussion that informs, influences, and drives our culture forward.

Join us at the Y for what promises to be the highlight of the season. Zachary Burns, beloved columnist for the newspaper of record, will lead a discussion on "How to Raise an Exceptional Child." With Zach will be best-selling author Mitzi Munitz, author of *Esteeming Your Child: How Even the Best-Intentioned Parents Violate, Mutilate and Desecrate Their Children;* Sophie Zee, best-selling author of *The Warrior Mother's Guide to Producing Off-the-Charts Children,* and Plato, best-selling author of *The Republic.*

BURNS: I want to thank all of you in the audience for coming out on this blustery, snowy evening. It's a real testament to the three outstanding panelists up here onstage with me that not only is there not a single empty seat in the room, but we've had to accommodate the large overflow audience in additional rooms where they're watching the discussion on closed-circuit TV.

But it's no wonder that so many would brave a blizzard to hear tonight's exchange. Not only are we privileged to have with us three internationally acclaimed authors who have written extensively and controversially on the subject of child-rearing, but the subject itself is one that's guaranteed to provoke visceral reactions. It doesn't require the meme of the selfish gene to remind us that every generation is heavily invested in wanting their children to do well in life, what my grandparents used to call *naches foon der kinder, naches* from the kids. Wait a minute. Do I need to translate *naches*? It's Yiddish and there's no exact English equivalent. Anyone?

ZEE: Doesn't it mean the pride that parents feel in the achievements of their children?

BURNS, *laughing:* You really *are* an overachiever, aren't you?

ZEE, *laughing:* Well, I'm married to a Jewish man. I have a Jewish mother-in-law. *Audience laughter. Zee smiles radiantly.*

BURNS: That's Sophie Zee, folks, and it's my privilege as moderator for tonight's event to formally introduce her and her fellow panelists. Of course, they're all so accomplished that if I were even to begin to do justice to their résumés we would have to camp out here all night, which, given the weather forecast, we may have to do anyway. Still, I know you're eager to hear from them, so I'll limit myself to the high points, particularly those that apply to tonight's discussion. I, by the way, am Zack Burns, a lowly journalist. *Scattered audience laughter.*

Dr. Mitzi Munitz, sitting immediately to my left, was born in Vienna, where she trained as a psychoanalyst, although after twenty years of practice she renounced psychoanalysis as colluding in the same vicious power structure of exploitative parenting as the rest of society does. Dr. Munitz's ideas have profoundly changed the culture of child-rearing, heightening our recognition of the more subtle forms that the victimization of children can take and encouraging people to look back on their own childhoods in search of the abuse they may have suffered. One critic has described Dr. Munitz as "the missing link between Freud and Oprah."* *Esteeming Your Child,* which has sold a whopping 800,000 copies—

MUNITZ: My apologies, Mr. Moderator, but here I must correct you. *Esteeming Your Child* has sold one million copies to date and been translated into more than thirty languages.

BURNS: *Esteeming Your Child,* which has sold a *truly* whopping million copies, will be on sale after tonight's event, together with Dr. Munitz's most recent book, *Post-Traumatic Distress Disorder: Another Name for Life.*

Seated to the right of Dr. Munitz is Sophie Zee. Professor Zee is a first-generation American who was born in Palo Alto, California, and attended Stanford University and then went on to Stanford Law School, where she is now the Fracas Professor of Law. Professor Zee

---

*This wonderful phrase comes from Daphne Merkin, "The Truth Shall Set You Free," *New York Times Book Review,* January 27, 2002.

is the author of several learned books and case studies, but the book that catapulted her to international fame is *The Warrior Mother's Guide to Producing Off-the-Charts Children*. Talk about controversy! This book ignited a firestorm, not just coast to coast but all around the world—

MUNITZ: Excuse me, but I must again interrupt, since, despite this international notoriety that you are describing, I remain unfamiliar with this author and her work and must ask whether the title was intended as satire. Because it does not seem possible that after my ground-breaking discoveries concerning the psychopathology of everyday parenting that anybody could, in earnest seriousness, put forth in a non-ironic mode such a thesis as seems suggested by a literal reading of this title.

BURNS: You see, ladies and gentlemen, I didn't lie to you! I promised you controversy, and we've got controversy before I've even finished the introductions! Let's hold off having Professor Zee answer your question, Dr. Munitz, at least until we get through with these preliminaries. Professor Zee was recently chosen as one of *Time* magazine's 100 Most Influential People in the World—

MUNITZ, *muttering:* Ach, too bad for the world.

BURNS: And of course her book will be available for sale and signing immediately after our event. And now I turn to our last panelist, Plato. It's just Plato, isn't it? Not Dr. Plato?

PLATO: That's right.

BURNS: But of course you've been teaching for most of your professional life. So I suppose it's only right to call you Professor Plato.

PLATO: Plato would be sufficient. Plato is what I'm called.

BURNS: Plato it is then! Plato has long been hailed as one of the most creative and influential thinkers in the history of Western thought. Indeed, some have argued that all of philosophy consists of footnotes to Plato, which is high praise indeed. He was born in Athens, Greece, a city where he has spent the bulk of his life and where he informally studied as a young man under the famous philosopher Socrates. Plato was so impressed with Socrates' thinking—even though, as I understand it, Socrates never published a single book or journal article—that he abandoned his hopes of becoming a dramatist and poet and instead became a philosopher. I think Dr. Munitz may have some questions to put to you regarding your parents' reactions to those two choices, poetry and philosophy.

MUNITZ: I do indeed have many urgent questions to put to Plato concerning his parents. Plato presents to me as a classic case of intellectualization in order to cauterize the deep wounds inflicted during his impressionable early life.

BURNS: In good time, Dr. Munitz, in good time. To continue, Plato is the author of at least twenty-six works—I know there are people who want to attribute even more works to you, but you've remained noncommittal. *Looks questioningly at Plato, as if hoping Plato will reveal whether* Alcibiades I *and* II, Hipparchus, Amatores, Theages, the Greater Hippias, Minos—*not to speak of any of the thirteen letters, including the tantalizing seventh—are apocryphal, but Plato maintains his inscrutable mien.* Ah, well, I suppose the mystery will not be solved at the 92nd Street Y this evening. All of Plato's works are written in his signature dialogue style, and a good number have been hailed as masterpieces, including his best seller, the *Republic,* which will be on sale, together with all of his other works to date, following our discussion and which I figure is on more college syllabi than any other book ever written. Any idea how many copies the *Republic* has sold, Professor Plato?

PLATO: Please, just Plato. And no, I have no idea at all how many copies the *Republic* has sold. But we could google it. I have my laptop right here. *Plato reaches for the computer shoved under his seat.*

BURNS: Oh, no, that's okay, Plato, no need to google right now. I was just curious. Any idea how many languages it's been translated into?

PLATO: I'm afraid I lack that knowledge as well. Into English and Latin, I know.*

BURNS: I should say so! I first read the *Republic*—and it was in English, I'm ashamed to admit, not the original Greek—when I was a freshman at Columbia, right across town, and it was on the syllabus for our mandatory course in CC, or Contemporary Civilization. In fact, the *Republic* was one of the first books we read—right after the *Iliad* and the *Odyssey*—during that amazing year of studying the classics that set the contemporary stage. If anyone would have told me back then that someday I'd actually get to speak to the author and have the opportu-

---

*The Greek title of Plato's work is the *Politeia*. The title of the *Republic* is derived from the Latin *res publica*, which means "public things" or "public affairs." The term came, during the Roman period, to refer to a form of government that wasn't headed by a king and in which at least some part of the populace—whether the people as a whole or an aristocracy—got to choose the government. Plato's ideal *polis*—what he calls his *kallipolis*, his beautiful city—is headed by a philosopher-king, and so doesn't qualify as a republic.

nity to pepper him with questions, I would have told him that was nuts. And you know what? It *is* nuts! *Audience laughter.*

Anyway, as successful as Plato has been in the publishing arena he's also a celebrated teacher. Not only did he found his own university, the highly ranked Academy, but he has continued to teach there, which is refreshingly at odds with tales we hear emanating from our own top-tier universities, where undergraduates often never lay eyes on the celebrity professors. I've heard tell that Plato's public lecture on "the Good" is legendary. And in these days when parents start saving for their kids' college educations when the student is in utero, the Academy somehow manages to charge no tuition. That's a grand total of zero dollars or, I suppose I should say, euros, assuming that Greece is still in the Eurozone—though does it make a difference whether it's euros or dollars when the amount is zero? That's the sort of conundrum made for a philosopher, I guess. Perhaps your colleague Parmenides?*

PLATO: Perhaps.

BURNS: Another interesting feature of the Academy is that it has only one entrance requirement, namely geometry.† Or to put it in terms familiar to an American, you don't require the verbal SAT, only the math SAT, which would certainly have put me at a disadvantage! But I don't hold that against you! Also—this was something that I was surprised to learn—your Academy is one of the only elite institutions that was co-ed from the very beginning. Co-ed, zero tuition, and only one SAT: you must be swamped with applications!

PLATO: Actually, we're not.

BURNS: Really? That surprises me. Some of you may have seen a recent article, published by my own esteemed employer, reporting the findings of two economists who did a long-range study which found

---

*The pre-Socratic philosopher Parmenides, born c. 515 B.C.E., was much obsessed with working out the logic of nothingness. From his fundamental tautology that what is not is not, Parmenides proceeded to ponder the question of what can truly be thought of the nonexistent. His conclusion: nothing.

† It is said that engraved on the doorway to the Academy were the words *mèdeis ageômetrètos eisitô mou tèn stegèn*, which translates: *Let no one ignorant of geometry come under my roof.* It was only the later Neoplatonists, coming a full ten centuries after Plato, who report this academic graffiti, for example Joannes Philoponus, a late Neoplatonic Christian philosopher who lived in Alexandria in the sixth century C.E., and Elias, another sixth-century Christian Neoplatonist, also of Alexandria. Aristotle, who studied at the Academy for twenty years, never mentions this engraving, at least not in his extant works (much of Aristotle is missing), even though, interestingly, Aristotle does use the word *ageômetrètos* in his writings. In his *Posterior Analytics*, I, xii, 77b8–34, the word is used five times within a few lines. See Bernard Suzanne, *Plato and His Dialogues*, http://plato-dialogues.org/plato.htm.

that good teachers make a lifelong impact on their students' lives that reach far beyond mere academics into all kinds of quality-of-life dimensions, including higher earnings.* Just one truly outstanding fourth-grade teacher could make a world of difference for her students, significantly improving their chances of going on to college, avoiding mistakes like drugs and teenage pregnancy, and earning an average of $4,600 more over their lifetime. I know, amazing, isn't it? And I think this finding demonstrates something I've been arguing for years, in my own small journalistic way, which is the fundamental importance of developing good character during childhood and the way in which good role models are essential in the development of good character.

MUNITZ: Yes, but how is one to say who are the good and who the disastrous role models, when every authority figure assumes the right to impose itself on the vulnerable young?

BURNS: A good question, Dr. Munitz, and one that I remember was close to the central question Plato set out to answer in the *Republic,* with his whole idea of the philosopher king—do I have that right, Plato? That the philosopher-king is held up as the best role model?

PLATO: Well, it's a little more complicated.

BURNS: Well, yes, of course, it is *very* complicated.

PLATO: I hope not as complicated as all that. Just complicated enough, and no more.

BURNS: Well, yes, of course, I didn't mean to suggest otherwise. The point I'm just trying to make, in these purely introductory remarks, is that Plato himself, in his own life narrative, presents a dramatic demonstration of what a difference the right mentoring makes in character development. Plato's teacher, Socrates, had a lifelong impact on him and he in turn mentored Aristotle, who, as I'm sure you all know, went on to phenomenal success and, in fact, even started up his own rival university just down the road from you. The Lyceum, right?

PLATO: Yes.

BURNS: Yes. And then Aristotle in turn had a lifelong impact on no less a figure than Alexander the Great, if I remember correctly.

PLATO: You remember correctly.

BURNS: Okay! Boy, what a pedagogical lineage! Maybe, I know we

---

*Annie Lowrey, "Big Study Links Good Teaching to Lasting Gain," *New York Times,* Jan. 6, 2012.

have so much to discuss this evening, but maybe this extraordinary tradition of mentorship speaks to another subject much in the news these days. I mean whether online teaching could replace classroom teaching. Did you know, Plato, that elite universities like Harvard and M.I.T. and Stanford have decided to create massive online open courses that students can take for free? Tests and quizzes will be embedded in the online material, and at the end of the course the students will get a certificate—not real credit, mind you, just a certificate—showing they've passed?

PLATO: Yes, and these MOOCs will be taught by the regular faculty, which means that I am able to have Nobel laureates teaching me cosmology and particle physics.

BURNS: Which, as several publications have pointed out, could create a campus tsunami.* Star professors, for which students—or rather their parents—shell out hundreds of thousand of dollars to sit at the feet of, will be accessible to anyone with Internet access.

PLATO: I love the Internet.

BURNS: Yes, I noticed. Do you always carry your Google Chromebook with you?

PLATO: Yes. Ever since I visited the Googleplex.

BURNS: Fascinating! Well, I wonder whether, despite your fondness for the Internet, you'd agree with me that massive online courses, even taught by the most famous professors of our day, being entirely impersonal, will never take the place of the intimate relationship between teacher and student—intimate, I mean, at least when the relationship is really working. I mean, maybe online courses are adequate when the subjects are practical—say, business courses—but what about your own field, philosophy? Even if they had you, Plato, delivering the online open-access lectures, I wonder if it would have the same impact as being in your presence.

PLATO: Of course it wouldn't.

BURNS: You agree with me!

PLATO: I do. In a field like philosophy, where understanding involves not so much the reception of knowledge but rather a transformation of the receiver itself, so that the receiver, which is to say the student, can generate the knowledge for him- or herself, then the physical pres-

---

* See, for example, "Get Rich U," by Ken Auletta, *The New Yorker,* April 30, 2012.

ence of the teacher is essential. This is the pedagogical paradox. The person of the teacher is required precisely because the knowledge itself is non-transferable from teacher to student.

BURNS: Absolutely fascinating. So you're saying the teacher can't transfer his knowledge into the student.

PLATO: Can't transfer his or her knowledge into the student.

BURNS: And that makes his or her presence all the more essential.

PLATO: Exactly. The fire for the subject and the fire for the teacher are intermingled in the receptive student. It's only by proximity to the beloved teacher, himself or herself on fire with love for the subject, that the fire can leap over and be kindled in the student in a self-generating blaze of understanding.[*]

BURNS: Or to put it another way, a brain isn't a computer, and we're not blank hard drives just waiting to be filled with data. People learn from people they love and remember the things that arouse emotion.[†]

PLATO: I agree with you.

BURNS: Well, I guess it stands to reason you'd agree with me, since I just realized that I got it from you! Every time I think I have an original thought I think a little more and realize I've gotten it from someone else, more often than not from you.

PLATO: There are no original thoughts. All knowledge is recollection.

BURNS: I've often thought that, too, and now I know where I got it! The *Meno*, right?[‡]

PLATO: I think rather you must have recollected it for yourself.

BURNS: Thank you! That's giving me a lot of credit! Well, I just can't express adequately what a thrill it is to have you—*all* of you—here participating in this dialogue. It's a once-in-a-lifetime privilege.

Okay, so let's begin. I might as well inform the audience up front that there has been some controversy about the title of tonight's discussion. In fact, one of our panelists was so dismayed that we called our

---

[*] Cf. *The Seventh Letter*, 344a–b.
[†] David Brooks, "The Campus Tsunami," *New York Times*, May 3, 2012.
[‡] In the *Meno*, Plato ponders an old sophistical puzzle, according to which nobody ever seeks to learn something new, since if they don't know it, they don't know to seek it. Plato solves the puzzle by his doctrine that all learning is recollection, or *anamnesis*, having Socrates propound: "All nature is akin, and the soul has learned everything, so that when a man has recalled a single piece of knowledge—*learned* it, in ordinary language—there is no reason why he should not find out all the rest, if he keeps a stout heart and does not grow weary of the search, for seeking and learning are in fact nothing but recollection" (81d). Right after this, Socrates singles out one of Meno's slaves and soon has him deducing a geometrical proof.

event "How to Raise an Exceptional Child" that she was tempted to cancel, and we're grateful that she overcame her deeply considered reservations and has joined us tonight. After all, it's in the spirit of dialogue that the best way to deal with disagreements is to bring the disputants together and let them go at it. Isn't that right, Plato?

PLATO: Just as thinking is the soul speaking to itself (*Theaetetus* 189e), forcing itself to articulate its reasons and exposing those reasons to evaluation as if to different aspects of its own self, so we enlarge our thinking by bringing others into the dialogue.

BURNS: Well, then, in the spirit of dialogue, which is nothing less, Plato has just told us, than the spirit of thinking itself, I'm going to begin by asking each of our panelists how they understand the title of tonight's dialogue. Dr. Munitz, why don't we start with you?

MUNITZ: How I understand this title is that it perpetrates and legitimates the most criminal of falsehoods, a falsehood not just of the mind but of the heart and of the soul, a falsehood that stands as the root cause of all the self-inflicted misery of humankind through the ages. How I understand this title is that it is premised upon a perversion.

BURNS: Strong words, Dr. Munitz. Could you say a bit more?

MUNITZ: I could say a great deal more. The desire to raise an exceptional child is a desire to sacrifice the integrity of the child, to nip the tender bud of the child's own nature before it has had a chance to poke itself above the ground. And for whom is this child's integrity sacrificed? Certainly not for the child itself, since nothing worse can befall a person than to be deprived of the possibility of ever becoming itself, that is, of coming into its rightful inheritance of autonomous personhood.

The sacrifice is therefore performed not for the sake of the child but for the sake of the parent. It is a criminal act of brutality, and what makes it even more criminally brutal is that it is brutally experienced by the child, who registers it in her psyche as an assault, and as such it is stored in the child's body for a lifetime, as all trauma is somatically stored, even if the memories are so repressed as to be irretrievable. The unknowing parent, who is operating without integrity of her own, having been subjected to precisely the same evisceration of autonomous selfhood, has, little wonder, no insight into the interior of her child, does not have a clue that such an interiority even exists, for she herself can retrieve no memories of what it was like to be a child. So she is fancy-free to regard the child not as a person but as a project, *her* project, to produce, to quote the title of this book I have not read, "an off-the-charts child" whom she can flaunt as a sign of her own superiority, since flaunting superiority is all that she knows how to do in order to assure herself that she herself exists, the legacy of her own brutal rearing. And so it persists, from generation to generation. The produced child, the project of its parent, is deprived of what ought to be its *own* project of self-actualization and autonomy, from which, in the healthy state of affairs, proceeds the stable assurance of its integral existence and inalienable individual worth.

BURNS: So if you were to modify the title of tonight's event, it would be . . . ?

MUNITZ: "How to Raise Your Child So That It Will Realize Its Own Personhood and Have the Strength to Resist the Criminal Authoritarianism of Others."

BURNS: Okay, well, I think we can see the outlines of a difference of opinion emerging here.

MUNITZ: This is not opinion, Mr. Moderator, but established scientific fact, and those who continue to deny it demonstrate, by their very denial, how deeply stored within their bodies is the abuse they cannot access. My conclusions are scientifically established from decades working with my clients, tracking their neuroses back to their childhoods, as well as my research into the childhood experiences of some of the most exceptional of people, the famous and the infamous, from the greatest artists to the greatest thinkers to the greatest criminals. And by the way, all such people can rightly be called "exceptional," which again exposes the fatal errors in the title of tonight's event.

BURNS: So it sounds as if you're almost suggesting that greatness is itself a symptom of a disorder, something we should cure rather than cultivate.

MUNITZ: Greatness is, by definition, an abnormality. The normal state of affairs is demonstrably *not* to be great. So the *production* of this abnormality demands that extreme measures be taken. In this regard I concur with the purely factual presupposition underlying the title of this book which I have not read. It does indeed take the brutal intervention of a warrior mother to produce off-the-charts children, though quite often their off-the-chartness will duplicate the brutality of the parenting and take on a monstrous form. But even when it does not, even when the greatness is of a non-monstrous nature, it is the *moral* presupposition of tonight's event, as opposed to its merely factual presupposition, which I am here to protest.

BURNS: And could you spell out what this moral presupposition is?

MUNITZ: That the raising of the—may the good Lord help us— *exceptional* child is something that a parent ought to want. This is a normative proposition, signaled by the word *ought,* and thrusts us into the moral dimension, a dimension of which warrior mothers are devastatingly oblivious.

BURNS: Well, on that minor note of agreement—right, there was some agreement in there, wasn't there?—why don't we turn to Professor Zee, and ask her what she makes of the title of tonight's event? I imagine you don't have the same difficulties with it that Dr. Munitz does.

ZEE: Well, actually I agree with just about everything that Dr. Munitz has said.

BURNS: You do?

MUNITZ: No, you don't. You can't possibly!

ZEE, *turning to face Munitz:* Yes, I do! I really do! I agree with you one hundred percent that it's the parents' obligation to see that their children's best potential is realized, so that the children can become the super-best people they can be, to realize their very own personhood!

MUNITZ: And what do you mean by "best" potential? Who decides what this "best" potential is? You, no doubt.

ZEE: Well, of course, at least when the child is very young, the parent has to be the authority on what's best. I mean, let's face it, a child is in no position to consider the choice between, say, playing video games for hours on end or, instead, memorizing her boring multiplication tables. I mean, what child is going to willingly stop playing video games and open up her arithmetic book? So of course a parent has to set the agenda. A child doesn't have the experience or knowledge to know what's best. She can't look ahead. She's thinking only of immediate satisfaction, of instant gratification, and can't see beyond the drudgery of memorizing her multiplication tables or practicing her scales on the piano or violin. She can't possibly see the benefits of the drudgery and boredom and can't imagine what mastery is going to *feel* like, how much *pleasure* it will eventually give her! How can she imagine that when she hasn't even experienced it yet? So this particular pleasure is just unimaginable to her.

MUNITZ: It will give her such pleasure because it will win for her the withholding love and approval of her parents, who have made this mastery the condition of bestowing any positive feedback.

ZEE: Well, maybe, at least partly, but that's not such a bad thing, since that's going to motivate her to do what she has to do. Since a child can't see the benefits for herself, her parents have to provide the motivation, get her to stop playing her fun video games and go memorize her multiplication tables, and part of the motivation is going to be the approval and disapproval that the parents hand out. And a little disapproval is not going to scar a child for life. I think it's a big mistake to think our children are little breakable porcelain dolls who will shatter to smithereens if we exert a bit of pressure on them, or show them disapproval when they're not living up to the standards that they can't possibly understand the importance of for themselves.

MUNITZ: Now, some of what you're saying right now sounds, on the face of it, reasonable to the point of utter vacuity. Obviously, a parent knows that there are certain skills that the child must master, such as multiplication, and so it's the obligation of the parent to make sure that those skills are secured—with great sensitivity to the child's own capacities and rate of learning and always, above all, a sense of the child's personal dignity, which will prohibit the tactics of disparagement and belittlement which a combatant mother, a *warrior* mother, if you will, so aggressively employs. But I suspect that putting forth such trivial claims does not account for the international notoriety of your book, and I am very curious as to why you are trying to evade your own controversial claims. I suspect that you yourself, having been raised by the draconian methods you advocate, are driven by a desperate desire to please people, most especially mother figures, a role I am temporarily assuming for you, precisely because I am displaying my disapproval of you. The more I reprimand you, the more you will try to appease me, even to the point of disavowing your own published thesis, thus enacting before us a dramatic demonstration of the lifelong dependence on the psychological narcotic of approval that results from the parenting techniques you peddle.

In fact, I strongly suspect that when you published your views you had no idea they were even so controversial, since it seems axiomatic to you that all parents desire highly achieving offspring who will attest to their own superiority as parents. You never once, in all the pages of your book, think to examine the fundamental premise of whether it is in the child's own interests, as opposed to the interests of the parents, to be "off the charts," especially given the methods that are required to produce such a child. A child is not a breakable porcelain doll indeed!

ZEE: Excuse me, I don't mean to be confrontational, but I have to ask you how you know that I never examine this premise if you haven't read my book? *Audience laughter. Zee smiles at the audience, and is answered with scattered applause.* I mean, it's been my experience that my most vociferous critics haven't bothered to read my book, or if they have, they've misunderstood it because they haven't gotten the context.

MUNITZ: In my line of work we learn to deduce a great deal. But it doesn't take great exertions of my analytic powers to perceive that your title is meant to hook like-minded parents who will never question the highly questionable premise of the desirability of raising an off-the-charts child. In other words, your readership consists of parents who are not interested in raising a *person* at all, but rather an organ grinder's little performing monkey that they can exhibit to win the clapping of hands and the tossing of coins.

ZEE: Not at all! Like any parent, I'm only interested in the future happiness of my children and, unlike so many parents today, I don't think it's in the interest of my children to have to win *their* approval and be *their* best friends. *That's* taking the easy way out. That's putting your own interests before those of the child. *Turning to the audience.* Isn't that what we're all interested in, the greatest future happiness of our children? Here's the missing premise, as far as I'm concerned. There's a deep happiness that comes from mastery and high achievement. Dr. Munitz says that greatness is an abnormality, and of course it is, at least in one sense, in the sense that it's rare. But "abnormal" doesn't mean "undesirable." Many things that are rare are also desirable, and high achievement is an example. And I would go even further and say that one of the ways that we evaluate our achievement is by the approval and adulation that it garners, so there's a deep happiness in the applause, too, and there's nothing wrong with that. It's really a form of humility—that we are not the proper judges of our own accomplishment, because of course we're biased to think well of ourselves, but we seek objective standards set by others. When others approve of us, it means we have met those standards, that we're not just stroking our own egos. I mean, you, Dr. Munitz, are an off-the-charts achiever! Why did you tell us, and with such evident self-satisfaction—such *justified* self-satisfaction—that your book has sold a million copies? Why have you pushed yourself to achieve all that you have? Why did you get

your advanced degrees and write all your many books, and face down
your critics and do all the other unpleasant things that achievement
requires, if you don't yourself know the greater pleasure that comes
from pushing yourself to your limits and doing something extraordinary and then having a million people buy your book in recognition
that what you've done *is* extraordinary?

MUNITZ: I do all this because I believe in what I am saying, and
believe that if I don't say the truth then nobody will. It's not for my
own pathetic need to assert my own importance in the world, to demonstrate my own superiority and have my name on many flapping lips
that I am writing, but rather because it is my firm conviction that I have
something important to say.

ZEE: And if you didn't have something important to say, then how
would you feel about yourself? Would you feel as good about yourself?

BURNS: Why don't we let Plato answer that question, since he certainly has had many important things to say, and on so many topics.

PLATO: The question seemed particularly directed to Dr. Munitz, and
I would like to hear her answer. Then, if you like, I will try to answer it
myself.

BURNS: Fair enough. Dr. Munitz?

MUNITZ: I prefer not to answer the question in the flagrantly personal terms in which it was posed. The question is not whether I, or
even whether most of us—who have been subjected to the psychopathology of everyday parenting—have a driving need to achieve, as an
alcoholic has a need for the bottle or an addict a need for the needle,
in order to feel, at least temporarily, the high of believing ourselves the
superior person our parents made us feel we were absolutely *required*
to be in order to be loved. That is not the question. The question is
whether that is how people *ought* to feel. Again, the normative, the
moral question. And here I answer unequivocally that it is not.

Let us consider the child abuse that is involved here. Yes, I do not
shrink from the charge of child abuse. The child's need for a parent's
love is overwhelming. A parent who stresses achievement is irresistibly imposing on the child the idea that its being loved is conditional
upon its performing well the tasks that the parent holds up before the
child's desperate eyes as the gateway to love. In this way, the child is
conditioned to consider only the parent's own desires for it, so urgent
is its need to win this love, and in time the child will even lose the sense
of any conflicting desires it has for itself, which, of course, is a situation

much to the parent's liking, since it makes its project for the child—the project that *is* the child—all the easier.

And I would add, by the way, that this projectification of the child has only increased with the liberation of women. Ambitious women, who have invested so much in their education and careers, are required to make sacrifices to their own advancement by the obstruction that is a child. And so these mothers will require of that child that it really be *worth* the sacrifice, worth the slowing down of their own scramble up the ladder of success. And so the pressure on the child to be off the charts is only intensified. I think it is no accident that the author of this book which I have not read is herself a practitioner in a fiercely competitive field. For such a woman, a child is so substantial a setback to her own ambition that, in order to offset the setback, the child must *earn back* its existence by being so exceptional as to add to, rather than subtract from, the ambitious mother's tallying of her successes. So feminism, I am sorry to say, has only intensified the projectification of children.

Now let me just say that this projectification will certainly get positive results in good grades, admission to brag-worthy schools, and other tricks of the little performing monkey, but at what price? A person raised by an organ-grinding mother—which I suggest as a more suitable epithet for the warrior mother, stripping her of the pseudo-heroics of your phrase—will forever conflate self-worth with surpassing others and garnering external signs of success, which will forever be confused with love, though of a most distancing and unsatisfying sort. One is raising a person who will forever be appraising others as possible competitors and so will feel profoundly isolated, knowing only the quick fix of personal achievement to temporarily dull the pain of being shut up in the small cold space of her own eternal need to justify her existence by excelling.

And this quick fix of achievement, mind you, is pursued not for the sake of the excellent work achieved, but rather for the sake of being *regarded* as excellent, whether there is true excellence there or not. Shortcuts to approval will be sought, methods of self-promotion will take precedence over the devoted hard work of true excellence, which, I might add, often goes unrecognized precisely *because* it is authentically superior. The author of this book, which has so much dictated tonight's agenda, says that there is a happiness in the approval of others because it is a signal that true excellence has been achieved, but let me assure you on the basis of my years of therapeutic work that this is not accurate. It is the approval that is desperately sought, not the achievement itself. In fact, such a person has little feel for the integrity of the work itself, just as she has no sense for the integrity of the self, and shoddiness will do just as well, if not better, so long as praise is attained.

In short, this is a prescription for lifelong anxiety, loneliness, and, in the end, mediocrity. So if that is what you would like for your child, then here is indeed the recipe to raise such an exceptional child.

BURNS: Sophie, I think you ought to have a chance to respond to that.

ZEE: It's hard for me to respond since, quite obviously, Mitzi—may I call you Mitzi?

MUNITZ: You may not. *Audience audibly gasps.*

ZEE, *laughs, a bit embarrassed, but quickly recovers:* Well, since Dr. Munitz hasn't read my book, it's hard to know how to respond. She certainly doesn't have to keep repeating that she hasn't read it, since that's obvious. As far as my abusing my children is concerned, that's just ridiculous, and in fact, even though they, of course, often get annoyed with me, they have actually *thanked* me—and every parent knows how rare that is—thanked me for pushing them as hard as I have, because it was the pushing that got them to accomplish what had seemed far beyond them. But it *wasn't* far beyond them, and too many parents never allow their children to discover how many goals they might have thought were beyond them are really in their power. It's really about empowerment! That's what child-rearing ultimately is all about, which is also what Dr. Munitz, I think, thinks, which is why I don't think that we're really, fundamentally, in disagreement.

MUNITZ: Well, about that you could not be more wrong. I do not claim to be an expert on everything, but I am the world's expert on what I think, and what I think is not within spitting distance of what you think. I very much hope for their sakes that your children are as happy as you seem to think them, and that this happiness will continue into their adult life, but my decades of research make me skeptical, to say the least. Of course, your children thank you, since you have incinerated any autonomous selves with which they could independently assess what it is you have done to them. You have made them success junkies for life. They will never be free, never. Empowerment! This is Orwellian. War is peace. Coercion is empowerment.

BURNS: Sophie, do you want to respond?

ZEE: Well, I think that being given more options in life is undeniably empowering. Going to the best schools gives you more options, opens doors that wouldn't otherwise be open, and more options means more freedom. Freedom is obviously a really complicated philosophical subject, and I feel embarrassed even broaching it in the presence of a philosopher of Plato's genius, but for what it's worth I think that freedom is measured by the number of options that a person can choose among. You're only free if you have multiple options. A less free person has limited options, maybe none at all that can make them happy. Give your child more options and you're giving your child more freedom, even if you have to resort to warrior-mother tactics in order to give them that freedom.

MUNITZ: And tell me, have your children the choice of being ordinary? Is that a choice you would allow them to consider? Is that, after they have been raised by you, a choice that they could ever allow themselves to consider?

ZEE: Why would they want to be ordinary, if they have it in them to be extraordinary?

MUNITZ: Nevertheless, if they did decide that's what they wanted, if that's what would make them happy, would they be free to choose it?

ZEE: Well, of course, they'd be free to choose ordinariness. That's the default position. You don't have to do much to prepare your children so that they can choose, if they want, to be ordinary. Almost everybody's born to be middling, which is where the job of parenting comes in.

BURNS: Well, I see that we've definitely strayed out onto some dense philosophical terrain here, conversing about the nature of freedom, no less! And fortunately, who should we have with us to help lead us into a clearing but Plato! What do you say, Plato? Do you agree with Professor Zee that the more options a person has, the freer they are?

PLATO: It seems to me that the free person has a severely restricted range of choices.

BURNS: That sounds rather paradoxical.

PLATO: Yes, and I will make my statement sound even more paradoxical. The free person's choices are completely determined.

BURNS: You're right. That does sound even more paradoxical. How can someone whose choices are completely determined be free?

PLATO: When it is the person's own better nature that is determining that person's choices. Imagine a two-horsed charioteer, with one horse unruly and unable to stay the course, and the other horse knowing his way even without the whip or goad (*Phaedrus* 253c–d). The charioteer has only to control the bad horse so that the better horse may lead him in order to be free. Freedom isn't the absence of control; rather, control is the essence of freedom.

Burns, Zee, Munitz all begin to speak simultaneously.

BURNS, *to Zee:* Go ahead.

ZEE: Well, I just wanted to say that I completely agree with Plato, and his point about freedom only reinforces what I was saying, since a child's nature is still very much in the making, which means that they don't really have a nature yet, certainly not a *better* nature, their bad unruly horse has total control, and so they can't possibly be free, so it's sort of conceptually incoherent for parents to try and give them freedom. I mean, given what Plato's just said, coercion of children shouldn't even be considered coercion, and what you're doing in perfecting their natures is ensuring that when they're grown-ups and actually have better natures that can determine their choices, those natures will be determining them to wonderful, off-the-charts results. Their freedom is going to express itself in excelling and achieving, as opposed to just taking up space. And I actually want to quote Plato here. *Zee turns to Plato.* Because I remember that in the *Republic* you used the expression "city of pigs"[*] (372c–d), which really made an impression on me when I was a freshman in college, which, like Zack, is when I read your masterpiece. If I remember correctly, the city of pigs is one where only the necessities of life are met, and the people are contented to just lie around and be, well, *content*, but there's no motivation to improve themselves, to get to the next stage of civilization. They're all equal and totally happy to be equal. It seems sort of idyllic, their lounging around on their rustic beds made out of leaves, and drinking wine and eating barley cakes and singing their simple little ditties together, probably most of them out of tune, but still it's a city of pigs, and there aren't any painters or poets or great musicians or scientists or philosophers, all the advancements that require people to get off their beds of leaves and practice their scales and study hard to get into the best schools and

---

[*] It's not clear that Plato shares Sophie's low opinion of the primitive communal city. Socrates never endorses Glaucon's judgment that it is a society fit for pigs. Plato's *Laws* ranks the primitive communal society the best—though unattainable.

push themselves to make something distinctive and distinguished out of their lives. So, yes, it's an idyllic city of pigs, but it's still a city of *pigs*. Maybe we'd all be more content, not always being haunted with the possibility of failure no matter how much we succeed, or not getting stabbed through the heart when our peers surpass us, if we were all just lying around living in the moment and toasting each other with cheap wine, but we never would have moved on in the history of our species, never would have achieved the glories of, say, a Leonardo da Vinci, or an Einstein or . . . a Plato! *Zee throws her arms open to Plato and smiles jubilantly.* I mean, what is it that moves us forward—and I mean collectively, as a *species*, a species we can all take pride in? What makes us worthier than those contentedly grunting pigs? It's the off-the-charts people who are constantly exerting themselves, driven by the high standards that have been instilled in them and who set the bar higher for all of us, and I think—*Zee turns to the audience, her arms wide in the same gesture she had directed toward Plato*—that's what we all want for our children!

Prolonged audience applause.

MUNITZ: Do you actually believe that your little performing monkeys are going to move us forward as a species? They are trained to be people-pleasers, looking for validation in applause. *Munitz glares at the audience, as if daring a single one of them to applaud. None does.* Those who move the species forward are, by definition, out of sync with the bulk of humanity and are not very likely to win its approbation for their efforts in their lifetimes. If it's approbation one seeks, then you do well to stay far back within the bounds of the "tried and true." And I, too, can avail myself of Plato's *Republic,* which I first read at the age of twelve and entirely on my own, to support what I am saying. It is to the famous cave that I am referring.

Recall how Plato describes the cave,[*] the prisoners chained so that they cannot move their heads and are staring at shadows that move across the rock wall at the back, cast by puppets they cannot see, which are being carried back and forth on top of a wall behind them, with a small fire burning behind it to provide the dingy light. For them, these shadows before their eyes constitute the world, and they have trained themselves to be experts in discerning any patterns to be found there,

[*] *Republic* 514a.

earning their bragging rights by predicting what shadows might follow next. Some of them will be off-the-charts shadowists, earning the
admiration of the other shadowists, expressed no doubt with prolonged
applause.

Then one prisoner manages to free himself* and see the shadows for
what they are. He sees the wall, the people behind it carrying the puppets, and the fire behind the wall, and pushes himself on, groping his
way to the cave's opening, suffering the pain of the bright light of the
outdoors entering his darkness-habituated pupils for the first time. But
he forces himself not to look away, not to slink back into the familiar
gloom in which he and everybody else were raised. Slowly, by himself,
in the grandeur of his solitude, he acclimates himself to the pain of illu-
mination until he can see clearly, until he is finally able to cast his eyes
on the radiance of the sun itself, the sustaining truth that gives the lie
to all that passes for truth within the cave. How he would love to ignore
the subterranean cave dwellers, to never go down into the sooty depths
of the darkness again, but he will feel an obligation to share his clarified vision with his former compatriots out of pity for their benighted
blindness and enslavement. How he will struggle with himself to turn
his head back to the awful cave, finally overcoming his resistance and
heading back down to the throngs in the darkness, speaking to them of
the undreamed-of radiance he's seen with his own eyes, urging them
to throw off their shackles and follow him out of the truth-swallowing
darkness.

And how do they receive him, he who has truly achieved the worthwhile in his lifetime? Is it with approval? Certainly not! His eyes now
accustomed to the light, he won't be able to make out the shadows,
and so they'll say that he has returned with his faculties ruined. For
his bumbling in the meaningless darkness, they will direct at him not
accolades but jeers. He will try to convince them that what they are
looking at is not worth the looking, and they will respond with hostility
and attempts at invalidation, throwing the pseudo-statistics of shadow-
ism at him. By challenging their erroneous version of the world, he will
provoke such a reaction formation in them that they will stop their ears
from hearing a single word of what he says about the actual reality that

---

*Dr. Munitz's rendition of Plato's Myth of the Cave is imperfect. So, for example, the one
prisoner does not free himself but rather is freed and almost involuntarily dragged out of the
cave. She is concerned to highlight the lonely heroics of the truth seer.

explains away their shadows. His achievement will be met with rejection and ridicule.

I cannot count the number of times I have gone back and reread Plato's account of the cave, understanding from it that jeers are the lot of the truth seers—at least until such time as the truth cannot be entirely ignored, and then it will be sanitized and turned into a Disney movie with a happy ending, which mothers who are raising off-the-charts children can safely take their little monkeys to see.

BURNS: Well, this is fascinating, ladies and gentlemen, because both Professor Zee and Dr. Munitz have appealed to Plato's work in support of their own divergent views. This highlights what I said in introducing Plato tonight, that everybody likes to trace their ideas back to him, happy to be counted among his footnotes. So after hearing Professor Zee cite you and Dr. Munitz cite you, let us finally turn to you, Plato, and let you speak for yourself.

PLATO: Where would you like me to begin?

BURNS: A good question! Well, why don't we begin with the projectification of the child that Dr. Munitz referred to. While she was speaking I couldn't help thinking of the intense program of child-rearing that you lay out in the *Republic*. You're as hands-on as any warrior mother, discussing everything from the kinds of bedtime stories little children should be told* to dictating what subjects ought to be studied in school and even which extracurricular activities should be encouraged. As I remember, you agree with Professor Zee that drama club is definitely out (*Republic* 398a) while music is in—though you two sharply disagree over sports. Sophie, you weren't at all enthusiastic that your kids participate in sports, while you, Plato, make athletics very central in raising the perfect child.

PLATO: Both music and athletics are essential, since they have opposing but ameliorative influences on the developing child (*Republic* 410b–412a).

---

* "You know, don't you, that the beginning of any process is more important, especially for anything young and tender? It's at that time that it is most malleable and takes on any pattern one wishes to impress on it. Then shall we carelessly allow the children to hear any old stories, told by just anyone, and to take beliefs into their souls that are for the most part opposite to the ones we think they should hold when they are grown up?" (*Republic* 377a–b). Among the storytellers whom Plato censures in the passages that follow are Hesiod, Aeschylus, and, above all, Homer. Plato takes Homer to task for presenting Achilles as indulging in behavior unbecoming for a hero—in particular his "slavishness accompanied by the love of money, on the one hand, and arrogance towards gods and humans, on the other" (*Republic* 391c).

MUNITZ: On all children?

PLATO: Well, I mostly confined myself to a certain subclass of children.

MUNITZ: Yes, the elite. The only ones who truly count in your estimation.

PLATO: I would not say so.

MUNITZ: No, I'm sure you wouldn't *say* so.

PLATO: I was trying to think about what was best for all in society, not just for one class (*Republic* 410b–412a). If there is an exceptional kind of person to be identified and then further trained in my just society, it is only for the collective good. Indeed my own admiration for Sparta stemmed from their promoting the collective good above those of any individual, even the most exceptional. It was somewhat different in my own Athens. The collective good ought to take precedence over all in matters of justice. And it is difficult for there to be justice in a society when groups are pitted one against another, so that occurrences that count as winning for one group count as losing for another.

BURNS: You're speaking about zero-sum conflicts.

PLATO: Yes, exactly. Zero sum. A society is a just one when zero-sum conflicts are minimized to the point of elimination (*Republic* 422a–423e; *Laws* 628b–e).

MUNITZ: But how can you avoid zero-sum conflicts when you give all the privileges of self-fulfillment to one group alone? Your entire social setup is about as zero sum as it gets.

PLATO: Self-fulfillment is, of course, a good for everyone, and the self-fulfillment of all is a positive-sum good. But there are different kinds of selves, with different abilities leading to different kinds of fulfillment. Insofar as fulfillment is a good, it is exactly the same, no matter if the particular fulfillment consists in philosophizing or in farming.

MUNITZ: Ah, so now we get at it, the very pinnacle of elitism, surpassing anything even the warrior mother would be prepared to say—or even to write—since she, at least, is an egalitarian in her ruthlessness, believing that her warrior-mothering can turn almost any child into a super-achiever! She, at least, would not have children tyrannically tracked like trolleys, but rather push them all out to try to compete to the death in the fast lane.

PLATO: And yet if you are truly to avoid the ruthlessness you decry, then precisely this specialization according to innate abilities is required. What is play and delight for one kind of child is coercion and

torture for another, and will not take no matter how much coercion is applied (*Seventh Letter* 341d–344a). And I agree with you wholeheartedly that children should not be subjected to torture in their education, since it is entirely counterproductive. In fact, as far as is possible, a child's education should not take the form of compulsion but the form of play (*Republic* 536d–e). In Greek, our word for play is *paidia* and the word for education is *paideia,* and it is very natural and right that these words should be entangled at the root, together with our word for children, *paides,* which gave you your words *pedagogy* and *pediatrician.**
What one tries to force into a child against its own nature will never come to good. A child's natural form of behavior is play, and in our aim to educate, play should be honored and preserved for as long past childhood as can be. So we may say, in fact, the sum and substance of education is the right training that effectually leads the soul of the child at play on to the love of the calling in its adult life (*Laws* 643d).

MUNITZ: Well, on that particular point I am, of course, in agreement with you. What I refuse to countenance is your arrant elitism. You lay out a program of enrichment only for your ruling class—your master race, as it were—as if the others, the merely average, do not concern you, since they are incapable of achieving the life of the mind you hold up as the highest ideal. You're just like our warrior mother in dismissing the average human life as possessing no value, no dignity, concentrating all your efforts on producing a class of the exceptional, the privileged, and the powerful.

PLATO: I have never been interested in producing the most exceptional class of person simply for the sake of their exceptionality. The city we want is one in which all the citizens are enabled to achieve the virtues and thus are happy (*Republic* 420b; *Laws* 630c3–6, 705d3–706a4). That would be a just city, since it does justice to all. And this requires rulers who are at one with the city and will never exploit it. So, for example, I do not think the rulers should be able to own substantive private property, for substantive property will immediately make them

---

* L. Brandwood in his *A Word Index to Plato* lists over sixty citations in the *Republic* to the noun variants of *paideia* and to the verb form *paideuein* in reference to education/culture and the educational process. The references to play/game(s) in its noun form—*paidia*—occurs over twenty-five times, and in its verbal form—*paidzein*—over eight times in the *Republic.* Both terms are linked with the education and activities of children—*pais* and *paides*—but also with the education of philosophers. The three terms *paideia,* the word for education/culture; *paidia,* the word for play/game/pastime/sport; and *paides,* the word for children, have the same root. See "Play and Education in Plato's *Republic*," by Arthur A. Krentz, http://www.bu.edu/wcp/Papers/Educ/EducKren.htm.

citizens of the city of the rich, with its own special interests to protect. And there are other privileges, besides wealth, that the citizens will enjoy but which should be denied to the rulers (*Republic* 416d–421c). The guardians of the just state should be the most underprivileged of all its citizens. It is an essential feature of the just state that the wealthy be kept away from political power and that the politically powerful be kept away from wealth.

BURNS: Fascinating. Absolutely fascinating. You must have some pretty strong views about campaign finance reform.

PLATO: The temptations of power are enormous. . . .

MUNITZ: Power corrupts, and absolute power corrupts absolutely.

PLATO: Yes, precisely, Dr. Munitz. Since absolute power corrupts absolutely, I am concerned that the ruling class—and I don't deny that that is what they are chosen to be and what they are trained to be, those whom I call the *guardians*—be, as far as is humanly possible, incorruptible, held in bond to the moral order. So long as they are, the just city will last.

MUNITZ: So the unguarded guardians will guard themselves. And how likely is that?

PLATO: Not very. Which is why I took such pains, in my beautiful city, to explain how the guardians should be trained to become such exceptional men and women as to be untouched—again, as far as is humanly possible—by the normal temptations of humankind. And yet even so I predicted, if you might remember, how even such a state will eventually unravel (*Republic* 546a–580b).

MUNITZ: Well, might I humbly suggest that, given how difficult it is to keep the power elite clean from corruption, it would be better to reconsider your entire social structure and allow citizens control over, and accountability from, their so-called guardians? Guardians, indeed! You may counter that you were only theorizing in a utopian fashion, but the values are real enough, and I think them profoundly offputting, first and foremost the paternalism that has free adult citizens requiring guardians assigned to them at all, as if they were orphaned children. Would it not be better to try and rear *all* citizens so that they can assume full power over their own lives as fully functioning grownups, according them the dignity and autonomy of responsible human beings, instead of putting them under the guardianship of those who would think and act for them?

PLATO: Thinking is very hard.

MUNITZ, *glaring:* Oh, so only your elite are allowed to think!

PLATO: All the citizens think, of course, to the best of their abilities.

MUNITZ: But some people's abilities will allow them to make all the important decisions.

PLATO: Just as some people's athletic abilities will allow them to compete at the Olympics.

MUNITZ: Oh, please, there's no comparison! You can deprive people of the opportunity to compete in the Olympics and you're not depriving them of the dignity of their humanity. But if you deprive your citizens of the right to call their rulers to account, then not only do you set the stage for tyranny but you diminish your citizens to the status of dependent children. And this is rendering them a grievous harm—even if the rulers have their best interests at heart.

PLATO: You are a true democrat, Dr. Munitz.

MUNITZ: Have you no faith in democracy?

PLATO: Very little, I'm afraid.

BURNS: I'm reminded of Winston Churchill's remark, that democracy is the worst form of government, except for all the others that have been tried from time to time.

MUNITZ: I wonder whether you would get Plato to agree to that.

PLATO: Your democracy is quite different from the one with which I am familiar from my own Athens, and I have been trying to understand it. The Internet is invaluable but what I have not been able to figure out yet is if the Internet itself strengthens your democracy or weakens it.*

BURNS: Well, when you figure it out, will you let us know? Anyway, as fascinating as all of this is, I'm afraid we're straying far from the topic at hand, which is, if I remember correctly, the rearing of children. Now then—

MUNITZ: Excuse me, Mr. Moderator, but if I might just be permitted to ask one pertinent question of Plato, quite apolitical. You do believe, don't you, that your method of child-rearing produces the best possible person, the—to use the odious phrase of tonight's event—most exceptional person?

PLATO: I do. I don't deny it (*Republic* 456e).†

---

* See pp. 358–359 below.

† "[A]re these not the best of all the citizens? And will not these women be the best of all the women? Is there anything better for a state than the generation in it of the best possible women and men?"

MUNITZ: And what is the measure of this person? On what scale is it decided that his exceptionality is the exceptionality that matters?

PLATO: Reality is the measure.

MUNITZ: And exactly *whose* reality would that be?

PLATO: Nobody's. Everybody's. That which simply *is*, the same for all of us, out there to be discovered.

MUNITZ: That which simply is because the ruling class *says* that it is.

PLATO: That is to get it exactly backward.

MUNITZ: Perhaps *your* way is to get it exactly backward. Who are you to say otherwise?

PLATO: Not only do I say otherwise, but so, too, do you, Dr. Munitz. You have already said otherwise, and quite forcefully.

MUNITZ: *I?* I hardly accept such a hegemonic vision of reality, which amounts to one more way for authorities to impose themselves on the powerless and deny them their rightful autonomy.

PLATO: And yet, in your eloquent restatement of the Myth of the Cave, you stressed the difference between the shadowists and the truth seers.

MUNITZ: The only realities I recognize are those embodied in the personal suffering buried deep in each person's history. And that reality is not for any authority to authorize or not; that, at least, belongs to the person whose reality it is.

PLATO: But it is a reality that few are able to see for themselves, according to your own account, but which you, Dr. Munitz, have seen and courageously try to get others to see, no matter that they reject the truth and make your life all the harder in their rejection.

MUNITZ: Yes, that much is very true.

PLATO: And there is far more of the reality that you have seen and the others have not, extending, as you yourself put it, into the normative sphere. Not only do you know of the personal suffering buried deep in each person's history, but you are outraged at this knowledge, knowing how wrong are both the child's suffering and also the stunted life to which it will lead. You know that it is injustice for one person to take away what rightfully belongs to another, and so you judge it unjust for the parents to deprive their children of the possibilities for joy and self-discovery that you know all people have coming to them. And when people who are chained so that they see only the flitting shadows on the back wall come to you for help with their lives, then you, knowing

far more than they do of the reality that led them to their false appre-
hensions of reality, help them to open their eyes to reality, to recover
their sight of that which *is*. You show them that which *is*, even though
all but you deny it to be that which *is*, because only in apprehending
that which *is* can they free themselves from their pain. Or have I not
understood the nature of your life's work?

MUNITZ, *almost whispering:* You have understood exactly my life's
work. I have never heard it better described.

PLATO: And when I, therefore, say, Dr. Munitz, that it is mistaken
to say that reality is whatever the most powerful say it is, I only concur
with you. And, conversely, when I say that it is right that the guardians should be those who are capable of apprehending reality, and most
importantly the aspects of reality that account for goodness and justice
and wisdom, then I would expect that you would concur with me. Let
it be reality that chooses the powerful, rather than the powerful who
choose reality. Isn't that less tyrannical?

MUNITZ, *still speaking uncharacteristically softly:* But then you enthrone
reality as the tyrant.

PLATO: It is a better tyrant than any one of us, certainly with more of
a right to impose itself on our minds than any human being possesses.

MUNITZ, *gathering steam:* But your guardians would still be human,
all too human, and, with no other humans to control them, they'll feast
on their power and fatten into fascists in no time.

PLATO: There is, of course, that danger, human nature being what
it is. And this is why I tried to impose over them the greatest clamp
of which I could conceive. Something far more forceful than other
humans is meant to limit the power of the guardians, and this is reality
itself. Any grandeur they might feel in their own position will seem ris-
ible to them compared to the grandeur of what *is*, awash in beauty and
goodness. Do you think that a mind habituated to thoughts of grandeur
and the contemplation of all time and all existence can deem this life of
man a thing of great concern? (*Republic* 486a). And what an enlarged
sense of responsibility the guardians will feel toward others, a tender
sense of caretaking for those who are deprived of that very vision that
gives blissful meaning to their own lives. It would not occur to them to
exploit those who are not capable of seeing what they see, no more than
it would occur to you, Dr. Munitz, to exploit the people who come to
you for your help in easing them toward the light. You yourself turned

away from the training you had undergone, even though it gave you so much power over others, precisely because you, seeing more of reality, understood that indulging that possibility of power is unthinkable. You yourself, in seeing what *is* and feeling the responsibility toward your fellow creatures that such seeing engendered in you, have demonstrated some of the qualities I would require of the guardians, and, I would even say, you have already attained in your life the position of a guardian.

MUNITZ: Are you saying that you would have me be a guardian in your utopia?

PLATO: The conclusion to be drawn is obvious.[*]

MUNITZ: I am overwhelmed.

BURNS, *grinning:* Well, Dr. Munitz, you got in the easy way, without going through Plato's extensive program of child-rearing.

MUNITZ: I assure you that it has not, any step of the way, been easy.

BURNS, *still grinning:* Okay, well, let's get back to Plato's recommendations for raising the exceptional child—raised, as he just pointed out, not so much for the sake of the exceptionality of the child but rather for the sake of society, to ensure that those with power don't abuse their power.

ZEE: But, Plato?

PLATO: Yes?

ZEE: What about the city of pigs?

PLATO: Yes?

ZEE: Well, isn't it also imperative, for the good of society as a whole, for the good of society in the future—and that's what we're really speaking about when we're talking about raising the best possible children—isn't it imperative that we form the habits of excellence, driving ourselves to gain more and more distance from the pigs, who just accomplish nothing with their lazy, lay-about lives?

PLATO: The real question you are asking is whether what we want from society is for it to protect us or to perfect us.

BURNS: Well put. And your answer, Plato?

PLATO: We must ask, first and foremost, for it to protect us—protect us from our outside enemies and also from the worst that we can do to one another.

---

[*] I'm not sure whether Plato is just managing Munitz here or is really implying that she's guardian material. Needless to say, he didn't have anything like psychotherapists in mind when he spoke about his guardians.

BURNS: Call me a cockeyed optimist—my wife often does—but why can't we ask both of those things from society, ask that it both protect *and* perfect us?

PLATO: I suspect your wife would call me a cockeyed optimist, too.

BURNS: Yes! That's what I wanted to hear from you! Because your beautiful city is meant to do both, isn't it?

PLATO: It is meant foremost to protect us. But by demanding that the best among us perform this task—and they are the best among us precisely because they would also protect us from themselves—it demands the program of perfecting.

BURNS: So the perfecting is the icing on the cake.

PLATO: Where I come from, we call it the honey on the baklava.

BURNS: The honey on the baklava: I like that! So now let's get to the honey. Wasn't that a line in some movie? Show me the honey! *Feeble laughter.* Okay, sorry. My sense of humor is something my wife has also commented on, to little effect, obviously.

Okay, let's get down to brass tacks here, your practical suggestions for how we can be perfected. You start very early in a child's life, and you single out a certain kind of child as having the potential for going all the way to the top. How can you see that kind of potential so early in a child's life? What is it that you're looking for? Is it a matter strictly of IQ, general intelligence?

PLATO: No, not strictly. A quick mind, yes, is part of what it takes to go, as you put it, all the way to the top, but there is also the all-important matter of character.

BURNS: Character, yes, exactly!

PLATO: And, in particular, mettle. Mere intelligence without mettle makes for a feeble material. There must be something of what we in Greek call *thumos.*

BURNS: *Thumos.* Can you give me an example?

PLATO: Well, I would say that Professor Zee presents a very good example of *thumos.* My warrior class is composed of those who are most distinguished by *thumos,* and Professor Zee is a warrior mother.

MUNITZ: I think you might have misunderstood the title of that book of hers, Plato. She's clearly using the word "warrior" figuratively.

PLATO: I understand. She is using the term "warrior" to signify a certain type of person, a certain type of nature, the type who has a great desire for recognition and glory.

MUNITZ: For self-recognition and self-glory.

PLATO: You say that as if there is something wrong with it. And yet this is the driving goal of the thumotic person.

MUNITZ: Yes, I do most vehemently say there's something wrong with an obsession with self-recognition and self-glorification. And I should have thought that you would agree with me on that yourself, and just as vehemently, you who paint a picture of the virtuous person as transcending his own petty self, subsuming his personal ends and desires under the requirements of the good.

PLATO: You cannot change human nature. You can only change the *polis* so that what is potentially dangerous is rendered innocuous or even, in the best-ordered society, beneficial. The desire for distinction is present in a great many, and in some natures it is a driving force, and it yields in such a person a vigorous spiritedness, characterized by a robust desire to distinguish him- or herself. Those who lack this vital spiritedness will never do much harm in the world, it is true, but they will never do much good either.

BURNS: So you approve of *thumos*?

PLATO: It isn't for me, or for any of us, to approve or to disapprove of human nature. It's only for us to try to work with it.

BURNS: And in fact, it's the children who are well endowed with *thumos* whom you single out for your enrichment program.

PLATO: Yes. Spiritedness shows up early, shows up in the manner in which children play. We should pay close attention to how children play, since the nature of their individuality is revealed in it (*Republic* 536). And a child who evinces great pride in its games, exerting itself with focus and passion, is the kind of person who will be dedicated to perfecting himself or herself. Mediocrity is not an option for such a soul. It recoils at the idea. Professor Zee presents a fine example of this spiritedness.

MUNITZ *mutters something inaudible, perhaps in German.*

BURNS: So the die is cast, you're saying, from an early age? You can't give spirit to an unspirited child?

MUNITZ: You can certainly go in the opposite direction and kill the spirit in a child.

PLATO: That would be a grievous thing to do, not only for the child but for the greater good.

BURNS: Because you make this spiritedness a requirement for the sort of perfecting of the child that you have in mind, the one you believe has the potential to do so much good in society.

PLATO: Yes, though again I would stress there is an equal potential for harm. *Thumos,* on its own, can lead to terrible excesses, to a brutal and savage manner of person, his or her pride and ambition crowding out all other values. Such people will pursue their drive toward self-aggrandizement with single-minded fanaticism. And when such thumotically endowed persons possess intelligence and charisma as well, the damage they can wreak on the world is profound.

BURNS: The sort of monsters to whom Dr. Munitz previously referred.

PLATO: Yes, sometimes monsters, sometimes simply unconscionably charming rascals.

BURNS: So even intelligence and spiritedness aren't enough. Is the rest all a matter of training then, the projectification that Dr. Munitz referred to before? Can you take any child who meets your requirements of intelligence and spiritedness and turn him into the kind of exceptional person you're looking for here?

PLATO: To turn him or her into such a person, there is one more character trait, which is also inborn and essential. It is a trait that demands both intelligence and spiritedness but is something additional, since I have certainly known those who lack nothing in intelligence and spiritedness who yet lack this quality (*Republic* 375e). And here I would point to Dr. Munitz as a paragon.

Munitz raises her eyebrows, which are pronounced and highly articulate, almost prehensile.

PLATO: I don't know whether to describe it as a desire or an antipathy, for it is equally both. It is an inborn horror of being deceived as to the nature of things, and it is an inborn desire to know the truth as to the nature of things. Perhaps the best name for it is love of wisdom, (ibid.), and it is something different from intelligence and different from knowledge. Those who have this trait love the truth not because it is like this or like that. They love the truth simply because it *is* the truth and are prepared to love it no matter what it turns out to be. They will stick to a view just so long as it seems to them the truth and will not be seduced away from that view no matter what others are telling them, or what flashier and more attractive options are dangled before them; but they are also the least reluctant among all people to abandon a formerly loved view, if once they become convinced that it is not true. They are always on the scent of the truth, like dogs, who are the most philosophical of animals (*Republic* 375d–e). And this trait is something different from intelligence and spiritedness, even though it enlists intelligence and spiritedness in its service. But it is surely something different since intelligence and spiritedness can exist without it. I have known certain types—notably poets*—who experience the aesthetic demands of their imagination far more keenly than they do the love of truth. If some proposition floods their sense of beauty, they will believe it with all their heart and soul, and express it with such touching beauty that others, too, will be induced to believe. Their sense of enchantment shapes their conception of the truth, rather than the other way round, as it is when the art is of a kind of which a philosopher can approve, the art which knows and imitates the forms.

BURNS: So then there is art of which you approve.

PLATO: Certainly. *Very softly.* Perhaps it can even be said that I aspired to it myself.

BURNS: But didn't you, in fact, deal quite harshly with the poets in your utopia? Didn't you banish them (*Republic* 398a–b, 606e–608b)?

PLATO: That is an overstatement.

BURNS: But you do advocate censorship of the arts? I know I'm violating my own rule that we should avoid politics tonight, but I've always wanted to ask you about that. It's always bothered me.

PLATO: Beauty has a profound effect on us, drawing out from us our love. It is the one thing that can capture our entire attention because of our love for it, that is to say, the one thing, besides our own selves, to which we naturally and obsessively attend. What is it that can break our natural enchantment with our own selves if not our natural enchantment with beauty? If it were not for beauty, there would be no hope of getting us to pay serious attention to anything that does not directly concern us, and no hope of getting us to see the things that truly matter outside the limited personal scope.† It alone can commandeer our dis-

---

* Plato's inner conflicts over art, and most particularly poetry, are flung across several of his dialogues, including *Ion, Phaedrus, Republic* (most especially Books III and X), and *Laws*. His conflicts form the theme of *The Fire and the Sun: Why Plato Banished the Artists*, by Iris Murdoch (New York: Viking, 1990). Murdoch's final verdict is that Plato thinks our response to beauty too ethically important for the artists, unformed by philosophy, to be permitted to manipulate it as they will. "Plato wants to cut art off from beauty, because he regards beauty as too serious a matter to be commandeered by art" (p. 17). Murdoch is not denying the important epistemological and metaphysical and ethical role that beauty plays for Plato—our sense of beauty leading us to the truth because beauty is embedded in the truth. But she thinks that Plato has no faith that artists will use their sense of beauty to get to the truth.

† See the *Symposium*, Diotima's speech, 209a–212e, which includes this passage: "And if, my dear Socrates, Diotima went on, man's life is ever worth the living, it is when he has attained this vision of the very soul of beauty." (Notice the language of "life-worthiness.")

interested passion. So beauty is a serious matter—too serious, I have often suspected—to be left in the hands of those who are drunk on its enchantments, and so will act irresponsibly in its presence, which is as good a way of describing the artists of whom I am forced to disapprove.

BURNS: It just seems so strange to me that, with all your talk about beauty and how central you make it in your philosophy, you'd show such little respect for the artists. And it's even stranger considering that you're a great artist yourself. Your dialogues are an art form, aren't they?

PLATO: I would hope that I have not shown disrespect to the artists simply by pointing out that a great artist is not, simply by virtue of being a great artist, a great philosopher, or even a passable one, and so our awe of his artistry should not give us any reason to pay attention to what he has to say regarding how things actually are and ought to be.

BURNS: So you don't think we should regard our artists—our painters and novelists and poets and movie and theater directors and actors—as public intellectuals?

PLATO: Not simply by virtue of their performing their art well, no. Sometimes they're public intellectuals. I considered Euripides to be one.

BURNS: He was quite critical of his society, wasn't he?

PLATO: He was.

BURNS: But perhaps no more than you?

Silence.

BURNS: I'm sorry to persist, but this business about art just sticks in my craw, and I guess this is my opportunity.

PLATO: Yes?

BURNS: I'm relieved to hear you're impressed with Euripides, but it bothers me how unimpressed you are by so much great art. Doesn't an artist have to be privy to at least as much wisdom as a philosopher to be able to move us so artfully?

PLATO: There are many ways that we are moved, and not all of them involve wisdom. How much more tractable those problems of democracy we were just now discussing would be were this not the case. There are those who know all too well how to move the people artfully even when it is very much to the detriment of the people.

MUNITZ: You mean the powers of the demagogue, no doubt.

PLATO: Who are often quite artful. Our public intellectuals must be driven by the quest for what is true and what is good and how lives should best be arranged in the light of what is true and what is good. Our sense of beauty is invaluable in leading us to the truth, because the truth, quite simply, is beautiful.

BURNS: So then what do you have against artists, since they're more devoted to beauty than anyone? You're not saying that artists are demagogues, are you?

PLATO: Those who are truly devoted to beauty are equally as devoted to truth and to goodness as well. Such a one will succeed in bringing to birth, not phantoms of virtue, because he is not grasping a phantom, but true virtue, because he is grasping the truth (*Symposium* 212a). And that is the art that I love. It is the art that is moved by the same love of truth of which we were just speaking when specifying the characteristics we will be looking for in our children and trying to cultivate further in them.

BURNS: Okay, I'm glad you've brought us back again on topic, which is raising our kids. So you think there's some kind of innate difference there, too, that not everybody is born to love the truth?

PLATO: There are people in whom this love of truth is a driving force, but not so very many. For them, the pleasures of learning are unmixed with pain, and they belong not to the general run of men but only to the very few (*Philebus* 52b). For them, there is pleasure in the truth, no matter its nature, simply in the thought that it is the truth. I think, perhaps, that Dr. Munitz is such a person.

MUNITZ, *clearly touched:* That is one of the finest compliments I have ever received.

PLATO: I do not say it to compliment you. I say it because it is true. Dr. Munitz's love of truth stands out prominently, but in children it is not so easily detected. Intelligence and spiritedness announce themselves in the manner of children's play, but this other trait, the love of truth, is more hidden. And so it was that I proposed a somewhat artificial test for detecting it.

BURNS: So instead of submitting your future leaders to a battery of IQ tests, you submit them to a truth-loving test, measuring their ΦQ, as it were.

PLATO: As it were. What I proposed was having our children be told glorious tales to stir their imaginations, very much stressing all the time that these tales were true, and then seeing which among the children can resist them, can see the logical inconsistencies within these tales, and see all their inconsistencies with other truths that they have been told (*Republic* 413c–414a).

MUNITZ: Sounds a cruel and unusual form of testing to me, Plato, to deliberately take advantage of a child's tendency to trust the adults in authority. Children, who have so much to learn in so short a time, have evolved the tendency to trust adults to instruct them in the collective knowledge of our species, and this trust confers survival value. But it also makes children vulnerable to being tricked, and adults who exploit this vulnerability should be deeply ashamed. It is altogether ironic that, just because truth-loving is so prized by you, you would traduce the brightest and most spirited of children with deliberate deception. Do you not see the logical inconsistency in your proposal?

PLATO, *smiling*: As I have said, Dr. Munitz, you are particularly well endowed with the love of truth. It throbs in you as Professor Zee's *thumos* throbs in her.

MUNITZ: Which is why I will not allow myself to be misled by your blandishments. Artful as your flattery is, it will not deter me from pointing out that exactly the same inconsistency that I have noted in your test for ΦQ permeates the fabric of your utopia, in which you valorize truth and yet authorize your guardians to put forth untruths, so long as they are what you dub "noble lies," by which you mean lies that serve the greater truth, which the guardians alone can see.* Putting aside the little matter of logical inconsistency here, as well as the infantilizing of citizens, who are being robbed of their dignity in being fed untruths, let us just consider such a proposal from a pragmatic point of view. I do not think we have to tax our imaginations very hard to come up with ways in which such a license to lie, on the part of political leaders, can lead to horrific abuses. History furnishes us with many examples of what can happen when a populace is bred to be passive and isn't provided the tools to see through their guardians' so-called noble lies, from the self-serving fabrications told through the ages by religious establishments to the equally self-serving deceptions told by oligarchs of capitalism, believing that their commitment to their truths of the free market justifies their noble lying. Not to speak of the dictators of totalitarian governments, who disseminate their mythologies, which

---

* "Then if it is appropriate for anyone to use falsehoods for the good of the city, because of the actions of either enemies or citizens, it is the rulers" (*Republic* 389b–c). See also *Republic* 414b–415d, for the famous, so-called "noble lie."

often demonize minorities to further the greater good of social cohesion. Even granting your exceptional specimens, culled from the crowd and cultivated to see the truth, how can you prevent their being subject to their own self-deception? If you think that you, or anyone else, can devise a program of child-rearing that will prevent self-deception in our guardians—*she drawls the word out sarcastically*—then I am afraid that you yourself are very much self-deceived, which is why it is only asking for the worst kind of trouble for leaders not to be held accountable by those whom they lead, who, I would suggest, have just as much moral authority to demand moral accountability as the guardians. It is the essence of the moral standpoint that each person has the authority to demand accountability. Anything less is an insult to human dignity.

BURNS: You're raising urgently important questions here, Dr. Munitz, and I think we're all grateful to you for bringing them so forcefully to our attention. In fact, I think everyone here would agree that the questions you raise, Dr. Munitz, are so important that they deserve a dialogue in their own right. *Zee is energetically nodding yes.* Maybe we can even reconvene this panel, only this time exclusively debating politics. But for tonight, I don't want the equally urgent issues raised by child-rearing to get short shrift.

And I'd feel remiss, Plato, if I didn't address your attitude toward parents in the whole business of child-rearing. We take it for granted in our society that these questions of how best to raise our children are for parents to ponder and decide, and the best-selling books on the subject are all directed at parents. But you, in effect, remove these decisions from parental jurisdiction and assign them to the state. In fact, your most radical proposal would take parents out of the picture altogether. It reminds me of the early kibbutz system, with the children raised communally. But that communal system didn't turn out well for the kibbutzim, and almost all of them have abandoned it and gone back to having the children live with their parents.

PLATO: I confess I know nothing about kibbutzim. I will have to google it. What went wrong?

BURNS: Well, I think the children longed for their parents, and the parents longed for their children.

PLATO: My proposal for the ideal society would circumvent that problem. What I proposed was that neither parents nor children would know who their blood relations were. Parents would not know which children were theirs and so would direct a more generalized love and sense of responsibility toward all the children who were their own children's age. And children, not knowing who their parents were, would feel a generalized love and veneration for the whole generation of parents. Of course, this was all put forth in an idealized theory, meant to answer the question of what perfect justice might look like. The sense of cohesion and unity in such a society, in which all would be acting for the sake of all, not out of compulsion but out of solidarity, seemed to me to promise a high level of justice.

BURNS: How does a warrior mother react to Plato's suggestion that in a perfectly just society you wouldn't know who in a whole generation of kids was your own?

ZEE: Well, speaking just as a mother—I don't think the warrior aspect is relevant—I'm sort of vehemently aghast! If justice consists in nobody being deprived of what's their own, which is what I understood Plato to be saying earlier on, then how can justice demand that a person be deprived of what's more her own, and what means more to her than anything else in the whole world, namely her very own children? I mean, take anything away from me—my shelter, my worldly goods, even my freedom—but not my children! You deprive parents of the privilege and joy of rearing their own children and you're removing the most meaningful part of our lives. That generalized love that Plato just mentioned would be a ludicrously impoverished substitute, and, forgive me, but I can't help thinking the proposal could only be made by a man who never had children of his own. That generalized love could never, not in a million years, take the place of the fierce-as-a-tiger attachment that every mother feels toward her own flesh and blood, and the sort of sacrifices that she's prepared to make for them to ensure that they live the best life possible for them to live.

Sustained applause from the audience.

BURNS: And you, Dr. Munitz?

MUNITZ: Well, I was also going to dismiss Plato's proposal as monstrous, but listening just now to the last speaker I feel myself warming toward his view. The fierce attachments she's speaking of here are, at heart, the projections of the narcissism that causes such a parent to see her children as extensions of herself. And since narcissism is rampant in our society—what Plato praises as spiritedness I would suggest is better described as runaway narcissism—the dangers to the developing child are legion. These fierce attachments are treacherous precisely because they're fierce, and they lead to the fierce projectification by which such parents try to produce children that will sustain their narcissistic fantasies. At least Plato wants his exceptional children to be exceptional for the collective good that they can render, whereas warrior mothers want their children to be exceptional simply because they are their very own, their flesh and blood, as she so graphically, if primitively, put it. No, Professor Zee, a child's flesh is its own flesh and a child's blood is its own blood, and it is a crime against the child to appropriate what is theirs for your own.

ZEE: I don't think that's exactly fair! Yes, a mother loves her child fiercely because it's her own, but that doesn't mean that that love ultimately reduces to narcissistic self-love! *Vigorous applause and even a few whistles, which Zee patiently waits out.* Are you really going down on record as being against a mother's love? *More applause.*

MUNITZ: No, I'm not going down on record as being against a mother's love. Though it's hardly relevant, I happen to be a mother myself, and it goes without saying that I love my children. And I truly mean that it goes without saying. I find it unseemly for a mother to repeatedly articulate to us that she loves her children. It's like a pig boasting it can root around in its filth. *Audience gasps. Munitz turns to the audience and smiles strangely.* Ah, you gasp. This is well and good, since it tells me who sits out there in cavelike darkness whose faces I cannot see. Can you contain your offended sentimentality and hear me out? I am simply trying to point out that a mother's love is a complicated thing, and to the extent that a mother cannot separate out her love for her child, who is an autonomous being, from her love for herself, then that love is as much a threat to the child's well-being as are all the external dangers that the mother so fears and tries to protect against with fierce-as-a-tiger vigor. It is the most pernicious of threats since it is by its very nature invisible to the mother. Plato's proposal, as extreme as it is, at least acknowledges the limits of parental omniscience on the matter of the child's well-being. That much I will say for his wild utopian proposal. But once again the lengths he is willing to go to in the direction of the authoritarianism of the state strikes me as naive at best, immoral at worst. The way to reduce the tyranny of the parent is not to transform the state and its educational system into the tyrant. One must find a solution that doesn't sacrifice the dignity and autonomy of all individuals. And you are entirely correct, Professor Zee, that this generalized love directed toward a generation cannot take the place of a personalized love, and the institutional attention cannot take the place of a more familial arrangement, whether conventional or not, that provides a highly individualized sense of attachment, commitment, responsibility, and yes, I would agree with you, intense, if not fierce, love. *Applause. Dr. Munitz grimaces.*

ZEE: Yes, I agree with you, Dr. Munitz, oh, I just completely agree with you! *Dr. Munitz stares fixedly at her, her eyebrows merging into a heavy line, as if trying to decide which category in the* Diagnostic Statistical Manual *to apply.* I think it might even be the end of the species! Why would parents even *have* children if they couldn't lay claim to them as their *own* children? The species would just end!

BURNS, *smiling:* What do you say to that, shall we say, *spirited* question, Plato?

PLATO: Yes, there is truth in what Professor Zee says, that the reason people are motivated to have children, and feel so strongly about their children, is bound up with themselves, with their own existence, and even with their desire to extend themselves beyond their limited existence into the future.

MUNITZ: Exactly. Narcissism.

PLATO: No, not exactly. Narcissus stared only at his reflection, an image, an *eidôlon* that partook of even less reality than he himself. What is of even more significance is that this love for the mere image did not pull forth anything from him, nothing that reached beyond himself. His was a love that gave birth to nothing.

MUNITZ: That was better. At least he did not bring people into the world—autonomous people with a right to their own existences—whom he would only regard as extensions of himself.

ZEE: Just because you love your children because they're yours doesn't mean you see them only as extensions of yourself! You're equating the two things, and they're just not the same!

MUNITZ: I am not equating them. Yes, one loves one's children because they are one's own, that much is trivial. If you'll remember, I wasn't even addressing my comment to you, but to Plato, to his statement that one's love for one's children is a desire to extend one's own existence. That is what I called narcissistic.

PLATO: Then, on that understanding of narcissism, which seems to me a perverse one, you are right. Mortal nature seeks as far as possible to be eternal and immortal, and this is one way, by producing offspring, that it is able to do so, through leaving behind another, a young one, in place of the old (*Symposium* 207d).

ZEE: You're not saying everybody has to have children, are you?

PLATO: Oh, no, not at all. There are many kinds of offspring. People are envious of Homer, Hesiod, and the other good poets because of the offspring they left behind, since these are the sort of offspring that, being immortal themselves, provide their procreators with an immortal glory and an immortal remembrance (*Symposium* 209d). And then there are those children left behind by the American founders, the laws they framed that continue to live to this day and bring justice to the state. These are children that bring far more glory to their parents than human children.*

ZEE: Exactly, that's exactly what I've been saying! What we want for our children is that they be people who bring forth these even greater children, which will bring far more glory.

MUNITZ: Glory to you.

ZEE: No, to them! What we want is for our children to live lives that are ultimately worth the living. That's what we're trying to figure out here. All parents want it for themselves, the life worth living, but they want it even more for their children. Or, at least, they feel like they have more control over creating the circumstances that will give it to their children.

MUNITZ: Yes, precisely. You are right. It is not narcissism. It is fascism.

BURNS: Fascism, Dr. Munitz?

MUNITZ: What would you call a point of view that distinguishes some lives as worth the living and others not. And these ones that are not: what shall we do with them? Shall we just round them up then and gas them? *An audible gasp from the audience. Burns is looking uncomfortable, sensing he might be losing control of the event.*

PLATO, *very quietly:* I think you misunderstand what Professor Zee

---

* Plato spoke of Solon rather than of the American founders: "You also honor Solon because of the laws that are his offspring, and there are other men in many places who are honored for other reasons. Among both Greeks and barbarians are men who produced many beautiful works, bringing forth *aretē* of every sort. Many shrines have been dedicated to men because of this sort of children, but none at all because of their human children" (*Symposium* 209d–e).

was saying, Dr. Munitz. She was not implying that the people who are living those lives which are perhaps not worth the living are themselves worthless. It is precisely because they are human beings, and therefore things of worth, that it is so important that their lives be worth the living.

ZEE: Exactly! Just what Plato said! I mean, imagine some child who's just raised in a tiny little cage, just kept alive in there but given nothing else. The reason that's so tragic for a child, as opposed to, say, for a chicken, is that it's a human being. I mean I'm not saying that factory farming is perfect, chickens have rights, too, but we just don't feel as awful when we hear that a chicken is living a life not worth living as when we hear that a person isn't living a life that's worth living.

MUNITZ: And I submit to you that to raise a child according to your methods, imposing on them the iron demands that they be off the charts, is to put them in a tiny little cage.

BURNS: Extracurricular activities! Why don't we talk about extracurricular activities? It's a question that every parent I know faces, and Sophie, you certainly put a lot of time and effort into your kids' extracurricular activities.

Zee nods her head vigorously.

BURNS: And we all know that colleges, at least here in the U.S.— I don't know how it is in your Academy, Plato—put a lot of emphasis on extracurricular activities, with the top-tier colleges, which you might think are geared more in the intellectual direction, looking—as every parent here who's been through this anxiety-producing process knows—for a certain ideal of the perfectly well rounded applicant. In fact, sometimes the ideal applicant seems so well rounded they remind me of that myth that you put into Aristophanes' mouth in your *Symposium*, Plato, that we'd all started out as sort of two people fused together so that we were perfectly round and could roll ourselves anywhere. *Audience laughs. Burns addresses audience.* Really, if you haven't read his *Symposium* I urge you to do so. Plato has Aristophanes trying to explain why it is that we fall madly in love with a particular person, and he comes up with this myth of a two-person fusion that gave us such completeness and hubris that the gods got angry and punished us by cleaving us down the middle. And so now we're all obsessed with finding our other halves—gays the person of their own sex they were once fused with, heterosexuals the person of the opposite sex—and when we find them we just want to fuse all over again with them, so much so that we'd rather stay, well, fused, physically fused, than eat or do anything else. *Audience laughs again.* Yes, it's a great dialogue, the whole thing, and I guess the source of our expression "Platonic love." But anyway, I'm getting off track. The point I was going after was that those colleges seem to want candidates that are so well rounded that they'd have to be two different people fused together with mutually exclusive characteristics! They have to be gung-ho athletes and sensitive artists, studious nerds and gregarious social networkers, future rulers of the universe and selfless altruists. You get the picture.

Now you, Plato, are, among everything else, the head of a famous university, really the prototype of all universities, and you've already mentioned how important you consider both athletics and music to be in raising children. Given what an intellectual you are—and let's face it, your idea of the exceptional child is basically someone who will grow up to be an intellectual—this emphasis on sports and music seems surprising. In particular, it's pretty amazing just how many pages of the *Republic* and also of the *Laws,* your later work, you devote to sports. Professor Zee, you certainly didn't encourage your children to pursue sports, as I remember.

ZEE: Well, no more than to keep them fit. I certainly didn't have any intention of ever becoming a soccer mom!

PLATO: Professor Zee's attitude seems reasonable to me. Children, even those who will grow up to be intellectuals, must be encouraged to acquire the habit of keeping fit, so that they possess a sound body to support a sound mind, but I also did not encourage, in my program of child-rearing, the extreme devotion to sports required of the professional athlete (*Republic* 407b).

BURNS: But you do spend a lot of time talking about sports. There seems to be something more than just sound minds in sound bodies that you're concerned with, or am I reading too much into it?

PLATO: You're right that I see something formative in sports, which fortunately also are, for most children, a natural form of play. But sports also provide lessons in the pleasures of self-mastery and self-discipline—something that Professor Zee, as a warrior mother, stresses. A child who is not naturally proficient in sports can usually attain a certain level of competence simply by putting in the hours. And because sports demonstrate that self-discipline is not incompatible with play, it's a model of what all learning ought to be.

BURNS: Even the highest learning? The kind of learning of a genuine intellectual?

PLATO: Even the highest. What is an intellectual but someone who has so disciplined his or her mind that he or she can take extreme pleasure in the free play of ideas?

BURNS: So you see this interplay of discipline and play as proceeding all the way up the scale? Instead of pay as you go, it's play as you go.

PLATO: The best thinking is always playful.

BURNS: That doesn't sound very warrior-like to me. What do you say, Sophie? Is Plato a little too frivolous for you?

ZEE: Plato ... frivolous? *She giggles and the audience laughs along with her.*

BURNS: No, but seriously, the tactics you employ as a warrior mother mainly consist in discipline. There's a lot of enforced practice involved, and there are even threats and punishments to get your kids to put in the hours of practice for their schoolwork and their music. And yet here's Plato telling us that true accomplishment is seriously playful.

ZEE: I agree with him entirely! It has to be fun and it has to be play. But nothing is fun if you're not good at it, and you can't get good at it unless you practice enough, and then it will be fun. The experience of mastering something that you thought you couldn't do is very empowering and it feels fantastic. But there's punishing work until you can get there.

MUNITZ: Which for a warrior mother like you justifies any extremes of enforced learning, including punishments and threats.

PLATO: I would hope not, and I do not think that Professor Zee really disagrees with me here—

ZEE: No, I don't! I agree with you entirely—

PLATO: Because of the important place she gives in the rearing of her children to music. Unlike you, Dr. Munitz, I did read *The Warrior Mother's Guide to Producing Off-the-Charts Children,* and I was impressed by how much time she devotes to the musical training of her children. For a warrior mother, raising her warrior children, music is essential. I, too, stressed that my future warriors must have a musical education. They must have sports to strengthen their natural spiritedness, and music to soften it, so that their pronounced *thumos* does not harden into something vulgar, harsh, and savage (*Republic* 410b–412a).

MUNITZ: But don't you see that a warrior mother just transforms music lessons into a form of competition, an *agon,* and so turns it into something that not only is vulgar, harsh, and savage but also likely a genuine agony for her kids? It's just one more way for her kids to beat out other kids—put their musical accomplishments on their résumés and use them to push themselves ahead of the crowd? For her, music is really no different from team sports, which is why she doesn't need any sports to strengthen her warrior children's *thumos*. She uses music, which is far classier and, in her social circle, brag-worthy, toward the same end.

PLATO: But the music will enter in just the same, and that is what is important. Musical training is a more potent instrument than any other, because rhythm and harmony find their way into the inward places of the soul on which they fasten (*Republic* 401e–402a). Music enters in and with it the sense of a beauty that stands in no need of any justification beyond itself. To be touched by this beauty which ends in itself, existing apart from that which can be made use of in the cause of self-advancement, is particularly beneficial for warrior children being raised by warrior mothers.

MUNITZ: Professor Zee is a lawyer, not a warrior.

PLATO: They are the same. *First time Plato gets a hearty laugh, which he ignores.* I mean by warriors simply those who respond most strongly to the sense of distinguishing themselves in victory and who revel in the thrill of competition. The teaching of the warriors, who constitute a far larger class than the class of intellectuals, must appeal, above all else, to the desire for recognition. And just as there is an important place in any society for the warriors, whether soldiers or lawyers or other specialists in fighting, so there must be pedagogical methods that quicken their *thumos*, appealing to their love of recognition and glory.

MUNITZ: And again, I repeat, *self*-recognition and *self*-glorification, which dubious goals are only strengthened if you quicken their *thumos*.

PLATO: All the more important, then, for them to have some music in their lives! For what should be the end of music if not the love of beauty? (*Republic* 403c). And what else is there to break the single-minded attachment to the self and its ambitions, occluding the sight of anything beyond the self and its ambitions, if not the love of beauty? If there is a tender spot within them, then music will find it and sink in. Any child who responds to music is responding to beauty, and any child who responds to beauty can be educated. Conversely, a child who is altogether indifferent to beauty cannot be educated, but fortunately there are few such children.

BURNS: It's surprising to hear you emphasize beauty so much. Most of the artists I know are too self-conscious to even utter the word "beauty." It's a word that's lost respectability in artistic circles.

PLATO: Artists embarrassed to appeal to beauty? I hardly know how to respond to such a situation. It seems beyond comprehension. My experience with the artists is that they are so besotted by beauty that they let it overwhelm them, and so lack that love of truth that Dr. Munitz so forcefully exemplifies. But artists who do not value beauty? What can be the good of them?

BURNS: Well, that's a topic for another dialogue as well. Meanwhile I'm intrigued how often you mention beauty in relation to raising the exceptional child. It's already come up a dozen times at least.

PLATO: The object of education is to cultivate the love of beauty. A teacher is charged with bringing his or her student into contact with the beauty that answers to that student's type of character and mind.

BURNS: So an educator is a kind of matchmaker. *Breaks into song:* "Matchmaker, matchmaker, make me a match, find me a find, catch me a catch." *Audience laughter.* Sorry, Plato. Maybe I'm getting carried away with your claim that thinking is playful.

PLATO: Quite right. Have you heard of the Myers-Briggs Type Indicator?

BURNS: No, I can't say that I have.

MUNITZ: Well, I have, of course. It's a psychometric questionnaire based on Carl Jung's personality typology, which, depending on the answers you give to some questions that probe the way you perceive the world and make your decisions, will categorize you as a certain type of personality.

PLATO: Yes, that is it exactly. I find it fascinating.

BURNS: You took the test?

PLATO: Of course, I took the test. You can do it entirely for free on the Internet.

MUNITZ: Let me guess: you came out an INTJ. *Turning to Burns.* That means an Introverted, Intuitive, Thinking, Judging Type.

PLATO: Dr. Munitz is right. I am an INTJ.

MUNITZ: The type which is characterized as a Mastermind.

BURNS: Well, that's hardly surprising! If Plato isn't a Mastermind, then who is? But tell me, Plato, are you bringing this up because you think that these personality types are correlated with different possibilities for learning, or I suppose I should say, putting it into your language, susceptibilities to different types of beauty?

PLATO: Yes. And what has furthermore fascinated me about these personality types is their degree of heritability. I had had a very dim—which is to say non-quantitative—grasp of the hereditary aspects of personality, for which the metallic composition of the three classes of people I referred to in my so-called noble lie was an exceedingly crude metaphor. According to the modern researchers, the hereditary input accounts for about half of the influence on variations in personality. The rest of the influence, they theorize, is due both to the environment, in which I should hope they include how they were educated by parents and teachers, and what they simply dub "randomness."

BURNS: So you're suggesting that there is an innate aspect about which beauty a particular child can love and therefore be educated about?

PLATO: Some are suited to be lovers of the beauty of sounds and of colors, of the words and the meanings of poets, of human faces and bodies, of the laws enacted by a just government or the laws that govern the celestial motions. Some are suited to be lovers of mathematical beauty and of moral beauty, to be lovers of the most abstract beauty that is inscribed in the necessity of being.

BURNS: And what you're saying is that you wouldn't force these various kinds of beauty on those who aren't innately suited to them?

PLATO: Certainly not. To what end? Forcing beauty is erroneous. Beauty is that which provokes desire and love. But we are not all the same in regard to the beauty that it lies within us to desire and love.

BURNS: So let's say you have a child who is impervious to, say, mathematical beauty. I think I probably qualified as such a child. *Audience laughter.* In fact, I don't know, I hear you say the words "mathematical beauty" and I'm sure you mean something by it, but really I have no idea what. *Audience laughter.*

PLATO: Nor should you.

BURNS: You really mean that? You wouldn't have required that I take algebra in the ninth grade, geometry in the tenth, trigonometry in the eleventh, and to tell you the truth I can't even remember what math I took in twelfth grade, but whatever it was, you wouldn't have made me take it?

PLATO: I see no reason why anyone should be force-fed information that does not agree with their cognitive digestive system. It will not nourish them. It will pass right through their system, just as soon as they have passed through the school system. How much of what got drilled into you do you remember now?

BURNS: I couldn't tell you what a sine, cosine, or tangent is if my life depended on it! *Audience laughter.* And remember those word problems they used to torment us with: if one train leaves New York for Boston at 11 a.m. going 75 miles an hour, and another train leaves New York for Boston at 11:30 going 100 miles an hour, at what time will the operator of the first train text his wife that he left his lunch bag on the kitchen counter?

PLATO: I do not understand the question.

BURNS: Wow! I stumped Plato! Anyone?

ZEE: At *no* time, since it's illegal for an operator to text while he's operating a train! *Audience laughter and applause.*

BURNS: And once again, ladies and gentlemen, Sophie Zee demonstrates the high achievement attainable by warrior mothers! But actually I want to ask you, Sophie, your take on Plato's views. I would imagine you might find Plato's easing up on certain requirements, like taking math in high school, somewhat soft.

ZEE: *Soft?* I don't know about *soft.* I was sitting here thinking how harsh his views are.

BURNS: Harsh, really? He wouldn't have made me take geometry. That sounds pretty sweet to me.

ZEE: And he also wouldn't have granted you admission into his top-tier Academy, since you wouldn't have passed his one requirement.

BURNS: Yes, that's true. I'd forgotten about that.

ZEE: I don't know, but I have to ask why Plato takes such a fatalist attitude toward human nature. He seems to be implying that some people have it and some people don't, and those who don't, well, there's nothing that they or their parents can do about it. That just seems such a defeatist attitude, and if people accepted it, a lot of kids who could achieve excellence are never going to.

I'll tell you a true story. When my daughter Mimi was in the third grade, she suddenly started doing badly in arithmetic. They were doing long columns of addition, having to carry, and she just kept making careless mistakes, one after the next. At report card time, before the teacher gave them out, she called Mimi over and had a private conference with her to prepare her for the shock, which was that she had gotten a C+, which of course for Mimi might as well have been an F− since she'd never gotten anything less than an A on her report card. Her teacher was just as sensitive to my little girl as she could be, stressing that this one bad grade shouldn't affect how Mimi feels about herself, she was still a very smart little girl, I mean just this great big fuss being made for fear that the C+ would take a major bite out of Mimi's self-esteem, which I think just ended up making Mimi feel a lot worse about herself. All those assurances just made her feel weak. When she brought that report card home I didn't make Mimi feel like a weak little thing who needed to be propped up and reassured that she was smart. Instead I was extremely stern with her about her carelessness, which was something that was completely within her control, and I let it be known that nothing less than an A was acceptable in our household. I made her practice and practice those columns until she was dreaming about them at night, and not even making careless mistakes in her dreams. And the next report card: an A in arithmetic!

MUNITZ: So you invaded the poor little mite's dreams as well. You didn't even give her autonomy to escape you in sleep.

ZEE: I was joking about her dreams. It was clearly a joke since I couldn't know whether she made mistakes in her dreams or not.

MUNITZ: I see. A joke. But it is still true that what you wished for this child was that your desires for her, of the high achiever that she must of necessity become for her to have a place in what you called your "household," were meant to invade the deepest recesses of her mind. And I ask you: why?

ZEE: Because what Plato just said about athletics, how it promotes a person's self-discipline and bolsters their sense of mastery, is true for other things as well, including the skills you're required to learn in school, even if you're never going to need those skills later on. Instead of trying to bolster a child's self-esteem by praising her in spite of what she didn't do, which is so artificial and pathetic, and kids are smart enough to see how artificial and pathetic it is, the right approach is to make sure the child can do what she thought she couldn't, and have her self-esteem be based on her own sense of mastery.

BURNS: Plato?

PLATO: When I spoke just now of mathematics as a subject that nobody need learn unless their minds held within them a love for mathematical beauty just waiting to be kindled, I was not thinking of adding up columns of addition. Few of us take delight in that.

ZEE: But if my daughter hadn't mastered her columns of addition, if she had gotten it into her head that she should feel good about herself even if she wasn't good with numbers, which was what her well-meaning teacher was telling her, she would never have had the confidence to do well in all her other classes in math, including geometry. She wouldn't have had the confidence to take her AP classes in mathematics, and if she had applied to your Academy, she would have been rejected!

PLATO: I would not have held it against your daughter if she could not add up columns of numbers without making careless mistakes. I am sometimes prone to careless mistakes myself. Unlike her teacher, I would have been able to see whether her mind, despite its tendency to be careless, was of a kind to take delight in abstract beauty. I once saw Socrates questioning one who came from the bottommost rung of society, a person who had enjoyed no privilege whatsoever, having never even been taught to read. But this unfortunate, led gently along by the teacher who knew how to ask the right questions and awaken in his mind a love for the beauty of logical connections, was able on his own to comprehend a subtle geometric proof (*Meno* 82b–85c).*

ZEE: I'm not saying that there's no difference among people in their natural talents. Some people have aptitudes that others lack. These inequities exist just as other inequities do, like class differences. But just like those class inequities can be corrected so can these inequities of talents, if a person is willing to put in the hard work. Mimi had to practice her sums hours more than some other kids, but in the end they performed the same, and that's what counts.

PLATO: Even if that were true, I still do not know why you would subject your children to long hours of grueling work, if it does not come naturally to them.

ZEE: How can a child feel truly good about herself if she has such glaring deficiencies?

MUNITZ: It's only you, the warrior mother, who instills such a sense of deficiency in your child. If children don't have parents who make

---

*The child was, in fact, a slave. Plato, fast study that he is, has learned to leave all allusions to slavery out of his conversations with our contemporaries.

them feel that if they are not exceptional, and exceptional in precisely those areas that the parents have designated as counting, then they are less than nothing, then it won't matter. They will be content with who they are, what their nature has stamped them to be. And again I'll quote Plato. Did you not write that the student who is not akin to the subject cannot be made so by any readiness to learn nor yet by any memory (*Seventh Letter* 341d)?

ZEE: No, Plato agrees with me on this—or rather I should say that I agree with Plato, or with Socrates, or whichever of you said that the unexamined life is not worth living. A life not worth the living is not exactly what I or any of us has in mind for our children!

MUNITZ: Your little performing monkeys are hardly in a position to live Plato's examined life. They're on a lifelong quest for applause and approval, not for the beautiful, the true, and the good.

ZEE: If the examined life is really the most perfect life, then I assure you that my children can achieve it! If it can be taught, then they'll learn it!

PLATO: And if it cannot be taught? If there is a subtle form of beauty that isn't expressible as other kinds of knowledge are, so that the only way for a mind to receive it is for it to keep constant association with the subject itself, living with it day and night, until suddenly, like the fire that is kindled by nearness to a fire, the soul itself bursts into the flames of knowledge of the beautiful (*Seventh Letter* 344a)?

ZEE: Yes, exactly! Living with it day and night! There's nothing that practice and hard work can't compensate for. If our kids aren't born with a natural aptitude for the beauty you're describing, then with enough effort they can overcome the deficiency!

PLATO, *gently*: And how do you know this?

ZEE: Because anything else is intolerable and ugly. You yourself mentioned moral beauty as a real thing, really existing, there to be loved for the soul who can love it. But how can the world be morally beautiful if the whole thing is rigged so that some people are going to live lives that aren't truly worth living no matter how hard they try? Where's the moral beauty in that?

PLATO, *somewhat sadly to Munitz:* Can you understand now my justification for proposing the noble lie? I didn't propose it for superficial reasons, but rather to prevent precisely such bitter recoiling against the state of how things are.

MUNITZ: No, Professor Zee's objection would still be pertinent no matter how noble you convinced yourself your lie to her was, telling her that she and her bright children are made of some lesser metal than gold and so she has not even the right to aspire to a life of dignity and responsibility and must leave the essential choices to her guardians. All that your noble lie would do is dishearten her so that she no longer had the spirit to protest the injustice of such a universe. You'd take the stuffing right out of her *thumos* so that it posed no threat to your social stability.

PLATO: Never underestimate the desirability of social stability. Its goodness emerges most fully in its absence. May you never live, Dr. Munitz, under circumstances that reveal to you how wrong you once were to have underestimated the desirability of social stability.

MUNITZ: But I have lived under such circumstances. I am a refugee to this country from such circumstances. When I protest that stability must not be purchased at any price, it is because of circumstances I fled.

PLATO, *quietly:* I understand.

MUNITZ: Putting those circumstances aside, have you never thought to draw a distinction between stability and stagnation?

PLATO: I have always been pessimistic, human nature being what it is, that perfection, in the unlikely event that it is achieved, can be sustained. There are internal forces that lead it to unravel (*Republic* 546).* I only propose safeguards to fend off its inevitable collapse.

MUNITZ: And those safeguards safeguard against all the citizens having equal access to the truth?

PLATO, *softly:* Not all share your love of the truth, Dr. Munitz. If they did, there would be no need to lie.

---

* "It is hard for a city composed in this way to change, but everything that comes into being must decay. Not even a constitution such as this will last forever. It, too, must face dissolution. And this is how it will be dissolved" (*Republic* 546a). The rest of Book VIII traces the progressive degeneration from a more superior government to a lesser one, always because the ruling class itself becomes degraded. So too much honor-seeking among the aristocrats degrades the government into a timocracy (548–550), in which people of *thumos* take precedence over truth-lovers, and the timocracy degrades into an oligarchy when its rulers fall prey to the lust for riches, finding their honor in being the wealthiest and then making wealth itself a condition of political power (550c) and creating a society in which the poor are not only powerless but despised, with the result that "such a city should of necessity be not one, but two, a city of the rich and a city of the poor, dwelling together, and always plotting against one another" (551d). An oligarchy encourages moneylenders, and many will lose their fortunes to these capitalists and will seethe with resentment. The next step after that is a tyranny, the worst of all governments.

MUNITZ, *first clearing her throat and then speaking so that her deep voice rumbles with her emotion:* I do not know if I am able to articulate what I am thinking.

BURNS, *smiling:* Oh, I think we all have faith in your powers of fully articulating what you are thinking.

MUNITZ: What I am thinking is that your faith, Plato, in a perfect state is incompatible with the finest moment in all your writings.

PLATO, *quietly:* And which moment, Dr. Munitz, would that be?

MUNITZ: When Socrates, about to die, says in his chipper, no-nonsense way, that those who fear death are only demonstrating our entrenched tendency to think we know what we don't. I apologize. I know that you—or Socrates—put it much better than that.

PLATO, *staring off into the noosphere:* To fear death, gentlemen, is no other than to think oneself wise when one is not, to think one knows what one does not know. No one knows whether death may not be the greatest of all blessings for a man, yet men fear it as if they knew that it is the greatest evil. And surely it is the most blameworthy ignorance to believe that one knows what one does not know (*Apology* 29a).

MUNITZ: Yes. Superb.

PLATO: I can take no credit for it.

MUNITZ: But you do agree, even until this day, with its point?

PLATO: I do.

MUNITZ: I do as well. I also think that the point it makes regarding our complacently thinking that we know what we do not know regarding death is equally true of life. Or to be blunter, this complacent ignorance is the error of anyone who tries to mold something, whether a child or a state, by brandishing some idea of perfection, and thinking that such an idea can cancel out such moral truths as always speaking the truth and according all humans the dignity to take responsibility for their own lives. This warrior mother, in her zeal to perfect her children, doesn't leave open the possibility that there is contained in the interior of her children creative possibilities of which she can form no conception and which she will mindlessly destroy in her attempts to mold those children to her own inflexible standards. You, Plato, seem to acknowledge these creative possibilities by emphasizing free play, but then you deny these same possibilities by trying to freeze your utopia in time, by having the guardians doing the thinking and deciding for the citizens, just as the warrior mother does the thinking and deciding for her children. No wonder the two of you have found such commonality. There among the populace, among your gold and silver and, yes, even in the iron and brass, there may be lurking creative possibilities of which your guardians can form no conception, precisely because they *are* creative possibilities. There was a superstition among the ancients in your own country that the gods punished man's hubris. But there is nothing superstitious about foreseeing bad consequences for the hubris of paternalistic utopianism. Humanity should never be frozen into a vision of the best. A creative society must be willing to tolerate some degree of instability because creativity is inherently unstable.

PLATO, *quietly:* How could I disagree with that?

MUNITZ: Of course, you couldn't. You're too marvelously creative yourself to disagree. But then we must be willing to tolerate instability in the political sphere, too. No look at reality can ever give it to us whole—the beautiful, the true, and the good. Maybe it's there whole. I'm ready to let you convince me that it is. But at no moment in time are we ever going to get it whole, or enough of it so as to be in a position to shape our society and freeze it in time. Are there things about our society that surprise you?

PLATO: Without a doubt. Too many for me to enumerate.

MUNITZ: I mean not just the scientific and technological advances, the computer to which you appear to be so oddly attached, but in the sphere of morality as well? Are there ways in which you find our *polis* more moral than your own, with a more evolved sense of the dignity and autonomy of all people than your own slave-keeping, misogynistic, war-mongering Athens?

PLATO: Again, without a doubt.

MUNITZ: Did you, the best of all Greeks, foresee these moral advances? Did you include them all in your *kallipolis?*

PLATO, *quietly:* I did not.

MUNITZ: Then I rest my case.

ZEE: But if I could just speak up here for Plato, whom you seem to be accusing of having been less than omniscient—

MUNITZ: Only because his moralocracy would demand such omniscience—

ZEE: Be that as it may, I'd just like to point out that if these moral advances are due to our reasoning out the implications of what human life is all about, for which I think Plato could make a good case, then

Plato shouldn't be put on the witness stand under your prosecutorial inquiry, Dr. Munitz, but rather praised as having gotten the whole process started. I read an article in *The New York Review of Books* that was called "Philosophy for Winners," which, for obvious reasons, is a title I love! *Audience laughs.* I wrote down a line from it because I thought it might be pertinent, and it is. "The moral philosophy of the ancients, much more than their science, was a living presence throughout the history of modern philosophy, and still is."[*] The very fact that Plato is surprised by how far we've gone beyond his best-reasoned conclusions, and how superior the laws of our *polis* are to those of his day, is proof of the contributions he's made to our progress. So rather than berating him for how much more we see now than he did, we ought to be applauding him for first pointing us in this direction!

BURNS: And this, ladies and gentlemen, is why, when you're in need of a good lawyer, you should always hire a warrior mother!

Vigorous applause, while a guy in T-shirt and jeans slinks onto the stage, hurriedly hands Burns a note, and then slinks off.

BURNS: I'm afraid, ladies and gentlemen, that we're going to have to let Professor Zee's spirited last statement stand as our summation since I've just been informed that the NYPD has declared a state of emergency out there. So unfortunately we're going to have to dispense with the Q&A. *Audience groans.* Yes, I know! I'm disappointed, too! So let's thank our illustrious panel for giving us such a lively evening of provocative dialoguing. Dr. Munitz, Sophie Zee, and Plato: thank you! *Audience applauds for several long moments. Zee jubilantly applauds, too, turning first to applaud Plato, who smiles and politely applauds back at her, then to Dr. Munitz, who grimaces and looks away.*

Afterword to Chapter δ: Plato's Responses to the Myers-Briggs Psychometric Questionnaire

1. You are almost never late for your appointments. YES

2. You like to be engaged in an active and fast-paced job. NO

3. You enjoy having a wide circle of acquaintances. NO

4. You feel involved when watching TV soaps. NO

5. You are usually the first to react to a sudden event: the telephone ringing or an unexpected question. NO

---

[*] M. F. Burnyeat, "Philosophy for Winners," *The New York Review of Books*, November 1, 2001.

6. You are more interested in a general idea than in the details of its realization. YES

7. You tend to be unbiased even if this might endanger your good relations with people. YES

8. Strict observance of the established rules is likely to prevent a good outcome. NO (so long as the established rules are reasonable)

9. It's difficult to get you excited. YES (about the things that concern most people)

10. It is in your nature to assume responsibility. YES

11. You often think about humankind and its destiny. YES

12. You believe the best decision is one that can be easily changed. NO

13. Objective criticism is always useful in any activity. YES

14. You prefer to act immediately rather than speculate about various options. NO

15. You trust reason rather than feelings. YES

16. You are inclined to rely more on improvisation than on careful planning. NO

17. You spend your leisure time actively socializing with a group of people, attending parties, shopping, etc. NO

18. You usually plan your actions in advance. YES

19. Your actions are frequently influenced by emotions. NO

20. You are a person somewhat reserved and distant in communication. YES

21. You know how to put every minute of your time to good purpose. YES

22. You readily help people while asking nothing in return. YES

23. You often contemplate the complexity of life. YES

24. After prolonged socializing you feel you need to get away and be alone. YES

25. You often do jobs in a hurry. NO

26. You easily see the general principle behind specific occurrences. YES

27. You frequently and easily express your feelings and emotions. NO

28. You find it difficult to speak loudly. YES

29. You get bored if you have to read theoretical books. NO

30. You tend to sympathize with other people. NO

31. You value justice higher than mercy. YES

32. You rapidly get involved in social life at a new workplace. NO

33. The more people with whom you speak, the better you feel. NO

34. You tend to rely on your experience rather than on theoretical alternatives. NO

35. You like to keep a check on how things are progressing. NO

36. You easily empathize with the concerns of other people. NO

37. Often you prefer to read a book than go to a party. YES

38. You enjoy being at the center of events in which other people are directly involved. NO

39. You are more inclined to experiment than to follow familiar approaches. YES

40. You avoid being bound by obligations. NO

41. You are strongly touched by the stories about people's troubles. NO

42. Deadlines seem to you to be of relative, rather than absolute, importance. NO

43. You prefer to isolate yourself from outside noises. YES

44. It's essential for you to try things with your own hands. NO

45. You think that almost everything can be analyzed. YES

46. You do your best to complete a task on time. YES

47. You take pleasure in putting things in order. YES

48. You feel at ease in a crowd. NO

49. You have good control over your desires and temptations. YES

50. You easily understand new theoretical principles. YES

51. The process of searching for a solution is more important to you than the solution itself. NO

52. You usually place yourself nearer to the side than the center of the room. YES

53. When solving a problem you would rather follow a familiar approach than seek a new one. NO

54. You try to stand firmly by your principles. YES

55. A thirst for adventure is close to your heart. NO

56. You prefer meeting in small groups to interaction with lots of people. YES

57. When considering a situation you pay more attention to the current situation and less to a possible sequence of events. NO

58. You consider the scientific approach to be the best. YES

59. You find it difficult to talk about your feelings. YES

60. You often spend time thinking of how things could be improved. YES

61. Your decisions are based more on the feelings of a moment than on careful planning. NO

62. You prefer to spend your leisure time alone or relaxing in a tranquil family atmosphere. YES

63. You feel more comfortable sticking to conventional ways. NO

64. You are easily affected by strong emotions. YES*

65. You are always looking for opportunities. NO

66. Your desk, workbench, etc., is usually neat and orderly. YES

67. As a rule, current preoccupations worry you more than your future plans. NO

68. You get pleasure from solitary walks. YES

69. It is easy for you to communicate in social situations. NO

70. You are consistent in your habits. YES

71. You willingly involve yourself in matters which engage your sympathies. NO

72. You easily perceive various ways in which events could develop. YES

*This response might come as a surprise, but not to those who have studied Plato's *Phaedrus*.

ε
<hr />

I DON'T KNOW HOW TO LOVE HIM

Jean-Baptiste Regnault, Socrate arrachant Alcibiade du
sein de la Volupté *(Socrates Tears Alcibiades from the
Embrace of Sensual Pleasure). Oil on canvas, 1791.*

One who is incapable of participating, or who is in need of nothing through being self-sufficient, is no part of a *polis,* and so is either a beast or a god.

—Aristotle, *Politics*

'Twere best to rear no lion in the state.
But having reared, 'tis best to humor him.

—Aristophanes, *The Frogs*

Inscribed on the Temple of Apollo at Delphi, where the god's own oracle sat on her tripod at the *omphalos,* the navel of the world, and issued the prophesies that people from all over Greece and even beyond came seeking, were two warnings. "Nothing in excess"—*mēdèn ágan*—warned one of the inscriptions. This was echoed by "Know yourself"—*gnôthi seautón.*

The presence of both warnings at such a solemn site, in the antechamber in which one waited before going in to keep one's appointment with Apollo, as channeled through the hierophants who "interpreted" the oracle's ravings, seems to show how central to the Greek worldview these two sentiments were. "Nothing in excess," in particular, is often offered as a summation of what is distinctively Greek—which is odd when you look at how they actually carried on.

Some scholars interpret the Delphic messages on the wall as having more limited application, as offering nothing more than instructions on how to behave in the presence of the god. Don't ask for anything in excess, but rather limit yourself to requesting exactly the information you need. "Know yourself" reaffirms the same point. Examine what you really need to know and then formulate your question extremely carefully.* The two together were practical instructions of just the sort meant to counteract the carelessness that provides the plot of fairy tales we tell our children, where three wishes are granted and the last of them must be used to cancel out the mistakes made in the incautious wishing of the first two.

Still, those two Delphic warnings do capture a truth beyond mere instructions on how to behave when seeking answers from an oracle. They are prescriptive, not descriptive. "Nothing in excess" is not a general observation on the nature of Hellenic behavior but rather a warning against the outcome toward which that behavior tended. And "know yourself" shouldn't be understood as recommending the kind of self-analysis that supports the mental-health industry and makes best sellers out of self-help books. Essentially a restatement of "nothing in excess," "know yourself" is a monition against self-delusion, which tends to take the form of our thinking too well of ourselves. That, at any rate, is how Plato reads the warning. In the *Philebus,* he has Socrates use "know yourself" to cite the three ways in which it is most often violated. People may delude themselves about how rich they are, and about their physical attractiveness, thinking themselves more handsome and tall than they really are. But by far the greatest number, he says, are mistaken as regards the "state of their souls," thinking themselves more virtuous and wiser than they are (48e–49a). This observation seems right. Shakespeare's Richard III may undeludedly declare in the privacy of his opening soliloquy, "I am determined to prove a villain," but most people, even the most villainous, have ways of presenting themselves to themselves in far more generous moral terms. We are our own best defense attorneys, determined to believe the best about ourselves. That, at any rate, is how Plato chose to interpret the famous warning, putting his own psychologically astute spin on the inscription.

Together, the Delphic messages captured an admonishment that the Greeks, perhaps knowing themselves, knew that they needed. Exhortations against excess and overreaching make sense in a culture in which excess and overreaching are a constant danger given the underlying ethos. So, too, does the horror of hubris make sense in such a culture,

---

*Petitioners didn't question the Delphic oracle, or Pythia, directly. Our best information is that people wrote or dictated their questions and requests and then gave them to priests who took them to the Pythia. There were also priests who then interpreted the responses—often ravings—of the oracle.

enforced by a religion that interprets misfortunes as a comeuppance earned by overreaching and intertwines the notion of hubris with a whole slew of others, such as *phthônos* (divine jealousy) and *nemesis* (divine indignation). All the restraining supernaturalism was necessary, as the story of Herostratus bears out. He was the nobody who tried to be a somebody by burning down the Temple of Artemis in Ephesus in 356 B.C.E. All he wanted was for his name to become known. *Kleos*. And it worked. Chaucer has Herostratus explain himself in *The House of Fame*:

> "I am that ylke shrewe, ywis,
> That brende the temple of Ysidis
> In Athenes, loo, that citee."
> "And wherfor didest thou so?" quod she.
> "By my thrift," quod he, "madame,
> I wolde fayn han had a fame
> As other folk hadde in the toun."

Contemporary analogues are unfortunately plentiful. As a recent book argues, the easiest way to become famous, if that is your most urgent goal and you are endowed with no particular assets or talents, is to "kill innocent people. The more random your victims the better because it sends the message that no one is safe. And when they're scared, people pay attention."[*] Obviously, the author isn't endorsing random violence as a course of action, but rather seeking to explain the mind-set that transforms too may ordinary settings into scenes of mayhem and tragedy. Perhaps a society that has slipped back into celebrating "acoustic renown" as an end in itself should be extra careful about its gun regulations. But back to ancient Athens.

If one aspect of the culture goaded a Greek to spare no efforts and to produce a life that would leave others gaping in astonishment, then excess and overreaching were exactly what was called for. It was as if the two clashing imperatives—one derived from the Ethos of the Extraordinary and the other from the advice inscribed on the Delphic temple—might somehow be reconciled in a life that would, at one and the same time, accept nothing less than the extraordinary and yet also be sensible, safe, and temperate, avoiding the wanton transgression of hubris.

---

[*] Adam Lankford, *The Myth of Martyrdom: What Really Drives Suicide Bombers, Rampage Shooters, and Other Self-Destructive Killers* (New York: Palgrave MacMillan, 2013), p. 108.

The tension can be transposed into the language of the gods. On the one hand, there is the desideratum, *Go forth and be godlike!* And on the other hand, the antiphonal response comes back: *In knowing thyself, O mortal, know first and foremost that you are no god!* It seems a dilemma, with no help forthcoming from Mount Olympus. There's never much help emanating from that particular sphere, most especially of a normative nature.

The Olympian gods and goddesses don't inhabit a higher moral plane, from which ethical standards can issue, together with divine assurance that our lives are worth something in their eyes. The Olympic denizens are only more powerful versions of ourselves[*] and, given their power, must be propitiated lest they lower the odds of our getting through life unscathed by tragedy. They can hold us back in our worldly endeavors or they can help us, demote us or promote us, and we must try to haggle and barter with them, which is what Greek prayers tended to do. The gods are less like the unspeakably holy Jehovah and more like idolized but bribable elder siblings—tantalizingly beyond us, yes, but not so unthinkably beyond that we can't emulate them, thereby living extraordinary godlike lives. But this achievement itself is fraught with danger. Just like those older siblings, the gods can turn mean when they feel us gaining on them, do something spiteful just to show us that they can. Even when they favor us, they can turn on a drachma, desert us at the very moment when we need them the most—just as Apollo did to Hector, leaving him to the blood fury of Achilles. Our reversals of fortune are to be read as signs of their disfavor, whether because of something we ourselves did or failed to do, or for some other reason entirely, a deal they'd hatched among themselves. The gods exist for their own sakes and not for ours and do little to make us feel at home in the cosmos, being unreliable at best and at worst downright hostile. The existential upshot is that, despite the Olympians' propinquity, we're left all on our own when it comes to solving the existential questions that emerged over wide reaches of the world during the Axial Age. In this sense, Greek society, as rife with religious rites as it was, prepared the way for a secular worldview.

---

[*] "[The Olympian gods] are more powerful than man but otherwise no different. Their immortality is no true immortality but an inability to succumb to death, which defeats their pygmy brother man. They are no wiser, no happier, and no more complete than those whose minds have contained their picture." David Grene, *Greek Political Theory* (Chicago: Phoenix House, 1965), p. 194.

It also prepared the way for the genius of Greek tragedy. Our being ultimately all on our own forms the background assumption of many of the masterpieces of Greek tragedy, none more so than *Prometheus Bound,* with its genocidal Zeus:

> When first upon his high, paternal throne
> He took his seat, forth with to divers Gods
> Divers good gifts he gave, and parceled out
> His empire, but of miserable men
> Recked not at all; rather it was his wish
> To wipe out man and rear another race. (229–234)[*]

If there was a certain strain in the culture that exhorted a person to become godlike—and Plato, in his own way, endorsed such a goal—then it wasn't in order to attract the attention of the gods. All things being equal, it's better if they just don't notice you. You must at once both impress your fellow mortals while also not attracting undue attention and possible resentment from the touchy gods.

It's odd for us—whether we adhere to a particular religion or not—to consider a religion that is silent regarding issues on which religions as we now know them are loquacious. I was brought up as a child to believe that all my deeds—a nibble from a friend's (not certified "kosher") Hostess Twinkie, a donation from my meager allowance to a charity for orphans, a spin-doctored rendition of a quarrel with a sister—get inscribed in a heavenly book, which, come the autumnal Days of Awe, will be scrutinized, tallied, and evaluated. Terrifying, yes, but also quite effective in inducing a robust sense of human consequentiality. No less than the Lord of the Hosts himself takes note of that Hostess Twinkie.

The Abrahamic religions powerfully address the problem of human worth, as do other religions that have demonstrated their millennia-long staying power. In the case of the Hebrews, the phrase *bi-tzelem elohim,* meaning "in the image of God," provides an answer to the question of human worth. The phrase is used three times in Genesis, all of them in what have come to be called the Priestly portions of the Torah, usually dated to the sixth or fifth centuries B.C.E., meaning relatively late for the Torah's authorship. It's in the last of the three passages (9:6) that the normative dimension of the phrase fully emerges. As the King James Version translates it: "Whoso sheddeth man's blood, by

[*]Translated by Edmund Doidge Anderson Morshead, http://www.sacredtexts.com/cla/aesch/promet.htm.

man shall his blood be shed: for in the image of God made he man."
The implication is that God's impressing his image on man confers
worth sufficient to make the shedding of blood prohibited. The Abrahamic religions have, in their turn, so impressed themselves on ethical
thought that it is sometimes hard for adherents, even now, to fathom
how humans could have worth *without* this divine impression.

In fact, the answers that religion, as we have come to know it, provides to the question of human worth have played so dominant a role
in the preceding centuries that believers often cannot conceive how
non-believers can muster sufficient commitment to their own lives to
get out of bed each morning, let alone the ethical wherewithal to regard
others as deserving of moral regard. Once one "comes out" as an atheist, these are the inquisitions to which one is often subjected.

But if we find Greek religion odd in its reticence regarding such
questions as had prompted my neighbor to twist himself into a metaphysical pretzel, it's only because we're forgetting that religion, as
we have come to know it, provides only one possible solution to the
question of human worth. In western culture, there is a tradition that
seeks to solve our existential and normative problems in strictly human
terms, and this is the tradition that goes back to the Greeks.

It goes back to the Greeks, even predating Socrates and Plato, an
aspect of the normative framework summed up by the notion of *aretē*,
as Nehamas explained it, with its significant social dimension, bundled
up with the idea of *kleos*. "From earliest times, the idea of *aretē* was
intrinsically social, sometimes almost the equivalent of fame *(kleos)*."
It was human sociality and its institutions—the family, the *deme*, the
phratry, the *polis*—that provided the context for understanding and
exhibiting *aretē*. The power of the *polis*, perhaps most especially in Athens, the sense of social identity and exceptionalism that it yielded during the years of political and cultural hegemony, made the institution of
the *polis* particularly prominent in the conceptualization of *aretē*. Aristotle's comment, quoted as one of the epigraphs of this chapter, bears
repeating: "One who is incapable of participating, or who is in need
of nothing through being self-sufficient, is no part of a *polis*, and so
is either a beast or a god." There was an intimate connection between
the notion of *aretē* and the *polis*—Plato, like Aristotle, acknowledged
it—and yet what that connection was could be interpreted in various
ways, with the primary concept tipping toward either the one, *aretē*,
or the other, the *polis*. Is the best state the one that maximally allows

aretē to exist and to flourish, where *aretē* is independently defined? Or is *aretē* to be defined in terms of the qualities that will allow a person to become *justifiably notable* in the *polis*, the qualities of an individual that best allow the values-setting *polis* to exist and to flourish? Plato (as well as Aristotle) will go in the first direction, though never so far as to cancel out the moral relevance of the *polis;* but the Athens that tried and executed Socrates tipped the balance toward the second direction. Plato moralized political theory, while the Athens to which he objected politicized morality—or at any rate it politicized *aretē*. And it judged Socrates to be severely lacking in the qualities that would conduce to the flourishing of his *polis*, which made him, though notable, not *justifiably* notable, and so deficient in *aretē*. Thus, once again we can say that, on its own terms at least, Athens was justified in judging Socrates guilty of normative heresy, since he rejected its values.

The politicizing of *aretē* made quite a bit of sense in a culture that so valorized extraordinary achievement and the striving that brings it about. How will a society of individuals striving to gain on the gods (only not enough to excite divine jealousy) attain political stability? Achilles didn't think about what might conduce to the greatest good of the greatest number of his fellow Greeks as he allowed his comrades to be slaughtered, their blood staining Trojan soil red. He sat in his tent, playing the lyre and singing of *kleos.** But one can't make a civil society out of citizens all of whom are going single-mindedly after their own personal *kleos*. The valorization of the extraordinary leads to an antisocial conclusion.

Thrasymachus, in the first book of the *Republic,* embodies that conclusion, a sophist who argues that an extraordinary person has a perfect right to do whatever he can get away with. (Callicles in the *Gorgias* argues along a similar line. Thrasymachus and Callicles, by the way, are among Plato's most arresting characters. They jump off the page. Just as in fiction, it's often the least laudatory characters who are the most lifelike and fascinating.) It is only the weak who seek to restrain him, and without any natural right on their side. Thrasymachus is a problem not only theoretical but practical. Plato has his ideas about how to dispatch a Thrasymachus, but then so did the *polis.*

The political configuration of Greece asserted a strong prescriptive normativity which offered a political solution to the Thrasymachus challenge. Duty to the *polis,* like duty to the family and the other social

* *Iliad* 9.189.

institutions, yielded behavior-regulating *oughts*. Participation in the collective life of the *polis* both restrains the extraordinary individual and enlarges the ordinary individual, allowing him (especially if he's lucky enough to be an Athenian citizen) to participate in the extraordinary. An individual can achieve participatory excellence via the accomplishments of the *polis* and need not always be caught in an agonistic struggle to outdo his peers. Even religious ritual was absorbed into civic duty, with the patron god or goddess of one's *polis*—Athena and Poseidon for Athens, Artemis and Ares for Sparta—largely regulating rituals, which, communally performed, strengthened the devotion to the *polis*. As Pericles puts it in his Funeral Oration, Athenians should feel *erōs* for Athens: they should fall in love with their *polis*.

The politicizing of *aretē* provides a solution to the Thrasymachus challenge. So long as the abundance of energy and striving encouraged by the Ethos of the Extraordinary is channeled into participatory extraordinariness, the free radicals of outsized ego can be soaked up and neutralized. Godlike striving can be deflected from the individual onto the *polis* and personal hubris avoided (though collective political hubris remained actively in play, as those non-Athenians staring up at the Acropolis might have had reason to reflect). Not surprisingly, the politicizing of *aretē* was most stressed by the two most extraordinary *poleis*, Athens and Sparta, with Sparta funneling its sense of exceptionalism into its collective martial superiority, whereas Athenians assessed their exceptionalism in far more varied terms—military, commercial, political, cultural, intellectual, psychological, ethical—as Pericles enumerates the Athenian virtues in his Funeral Oration.

But what of an extraordinary person who refused to be dictated to and defined by the *polis*, who defied the politicizing of *aretē* and insisted on his singularity remaining exclusively his own, unwilling to deed it over to any one *polis*? Such a person might be expected to arouse violently contradictory emotions. His fellow citizens couldn't help but revile him for violating the politicizing of *aretē*, while at the same time they could not help admiring him—especially if he also happened to possess a beauty so extravagant that a contemporary might write of him that "if Achilles did not look like this, then he was not really handsome."[*]

---

[*] See Charles H. Kahn, "Aeschines on Socratic Eros," in *The Socratic Movement*, ed. Paul A. Vanderwaerdt (Ithaca, NY: Cornell University Press, 1994), p. 90.

I am speaking now of none other than Alcibiades, who provided fodder for the Athenian version of Gawker throughout his whole life—from his obstreperous childhood,* through his attention-grabbing career, to his violent death at (perhaps) the hands of hired assassins. Alcibiades was never in love with Athens; he was, rather, intent that the Athenians should be in love with him. At every stage of his life, Plutarch tells us, he exemplified the physical beauty appropriate to it, and "never did fortune surround and enclose a man with so many of those things which we vulgarly call goods." It was as if Athens, in the reclamation of Greek-speaking post-Homeric glory that reached its height in the time of Pericles, had produced an avatar of Achilles. Everything about him conspired to make him a person who lived out a life that was abundantly worth the telling—and also to make a thoughtful person question whether living a *kleos*-worthy life was necessarily a good thing, if not for the person himself then for everyone else involved.

In his own time, there was nobody like him, and there haven't been many since, a point forcefully delivered by a chart I found on the comedy site Cracked.com:†

	Hilariously Rich?	Brilliant Orator?	Unstoppable War Machine?	Lady Killer?	Violent Lunatic?
Justin Bieber	✓	X	?	✓	X
O.J. Simpson	✓	X	X	?	✓
Leon Trotsky	X	✓	✓	X	X
Genghis Khan	X	X	✓	✓	✓
Alcibiades	✓	✓	✓	✓	✓

* One story told by Plutarch is that, when Alcibiades was fighting with another little boy who had wrestled him to the ground, he bit his finger. "You bite, like a woman," the boy said. "No," answered Alcibiades, "like a *lion*."
† http://www.cracked.com/funny-5516-alcibiades/. I recommend the whole Cracked article on Alcibiades.

His uniqueness grew out of the uniqueness of Athens. It took the special nature of Athenian democracy and Athenian hegemony to produce the mix that was Alcibiades. The nature of its democracy shaped how he honed his intelligence into rhetorical brilliance. The people adored him, and he adored their adoration. And because he was, as Cracked.com put it, "hilariously rich," he could be lavish, a trait cherished by Athenian citizens, who were often not above being bought. (Pericles did plenty of buying.) Wanting to be certain that he would go home with an Olympic prize one year, Alcibiades extravagantly entered seven chariots—everybody couldn't stop talking about it—especially when he then won first, second, and fourth prize.* He was, of course, an aristocrat, a scion from one of the most ancient families, the Alcmaeonidae. (This also meant that he carried the famous curse of the Alcmaeonidae.)† But, like Alcmaeonidean Pericles, he threw his lot in with the *demos* of Athens—with the people—whom he would by turns delight, outrage, thrill, provoke, torment, charm, shock, mystify, and seduce. But he always kept their attention, kept them talking of him. Imagine John F. Kennedy, Donald Trump, David Petraeus, Muhammad Ali, Julian Assange, Johnny Knoxville, Bernie Madoff, and Jude Law all combined in one.

Pericles, who was a cousin of Alcibiades' mother, became Alcibiades' legal guardian when the ten-year-old's father was killed in battle, so that Alcibiades was raised in Pericles' house, which makes the difference between them the more dramatic. Pericles, more than anyone, articulates the politicizing of *aretē*, of which Alcibiades would have little.‡ Alcibiades refused to yield his prodigiousness to the *polis*, claim-

---

* In the chariot races it was the person who paid for the chariot and driver, rather than the driver, who won the prize.

† Herodotus writes of this famous curse in *Histories*, Book V. In 632 B.C.E., an Olympic victor named Cylon decided to parlay his fame into political power, attempting to become the tyrant of Athens. He and his followers attempted to seize the Acropolis, but their attack failed and so they sought sanctuary at the goddess's temple, which meant that they could not be attacked so long as they were suppliants of the goddess. But a member of the Alcmaeonid family violated their sanctuary, telling them to come down and be judged and they would be safe. The story is that they came down, but kept themselves attached to the sanctuary by a rope or thread, but when that broke, it was taken by the Alcmaeonidae to be a sign from the goddess that she did not offer her protection, and they killed the failed tyrant and his supporters, thus incurring the curse. For some time, they were banished from Athens.

‡ Pericles had explicitly shifted the focus of exceptionalism to the *polis*, collectivizing both the burden and the glory of superiority. "To be hated and to cause pain is, at present, the reality for anyone who takes on the rule of others, and anyone who makes himself hated for matters of great consequence has made the right decision; for hatred does not last long, but the momentary brilliance of great actions lives on as a glory that will be remembered forever

ing his superiority as supremely and inviolably his own. Still, his recklessness could not be contained in the personal sphere, and the ruin
that he made of his life would be Athens' ruin as well.

Here are some highlights from the extravaganza that was his life:

He convinced the *ekklêsia* to undertake a risky attack on Sicily in
the middle of the Peloponnesian War, in 415 B.C.E., and then talked
them into assigning him as one of the co-generals.* Having gotten
his way in regard to the Sicilian venture, he and some of his high-
spirited companions may or may not have chosen the night before the
fleet was to sail to go on a spree of vandalism, defacing the faces and
genitalia of the herms—stylized statues of Hermes that were used as
property-boundary markers throughout the *polis*. Hermes was the god
of travel, which would have made the sacrilege a bad omen for the voyage to Sicily. It is unlikely that Alcibiades was involved, but such was
his reputation that many Athenians, according to Thucydides, blamed
him. Alcibiades and his lark-loving friends were also rumored to have
staged a blasphemous charade of the solemn initiation rites of the
Eleusinian mysteries, with Alcibiades officiating as high priest. (A
slave named Andromachus brought evidence against Alcibiades.)† He
set sail for Sicily the day after the herms debacle under a cloud of misgivings. Knowing his power to turn public opinion his way, he pleaded
to be allowed to clear himself of the charges first. It was his enemies
who hastened his departure. And sure enough, while he was sailing
to Sicily, the hysteria over the impiety *(asebeia)* grew, with Alcibiades'
enemies arguing that the blasphemy bore all the marks of outrageous
Alcibiades. Obviously, he couldn't at the same time be leading an Athenian force on a major expedition, and so a trireme was dispatched to

---

after." (Thucydides, *History of the Peloponnesian War*, ii, 64, trans. Woodruff. See Appendix
B.) Alcibiades just as explicitly shifts the focus of exceptionalism back to the individual—
specifically to *him*—translating the burden and the glory into personal terms. The short-lived
enmity of the envious is focused on him: "What I know is that persons of this kind and all
others that have attained to any distinction, although they may be unpopular in their lifetime
in their relations with their fellow-men and especially with their equals, leave to posterity the
desire of claiming connection with them even without any ground, and are vaunted by the
country to which they belonged, not as strangers or ill-doers, but as fellow-countrymen and
heroes" (Thucydides *History of the Peloponnesian War*, vi, 61, trans. Woodruff). The reverse
mirroring of the language Pericles had employed is noteworthy.
* Alcibiades, with all his daring, and Nicias, who was very cautious, had taken opposite positions on the subject of invasion, so they were appointed co-generals. To break any ties on the
field, Lamachus was appointed co-general with authority equal to that of the other two.
† See Debra Nails, *The People of Plato*, pp. 17–20, for a list of sources and a summary of the
events.

fetch him back to Athens for trial.[*] The full extent of the trouble that
Alcibiades was in was minimized for fear that his men would mutiny
if they suspected he was being taken back to be treated harshly. Alcibiades wasn't taken prisoner but was allowed to follow under his own
sail. But he quickly caught on to the situation, and, miffed at how little
faith his *polis* had in him, he promptly vindicated the lack of faith by
defecting to Sparta, offering invaluable advice on how they could use
Athens' attack on Sicily to their advantage. The attack did indeed prove
disastrous to Athens. Athens sentenced him to death in absentia.

Being now a Spartan, he ostentatiously adapted to Spartan ways,
at least as Plutarch reports it: "People who saw him wearing his hair
close cut, bathing in cold water, eating coarse meal, and dining on
black broth, doubted, or rather could not believe, that he ever had a
cook in his house, or had ever seen a perfumer, or had worn a mantle
of Milesian purple."[†] (Plutarch's account of Alcibiades is fun, but perhaps can't be counted on for every detail. As an indication, he here
gets the hair wrong. The Spartans wore it long.)[‡] He delivered a speech
in non-democratic Sparta in which he trivialized Athenian democracy
as "acknowledged lunacy." The Spartans found his transformation
convincing, especially as he had sought to renew his paternal grandfather's role as *proxenos* to Sparta; the name itself is of Spartan origin; the speech he gave (albeit Thucydides' invention) is excellent. But
tucked safely within his outward changes, the willful perversity of his
character survived intact. No *polis* was going to define and contain him,
not Athens and certainly not Sparta. As might have been predicted,
he went on to betray Sparta, too, and then betrayed both Athens and
Sparta to Persia (perhaps).

And through all the twistings and turnings he remained Athens'
own prodigal darling. In 411 he was recalled by Athens to the Helles-
pont to take command of the fleet, though he avoided actually setting
foot in Athens. In 407, he reentered Athens, throngs greeting him at

---

[*] Thucydides vi, 61.

[†] Plutarch, *Parallel Lives,* "The Life of Alcibiades," 6.2, vol. IV of the Loeb Classical Library
edition, 1916.

[‡] Plutarch (ca. 45–120 C.E.) was a Platonist who had studied at the Academy, though he
lived centuries after the events of the fifth century B.C.E., a reason that he can't be altogether
trusted. He includes many salacious details about Alcibiades' life, which I have left out as
unreliable, though, again, they're a lot of fun. He had a truly wonderful job: he was one of the
two priests on duty at the Delphic oracle, who interpreted the Pythia's ravings. Not only was
this interesting work, but it seemed to have left him a lot of time for writing.

the harbor and throwing garlands over his godlike head. He reiterated his innocence of the charges brought against him in 415, and the Athenians threw the bronze stele, which had recorded the death verdict against him, into the sea. Xenophon reports that the opposing voices in the crowd were intimidated into silence.* But he fell under suspicion—unfairly this time, but you can't really blame the Athenians—when one of the military expeditions he was leading failed. Plutarch remarks, "It would seem that if ever a man was ruined by his own exalted reputation, that man was Alcibiades. His continuous successes gave him such repute for unbounded daring and sagacity, that when he failed in anything, men suspected his inclination; they would not believe in his inability. Were he only inclined to do a thing, they thought, naught could escape him."

In Aristophanes' *The Frogs*, which was performed in 405 at the Athenian Lenaean Festival, where it won first prize, the god Dionysus, who has traveled to Hades in despair about the decline in tragic poetry, asks the two dead tragedians, Aeschylus and Euripides, the latter having only died the year before, what the suffering city, in the direst of straits after the decades-long war with Sparta, must do to save itself. But first another question must be settled: whether Alcibiades, now ensconced in a castle in Phrygia, should be welcomed back:

> Now then, whichever of you two shall best
> Advise the city, he shall come with me.
> And first of Alcibiades, let each
> Say what he thinks; the city travails sore.

Euripides asks the god what Athens itself thinks about Alcibiades and is given an answer that conveys the conflicted fixation that fastens Alcibiades and the Athenians together:

> What?
> She loves, and hates, and longs to have him back.

This line of Aristophanes', written only a few months before both the capitulation of Athens to Sparta and Alcibiades' violent death, shows

---

*Xenophon, *Hellenica*, 1.4.20.

how riveted on Alcibiades the *polis* remained. It was Alcibiades' nature to try to get away with everything, and it was Athens' nature to—more often than not—indulge him.

Time and again his brilliant intrigues had been motivated by what he alone could get out of the situation, deriving from his assessment of how his power could be augmented and his own image magnified so that it blotted out all others. He talked the *polis* into rejecting a peace treaty that would have been to Athens' advantage, since he was annoyed that two rivals of his, Nicias and Laches,[*] had arranged it, while he had been ignored because of his youth. When the Spartans were to appear in the *ekklêsia* to try to arrange a lasting peace, Alcibiades engaged in ingenious double-dealing. He presented himself as a sympathetic negotiator to the Spartans, who were at a loss about how to address the singular institution of Athenian democracy and submitted themselves to Alcibiades' coaching. Then he worked the crowd up into a fury over what the Spartans had—at his prompting—said. An incensed Thucydides gives us the details, showing that in such a way was the fate of Athens determined by one man's drive to make himself the focus of everyone's attention, his name on everyone's lips. "Meanwhile I hope that none of you will think any the worse of me," he tells the Spartans after defecting to their side, "if, after having hitherto passed as a lover of my country, I now actively join its worst enemies in attacking it, or will suspect what I say as the fruit of an outlaw's enthusiasm." He ends with a rejection of participatory exceptionalism as explicit as the endorsement of it in Pericles' Funeral Oration. It is the supremacy of the self over the city that he asserts—not just anyone's self, but rather his: "Love of country is what I do not feel when I am wronged."[†] Nietzsche's *Thus Spake Zarathustra* could as easily have been named *Thus Spake Alcibiades*:

> I teach you the superman. Man is something that is to be surpassed. What have ye done to surpass man?

Alcibiades held that what he had done to surpass man was to be Alcibiades. And there's no denying the man was extraordinary.

---

[*] Plato's dialogue *Laches* features both of these generals in conversation with Socrates and deals with the nature of courage.
[†] Thucydides, vi, 92.

It wasn't only Athens that was undone by Alcibiades' uncontainable assertion of self. Thucydides, who despises Alcibiades, makes him the central figure in prolonging the Peloponnesian War. Such instability provides good material for the making of an extraordinary life. And the Peloponnesian War weakened all the *poleis* of the Hellenes and made them vulnerable to attack by an external empire. This time, it wasn't the Persians to the east who came to conquer Greece but rather the Macedonians to the north. In 357 B.C.E. Macedonia broke its treaty with Athens, and by 351 the famous Athenian orator Demosthenes was delivering the first of his Philippics, warning of the designs on the independent Greek city-states that the expansionist king Philip of Macedonia presented. The Macedonians, although ethnically not dissimilar, were regarded both by themselves and by the Greeks as non-Greek.* In his third Philippic, Demosthenes described Philip as "not only no Greek, nor related to the Greeks, but not even a barbarian from any place that can be named with honors, but a pestilent knave from Macedonia, whence it was never yet possible to buy a decent slave."

The protracted Peloponnesian War, which finally ended in 404, had depleted the city-states, most assuredly Athens, and a mere sixty-five years later, in 338 B.C.E., Greek autonomy was, at least officially, over.† The ancient Roman and Greek historians consider the battle of Chaeronea, fought on August 2, 338, between Philip II and the amassed forces of the Greek *poleis,* as signaling the end to Greek liberty and Greek political history. In those sixty-five years Athens, though never again an empire, enjoyed many of its cultural successes. Plato's philosophical life is lived out during this time. This is the period during which

---

*Thucydides, too, regarded the Macedonians as barbarians, not Greeks. They were, in Greek eyes, *barbaric,* as we use the term, savage and uncouth. They hadn't gone through the civilizing steps that not only Athens but all the *poleis* had worked out for transfers of power. In Macedonia those transfers were regularly effected by way of assassination. (It's still a matter of speculation how far down the line of Macedonian kings this practice continued. Philip II's suspiciously sudden death might have been the handiwork of his son, Alexander.) The Greek system of independent *poleis,* the very context for the Greek notion of *aretē,* was a thing as foreign and unintelligible to the Macedonians as the Greek language itself.

† This is not to say that the Greek *polis* ceased to function, however. "The Macedon conquest of Greece put an end to the oppression of one Greek city by another, both because of Macedon's overwhelming military power and because Macedon conquests outside Greece removed the potential for using international power politics to cajole cities into preferring subordination for the known quantity of another Greek city, rather than to the uncertain quantity of a foreign power. But for many Greek cities, loss of an independent foreign policy was nothing new, and the characteristic life of the Greek city-state continued long after Philip of Macedon's establishment of the League of Corinth in 338 B.C." Robin Osborne, *Greek History* (London: Routledge, 2004), pp. 3–4.

he founds his Academy and writes his dialogues. This is the period, too, when Aristotle comes to Athens, first to study at the Academy, eventually to found his Lyceum. It's arguable that in the sixty-five years between Athens' defeat in the Peloponnesian War and its absorption into Macedonia, it enjoyed its period of fullest democracy. So one can hardly say that the Athenian experiment with the extraordinary died with its defeat by Sparta. But in terms of long-term damage, not only to Athens but to Hellas as a whole, the protracted Peloponnesian War played a pivotal role, and so then did the mischief of Alcibiades.[*]

But the Athenians who longed for their prodigal son's return couldn't know any of this. In 404, in the misery of their defeat, their democracy itself (temporarily) toppled and replaced by the brutal oligarchy of the Thirty, in cahoots with Sparta, still Athens didn't cease in its longing for Alcibiades. He had taken himself off to Asia Minor and was living in luxurious decadence, at least according to Plutarch, in Phrygia, in cahoots now with the Persians. But the Athenians had not forgotten him nor been able to bring themselves to believe that he had forgotten them. "And yet in spite of their present plight," writes Plutarch, "a vague hope still prevailed that the cause of Athens was not wholly lost so long as Alcibiades was alive. He had not, in times past, been satisfied to live his exile's life in idleness and quiet; nor now, if his means allowed, would he tolerate the insolence of the Lacedaemonians and the madness of the Thirty."

It was perhaps for this reason that he was killed, someone among the Thirty having convinced Sparta that the Athenians would never give up hope of regaining their hegemony so long as Alcibiades lived. (It may even have been Critias, Plato's mother's uncle, so powerful among the Thirty.) Secret assassins were perhaps dispatched, either from within the Thirty in Athens or from Sparta itself. Plutarch cites such a political intrigue as one possible scenario for how Alcibiades died. But Plutarch offers another scenario as well, in keeping with the operatic life of Alcibiades. In this version Alcibiades had seduced a country girl, and it was her brothers who came to avenge her stolen maidenhood.

---

[*] Lord Byron, who is in so many ways reminiscent of Alcibiades, took up the cause of Greek autonomy, and spent £4,000 of his own money to refit the Greek fleet to attack the Ottoman Empire, planning, despite his lack of military experience, to lead the charge together with Alexandros Mayrokordatos, a Greek politician and military leader. He became ill before his plans could be fulfilled, but he's still revered in Greece, with Vyron a popular name for boys and Vyronas, a suburb of Athens, named after him.

In this telling, he died a farcical death, sacrificed to no greater cause than his own inability to keep it in his *chitōn*. Whatever the truth of his death, Martha Nussbaum remarks: "His story is, in the end, the story of waste and loss, of the failure of practical reason to shape a life. Both the extraordinary man and the stages of his careening course were legendary at Athens; they cried out for interpretation and for healing."*

I've left out many of the details of this story, including one of the most important, certainly from the point of view of Plato and philosophy. In his youth, Alcibiades had, like so many aristocratic young Athenian men, attached himself to Socrates, and their relationship had been intimate and passionate. Socrates, too, had been unable to resist the irresistible boy and had loved him in his fashion, which meant focusing his attentions on trying to stimulate not his erogenous but his philosophical zones.

But what is even stranger is that Alcibiades, from all accounts, had been unable to resist Socrates, at least for a while, when he was of that age when many young aristocratic men fell under Socrates' spell. Baffled by his inability to have his way with Socrates—he who could have his way with one and all—Alcibiades had felt, perhaps for the only time in his life, the force of a personality more powerful than his own. Socrates alone had been able to flip the vectors of Alcibiades' attention so that they weren't pointing exclusively at Alcibiades himself, and the disorientation could bring the young man to tears by making him see his own mediocrity when measured by the entirely separate and self-contained standards that Socrates embodied. Plutarch writes:

> Alcibiades was certainly prone to be led away into pleasure. That lawless self-indulgence of his, of which Thucydides speaks, leads one to suspect this. However, it was rather his love of distinction and love of fame to which his corrupters appealed, and thereby plunged him all too soon into ways of presumptuous scheming, persuading him that he had only to enter public life, and he would straightway cast into total eclipse the ordinary generals and public leaders, and not only that, he would even surpass Pericles in power and reputation among the Hellenes. Accordingly, just as iron, which has been softened in the fire, is hardened again by cold water, and has its particles compacted together, so Alcibiades, whenever Socrates found him filled with vanity and wan-

*Martha Nussbaum, *The Fragility of Goodness: Luck and Ethics in Greek Tragedy and Philosophy* (Cambridge: Cambridge University Press, 1986), p. 166.

tonness, was reduced to shape by the master's discourse, and rendered humble and cautious. He learned how great were his deficiencies and how incomplete his excellence.[*]

For a brief period, Alcibiades had been susceptible to the kind of beauty that Socrates, that ugly little man who looked like a prancing, lubricious satyr, was able to make men see. Alcibiades tells us, at least in the dramatically and philosophically rich account which Plato has given us in the *Symposium*, of his humiliating attempts to seduce Socrates in the more conventional fashion, a failed seduction all the more humiliating since it required Alcibiades, though younger, to assume the role of the importunate lover.

This reversal constituted a violation of *paiderastia*, "a general pattern of feeling and conduct which is unique in the history of Western society: a code of male homosexual love openly practiced and socially acceptable."[†] *Paiderastia* not only was socially acceptable but was considered by many Greeks, most especially the more aristocratic among them, as essential to the cohesion of the *polis*, and therefore something that went into the making of a citizen's *aretē*. And the fact that our word for perhaps our most despised form of criminality is derived from practices that went into the Greek idea of *aretē* ought to heighten attention to a point the classicist Robin Osborne makes, that "in the comfortable analysis of a culture so like our own, we come face to face with the way the glory that was Greece was part of a world in which many of our own core values find themselves challenged rather than reinforced."[‡] Changing views on sexual morality have, in many ways, brought us closer to the sexual attitudes of the ancient Greeks, who never thought that the gods cared a fig leaf about what we did with one another sexually, especially if no harm came to any participant. (Actually, Greek gods were far less sensitive than mortals on the matter of sexual predation.) We have been brought closer to the Greek acceptance of homosexuality and bisexuality as completely natural ways for people to relate to one another. But it has perhaps carried us away from their relaxed attitude toward sexual relations between the powerful and the powerless, especially as this imbalance of power is reflected in age. Even this is not

---

[*] Plutarch, *Parallel Lives*, "The Life of Alcibiades," 6.2.
[†] Bernard Knox, "The Socratic Method," *The New York Review of Books*, January 25, 1979.
[‡] Robin Osborne, *Greek History* (London: Routledge, 2004), p. 22.

entirely clear, however, due to our continuing uncertainty as to what was meant by a *pais*—a boy.

The age at which a boy was initiated into sexual relationships remains, for us, undefined. Various scholars have estimated the youngest suitable age, and their answers range between twelve and eighteen. That is, from our point of view, a significantly wide gap (though it was not so many years ago that, even in the West, girls as young as twelve were married off, and often to men far older than they). There are many allusions in Greek literature to the first facial hairs sprouting. Peach fuzz was desirable, but a beard was not.* Sexual mores varied from *polis* to *polis*.† In Athens, seventeen may have been the operative age of consent for boys, which was the age of citizenship, but since the Greeks didn't have a zero, that actually translates to sixteen (except that it was the year of birth, not the day, that was used in the calculation). What is clear was that there was far greater care expended on protecting boys than girls. Well-to-do Athenian families hired a special slave—a *paidagōgos*—to accompany an underage boy and see that nobody approached him with indecent intent, but these special precautions are an indication that, whatever the age considered appropriate for a boy, younger boys were eroticized. In Plato's *Charmides*, Socrates, eager to engage in conversation with the exceptionally beautiful Charmides (Plato's uncle), self-consciously says that "there could be no impropriety" in talking to him, since Critias, Charmides' uncle and guardian, is present. Still a subterfuge is required, so highly eroticized is the mere sight of Charmides. Critias says that Charmides has been suffering from headaches, and Socrates should make the boy believe he knows a cure. One might think that Plato was subtly indicating the character deficiencies in Critias, future leader of the Thirty, if he didn't

---

*In the *Symposium*, when Pausanius distinguishes between the good *erōs* and the bad, he argues that the good is always directed at "boys" rather than females, "cherishing what is by nature stronger and more intelligent" (181c). However, good love is not directed toward boys until they begin to show some intelligence, which starts happening when "their beards begin to grow" (181d). Pausanius then goes on to explain why it is wrong to love those who are prepubescent, and it has nothing to do with protecting the boys but rather the lover of boys: "Actually, there should be a rule against loving young boys, so that a lot of effort will not be squandered on an uncertain project. It is unclear how young boys will turn out, whether their souls and bodies will end up being bad or virtuous" (181e). We have no way of knowing when puberty hit in ancient Athens. James Davidson in *The Greeks and Greek Love* (London: Phoenix, 2007) argues that it was far later than in our day, though scholars have challenged his arguments.

† Plato has Pausanius make this point in the *Symposium* (182). He concludes, not surprisingly, that the way of loving boys in Athens is the most noble.

also make Socrates eagerly respond, "Why not, if only he will come" (155b).

But though boys who were too young—whatever that means—were, if desired, still off limits, when they did come of age then a relationship with an older male was expected. The lover might be only a few years older, but still he would be at a different stage of political engagement. The difference in their political stages—their age-appropriate duties to the *polis*—was important, since it was by way of the intense relationship between the older lover and the younger that the latter came to learn the politicized *aretē* that was to govern his life. There is, as Kenneth Dover memorably puts it, "a consistent Greek tendency to regard homosexual erōs as a compound of an educational with a genital relationship."[*] And this was true not just in Athens but throughout the Greek *poleis*. "Such relationships were taken to play such an important role in fostering cohesion where it mattered—among the male population—that Lycurgus even gave them official recognition in the constitution for Sparta."[†] Since the publication of Dover's groundbreaking book, the standard term adopted among scholars for the older lover is *erastês*, the lover, and for the younger, *erômenos*, the beloved, both words deriving from *erōs*.

The earliest scenes of homosexual courtship on Attic black-figure vases are contemporary with the Solon fragment: "When in the delicious flower of youth he falls in love with a boy *(paidophilein)*, yearning for thighs and sweet mouth."[‡] Artistic renditions of homoerotic couplings are most prolific around the time that the Greeks defeated the Persians, at the height of Hellenic self-confidence teetering on self-exaltation. Homosexual relationships are often praised as more desirable, more manly, than heterosexual relationships, which are mired in mere carnality or reproductive goals, with little to contribute to the cultivation of *aretē*, since females didn't participate in the political life of the city. All the partygoers in Plato's *Symposium*, except for the comic poet Aristophanes, don't deign to extol any *erōs* but the homoerotic.[§] And

---

[*] K. J. Dover, *Greek Homosexuality*, rev. ed. (Cambridge, MA: Harvard University Press, 1989), p. 202.
[†] George Boys-Stones, "Eros in Government: Zeno and the Virtuous City," *Classical Quarterly* 48 (1998): 169.
[‡] Fragment 29. See K. J. Dover, *Greek Homosexuality*, rev. ed., p. 195.
[§] Ancient Greek has several words for love. *Agapē* is chaste enough that the word and its cognates are profusely used in the New Testament. *Pothos* is a sort of longing, usually for

Aristophanes' speech isn't in an extolling vein at all. Rather it details the more undignified and embarrassing aspects of *erōs*, although the speech keeps itself clean, with no gratuitous vulgarities. Considering that this is Aristophanes, whose plays revel in obscenities, this daintiness is out of character. But it's Plato who is writing Aristophanes here, and he is permitting himself to go as far in the direction of vulgarity as his sense of dignity allows. So, for example, Plato has Aristophanes first miss his turn at giving his speech in praise of the god Erōs because he is temporarily overcome by, of all things, the hiccoughs, indicating the absurd indignities to which our bodies subject us when they overcome us. A famous writer can't even get the words out, and the godlike gift of speech is rendered ridiculous by animal-like eruptions. When the poet finally recovers, he tells the myth, to which the moderator Zachary Burns of chapter δ briefly alludes: of how humans were once physically coupled in pairs, spherically rolling around, so that each sphere was a complete entity. There were some spheres that paired a man with a man, some a man with a woman, and some a woman with a woman. Complete in themselves, mortals were too godlike for the immortals' tastes. "They had terrible strength, and power, as well as grand ambitions" (190b). The gods powwowed and instead of ending the race— always a live option for the Olympians—Zeus hit on the plan of slicing the self-contained mortals down the middle, like a hard-boiled egg sliced by a wire, so that each half was separated from its other half. (Dover attributes the origins of this erotogenic tale not to Plato, nor to Aristophanes, but rather to an Orphic myth.) Thus in this gathering, Aristophanes alone acknowledges the common sources of homosexual and heterosexual love, implying that they are of the same nature: *erōs* is, in all of us, a matter of our running around in desperation, trying

---

someone who is absent. *Himeros* comes over a lover almost like a physical sensation, from either the sight of the beloved or the thought of him. Plato's *Phaedrus* is filled with images of *himeros* streaming into the lover at the sight of the beloved, and Plato describes this stream as reversing itself and flowing back to the beloved until both of them are drowning in *himeros*, this exchange of himerotic fluids not necessarily involving any actual bodily fluids—in fact, from Plato's point of view, far better if not. *Erōs* is the full-on obsessional "in love" experience, the kind that makes a person do crazy things, like move from New York to Boston. It is very heavy on the longing. Davidson, who nicely distinguishes between these various Greek love terms, remarks, "*Erōs* is, with only a few exceptions, utterly one-sided. You can be longed for, loved *(philein)*, desired 'in return' . . . with no problem, but for the Greeks there can be no mutual *erōs*, not concurrently. *Erōs* doesn't work like that. He is a vector, a one-way ticket from A to B" *The Greeks and Greek Love*, p. 23.

to find our one and only lost half, and so complete ourselves.* His is the tale that emphasizes the contingency of erotic satisfaction, placing *erōs* outside the domain of our own choice and making it a matter of pure chance whether you'll ever find the one for whom you were, so to speak, destined from the moment of your birth. You leave that bar five minutes too soon, and instead of endless bliss it's endless frustration. The gods knew exactly what they were doing, too, according to the Aristophanes of the *Symposium*, since it's our erotic desperation that prompts our piety. "If we are friends with the gods and on good terms, we will find and establish relationship with those darlings meant for us, which few do now" (193b).

There is no evidence that the privileging of homosexual relationships, particularly within the aristocratic classes, existed during the Archaic period when the Homeric poems were being composed, let alone during the earlier Bronze Age period, in which the poems were set, though the men of the classical period read such relationships *into* the Homeric tales. The all-important friendship in the *Iliad* between Achilles and Patroclus is a case in point. Though the love between Achilles and Patroclus is a special one, sufficient for the death of the latter to rouse the former out of his inanition and take the life-terminating action that makes him the greatest of the legendary heroes, Homer provides no hint that this love is erotic. But the Greeks of the classical age nevertheless read the erotic into love, interpreting the intensity of the love—Achilles' stated wish that when he dies his ashes should be intermingled with the ashes of Patroclus—as leaving little question.† The mainstay of the erotic interpretation of the central motif of the *Iliad* was Aeschylus' lost trilogy *Myrmidons, Nereids,* and *Phyrgians,* of which only fragments remain. In our own day, a genre known as slash fiction, written by fans of such shows as *Star Trek,* fancifully eroticizes the same-sex relationship of characters such as Captain Kirk and Mr. Spock. (As the voluminous academic literature on the genre helpfully explains, the term "slash" comes from the punctuation mark used in the dyads which identify the subgenres, such as Kirk/Spock and Starsky/Hutch.)

---

*If the erotogenesis is the same, in Aristophanes' tale, for homosexual and heterosexual love, the former is still superior. "Only men of this sort are completely successful in the affairs of the city. When they become men, they are lovers of boys and by nature are not interested in marriage and having children, though they are forced into it by custom. They would be satisfied to live all the time with one another without marrying" (*Symposium* 192a–b).
† Dover, *Greek Homosexuality,* p. 197.

The lost trilogy of Aeschylus can be regarded as the original slash fiction. Plato, too, has his character Phaedrus in the *Symposium* read an erotic relationship into Achilles and Patroclus (180a), differing only in making Achilles rather than Patroclus the younger *erômenos*.

Why did the Greeks come to have the attitudes they had toward sexuality, mingling, as Dover put it, the educational and the genital in homosexual relations that were cherished above heterosexual relations (at least by the aristocrats of Plato's class)? The complexity of the problem, the morass of nature and nurture questions it asks us to take on and sort out, has put the question beyond the reach of all but the most daring speculations.* But if we can't say quite *why*, we do know much about *how* they practiced homosexuality, the norms that dictated what was good love and what was bad—unseemly and unmanly. And we know that the respective behaviors of the *erastês* and the *erômenos* were central to these norms.

Laws protected boys below the accepted age from being shown the wrong kind of attention. The *erômenos* must be of an age at which he can obtain some benefit from his relationship with the *erastês*, suitable for someone who is soon to enter, or has just entered, into the life of the *polis*. What he is not supposed to derive from the relationship is sexual gratification.† The *erômenos* is supposed to assume a position of non-aroused passivity in erotic activity, maintaining a pose of such indifference as to seem barely aware of what is being done with his body, which tended to be, at least ideally, intercrural (between the thighs, at least according to most scholars). Penetration was considered crude. In fact, it was illegal for citizens, but practiced on catamites. There was a complex ideology at play here, including the perceptions that for a man to allow himself to be penetrated is to denature himself by assuming the woman's role.

Given these norms, we can appreciate all the more the radical rever-

---

* Georges Devereux, in "Greek Pseudo-Homosexuality and the 'Greek Miracle' (*Symbolae Osloenses* 42 [1967]: 69–92), speculates that Greek society cultivated prolonged adolescence and that this explains both their polymorphous sexuality and the explosiveness of their genius. Devereux is important to Greek scholarship, according to Davidson, because of his "seminal influence" on Dover: "indeed he," [that is, Dover] "began the project," [that is *Greek Homosexuality*], "with Devereux as co-author." *The Greeks and Greek Love*, p. 84.

† "Since the reciprocal desire of partners belonging to the same age-category is virtually unknown in Greek homosexuality, the distinction between the bodily activity of the one who has fallen in love and the bodily passivity of the one with whom he has fallen in love is of the highest importance." Dover, *Greek Homosexuality*, p. 16.

sal that is effected in the story that Alcibiades tells about his love for Socrates in Plato's *Symposium*. In this story, the roles of *erômenos* and *erastês* are toyed with. It's Socrates who acts the part of the young and aloof *erômenos*, spending the night turned away from Alcibiades, while beautiful Alcibiades takes on the role of the turned-on *erastês.** One scholar remarks that the very inversion of this relationship that Plato presents in the *Symposium* as taking place between Socrates and Alcibiades in itself reflects the dangerous way in which Socrates reverses *paiderastia.*† And since *paiderastia* is intimately related to politicized *aretē*, the reversal indicates the deeper ways in which Socrates challenged the values of his culture.

First Alcibiades lured Socrates into spending the night, then he declared that Socrates was the only man worthy of being his lover, to which Socrates had jokingly replied that "you must see in me a beauty that is extraordinary and quite different from your own good looks." And then Alcibiades, convinced that he had "as it were let loose my arrows" (all Alcibiades' metaphors of *erōs* are violent), makes a definitive move.

> I stood up, and not letting him say anything further, I put my own cloak over him, since it was winter. Then I lay down, getting under his own worn garment, threw my arms around this truly daimōnic and amazing man, and lay there the entire night. (And you can't say that I am lying about this, Socrates!) After I had done these things, he acted far better than I had; he disdainfully laughed at my youthful good looks, in a quite outrageous manner—and this was about something I thought was of real importance! . . . By every god and goddess, I swear I got up after having slept with Socrates in a way that had no more significance than sleeping with a father or an older brother. (219b–d)‡

---

* "The inner experience of an *erômenos* would be characterized, we may imagine, by a feeling of proud self-sufficiency. Though the object of importunate solicitation, he is himself not in need of anything beyond himself. . . . For Alcibiades, who had spent much of his young life as this sort of closed and self-absorbed being, the experience of love is felt as a sudden openness, and, at the same time, an overwhelming desire to open." Nussbaum, *The Fragility of Goodness,* p. 188. Nussbaum's superb readings of both the *Symposium* and the *Phaedrus* in *The Fragility of Goodness* have influenced me a great deal.

† C. D. C. Reeve, "Plato on Friendship and Eros," *The Stanford Encyclopedia of Philosophy,* ed. Edward N. Zalta. Spring 2011 edition. Online at http://plato.stanford.edu/archives/spr2011/entries/plato-friendship/.

‡ *Plato's Erotic Dialogues, The Symposium and Phaedrus,* translated with introduction and commentaries by William S. Cobb (Albany: State University of New York Press, 1993).

It's significant that Plato has Alcibiades describe Socrates as "dai-mōnic." A *daimōn* is, in the Greek conception of the continuum progressing from mortals to gods, an intermediate sort of creature, partaking—just as some of the legendary heroes do—of both mortal and divine properties.* (Our word "demon" derives from the Greek.) Socrates notoriously claimed that he had his own private *daimōn* who whispered to him whenever he was about to do wrong.† But here it is Socrates himself who is likened to a *daimōn*. In Alcibiades' telling—or in the words that Plato gives him—it's Socrates, not Alcibiades, who is the creature whose metaphysical placement is beyond that of normal mortals. Alcibiades is no latter-day Achilles at all, but rather one who stands in awe of the superhuman capacities of ugly old Socrates.

A number of Plato's dialogues, in addition to the *Symposium,* make allusions to Socrates as the *erastês* of Alcibiades, though this shouldn't be interpreted as implying a sexual relationship between them. Plato often uses the word metaphorically, as in the *Gorgias* (481d), where Socrates describes himself as the *erastês* of two things, Alcibiades and philosophy. What he's saying is that he has fallen in love with both of them. Both loves induce transformative longing. Plato begins the *Protagoras* with an unnamed friend asking Socrates where he has come from, then supplying the presumed answer: "no doubt from pursuit of the captivating Alcibiades," which Socrates doesn't deny. Plato seems to go out of his way to call attention to Socrates' relationship with the controversial figure, which also calls attention to Socrates' abject failure in impressing any semblance of his own *aretē*—here meaning something much closer to what we mean by "virtue"—into his *erômenos.* Whether or not virtue can be taught is a question that comes up repeatedly in the

---

* Cf. *Cratylus* 398c, a dialogue that investigates language and tries, albeit playfully and tentatively, to come up with possible explanations for why things are called (in Greek) as they are. When the word *daimōn* comes up, Socrates connects the word with *daimones* (knowing or wise). "And I say, too, that every wise man who happens to be a good man is more than human (*daimonion*), both in life and death, and is rightly called a *daimon*." Translation here is by Benjamin Jowett (Princeton, NJ: Princeton University Press, 1961). He then goes on to connect the word *hero* with *erōs*, and to connect the word *eros* with *erôtaô*, which means "I question," concluding that heroes are those who know how to question, meaning philosophers, while at the same time connecting the erotic with questioning, and also claiming an essential connection between philosophers and the erotic, which last claim accords with *Lysis,* the *Symposium,* and the *Phaedrus.* All of this etymology is intimately connected up with the transformation of the Ethos of the Extraordinary wrought by Plato, by means of which the many ways of becoming extraordinary are reduced to only one, consisting in the extraordinary exercise of the faculty of reason that is, as Plato would have it, philosophy.
† See chapter θ.

dialogues, centrally in the *Protagoras* and the *Meno*. It's a question that's intimately connected with the vexed proposition of whether virtue is knowledge or not. If it is, then it should be teachable. But if it isn't, then what does philosophy have to do with any of it? We're back to the sort of question Stanley Fish raised in the *New York Times,* a question that Plato is always alert to. If philosophy doesn't end in knowledge of some sort, including knowledge of virtue, then perhaps it just reduces to a sort of game that "doesn't travel." And the question of whether virtue can be taught is one that we could well imagine the historical Socrates pondering in a rather personal way, considering his own failures with the gifted but wayward Alcibiades. Homosexual relations could be looked on as serving a higher moral value—as opposed to the mere sensuality and reproductive goals of heterosexual relations—because the Greeks regarded homosexual *erōs* as "a compound of an educational with a genital relationship." Though Socrates may have chastely excised any bodily aspect from his eroticized relationships, there was no one who more embraced the educational aspects.

Did Plato regard the moral failure that Alcibiades undoubtedly was as a pedagogical failure on Socrates' part? Does he spare no reference to the special relationship between these two extraordinary characters in order to indicate Socrates' limitations as a teacher of virtue? Or are the copious allusions rather there to make the point that virtue cannot be taught, that *aretē* can't be transmitted from the knower to the ignorant as mere information can be? Alcibiades, smart boy that he was, had mastered the purely formal aspects of Socrates' methodology. Xenophon relates an amusing story of a young Alcibiades, still the ward of Pericles, trying to show off the sort of dialectics that Socrates wielded. When he backs Pericles into a corner, Pericles remarks, "At your age we were clever hands at such quibbles ourselves. It was just such subtleties which we used to practice our wits upon; as you do now, if I mistake not." To which Alcibiades cheekily replies: "Ah, Pericles, I do wish we could have met in those days when you were at your cleverest in such matters."

Plato was nine years old in 415 when the fleet sailed to Sicily, so he would have had only one opportunity to lay mature eyes on Alcibiades. This was when Alcibiades returned to a forgiving Athens in 407 and stayed in the *polis* for four months. Nevertheless, though he probably never knew him intimately, Plato's sense of the man in all his contradictions must have been unusually vivid, allowing him to create the dramatic presentation we get in the *Symposium*. Alcibiades had insisted on his individuality above all else, and Plato gives a sense of that individuality in the few short pages of Alcibiades' tumultuous appearance. It seems important to Plato to do full justice to who and what Alcibiades was, his charm as well as his dangerous recklessness, and one of the reasons might have been to exonerate Socrates of any responsibility for who and what Alcibiades was. After all, one of the charges brought against Socrates in 399* was that he had corrupted the young, and nobody's corruption had unleashed more dire consequences for Athens than Alcibiades' (unless of course it is the corruption of Critias, who also had once been close to Socrates and who, you'll remember, became a leader of the notorious Thirty, who ruled Athens for eleven months after its defeat by Sparta in 404).†

Alcibiades makes his entrance late in the *Symposium*, when it feels as if it's almost over, that we have reached the climax and all is winding down. The party had been an unusual one in that no wine had been drunk (only enough to disguise the taste of water), at least not until Alcibiades shows up. The party that night takes place in the middle of the Dionysia, the festival of theater, and is attended by men prominent in the *polis*. They are so hungover from their previous night's celebration that they decide to make this symposium a dry one. Instead of toasting with wine, they'll compose, each one in turn, a paean to the god Erōs. A good deal of flirtation has gone on as the men, some of them lovers, recline together on their couches and sing the god's praises. Socrates has been the last to speak and he has just delivered

---

*The internal dating of the *Symposium* presents complicated and contentious issues. The party is in celebration of the young poet Agathon's having, for the first time, taken first prize, so that allows scholars to date the party as having taken place in 416. But the party is recounted from a distance of several years, and the question is *how* many years? Why are people suddenly running around Athens trying to get the details of that party straight, as the prologue to the dialogue suggests? The date is traditionally put at 400. David O'Connor, in his edition of the poet Shelley's translation of the *Symposium*, argues that, on the contrary, it is sometime in the spring of 399, and the flurry of interest in Socrates is occasioned because word has leaked out that he is to be put on trial. Martha Nussbaum (in *The Fragility of Goodness*) argues for an earlier date, 404, on the grounds that the flurry of interest is actually centered on Alcibiades, with all of Athens first fixated on whether Alcibiades will return to save a languishing Athens, soon to lose in the protracted Peloponnesian War, and then learning—during an interim indicated in the prologue—that Alcibiades has been killed.

† Xenophon, in fact, links Critias with Alcibiades: "Among the associates of Socrates were Critias and Alcibiades; and none wrought so many evils to the state. For Critias in the days of the oligarchy bore the palm for greed and violence: Alcibiades, for his part, exceeded all in licentiousness and insolence under the democracy. Ambition was the very life-blood of both: no Athenian was ever like them. They were eager to get control of everything and to outstrip every rival in notoriety" *Memorabilia*, Book I, Chapter I.

an exquisite disquisition on *erōs,* telling us that it is the one thing that can save us, since it alone can break the spellbound fascination that our own self casts over us. *Erōs* alone can make us take passionate notice of something outside of ourselves, turn the vectors of our attention so that they are facing outward. *Erōs* is our longing to possess the beautiful, and it is this aching longing that draws our very selves out of ourselves. But the longings of *erōs* can't be left in their natural state or they will wreak havoc on our lives. And besides there is so much greater beauty to be discovered than that which resides in beautiful boys. Our love for the beautiful must, by stages, be turned from our fixation on individual beautiful bodies to the more general idea of embodied beauty, and from there to ever more impersonal and disembodied forms of beauty—the beauty of Athenian laws, for example, and of the abstractions that go into genuine knowledge. And as the lover "becomes more capable and flourishes in this situation, he comes to see a knowledge of a singular sort that is of this kind of beauty . . . The person who has been instructed thus far about the activities of Love, who studied beautiful things correctly and in their proper order, and who then comes to the final stage of the activities of love, will suddenly see something astonishing that is beautiful in its nature. This, Socrates, is the purpose of all the earlier effort"(210e).

Socrates pretends in this passage that he is only reciting what he had heard, long ago, from a priestess named Diotima, of the city of Mantinea, from whom he learned everything he knows about *erōs.*[*] Socrates introduces Diotima by telling us that she had delayed the plague from arriving in Athens by ten years; the suggestion is that erotic love has pestilential possibilities, capable of killing off men's bodies, as well as their souls, as the sudden appearance of Alcibiades will make clear.

---

[*] "We have no evidence outside the *Symposium* for a female Mantineian religious expert named Diotima, and in any case it is unlikely that any such person taught Socrates a doctrine containing elements which, according to Aristotle, were specifically Platonic and not Socratic. Plato's motive in putting an exposition of Eros into the mouth of a woman is uncertain; perhaps he wished to put it beyond doubt that the praise of *pederastia* which that exposition contains is disinterested, unlike its praise in the speech of Pausanias." Dover, *Greek Homosexuality,* p. 161, n. 11. Some commentators have speculated that Diotima might have been based on the historical figure of Aspasia, who was Pericles' brilliant mistress, a woman who actively participated in the circle of intellectuals surrounding Pericles. Plato mentions her by name in *Menexenus,* having Socrates there declare that not only he learned the art of rhetoric from her, but so did Pericles. "She who has made so many good speakers, and one who was the best among all the Hellenes, Pericles, the son of Xanthippus" (235e). Commentators have pointed to this passage as an indication of the high regard in which Plato held Aspasia, although other interpretations—insulting to Pericles—are also possible.

Alcibiades embodies the dangers of *erōs,* of letting our helpless adoration of unworthy love objects, as extraordinary as they might be, destroy not just our loving self, not just the love object, but even—given the role of *erōs* in teaching the duties of a citizen—the *polis.*

In the famous Socratic speech in the *Symposium, erōs* is not to be denied, but is rather to undergo a process of education. That is what philosophy is: the education of erotic desire. The purpose of our erotic longings turns out to be the same as the purpose of our cognitive longings—to get us outside of ourselves, to allow us hard-earned contact with what is, *to on.* This longing of ours to possess the beautiful can only be slaked in our knowledge of the astonishing beauty immanent in the structure of the world, the source of all instantiations of lesser beauty. Knowledge effects a possession that nothing else can achieve. Those unquenchable longings associated with erotic love, accompanied always by a touch of sadness, of disappointment, of the mournful sense of unfulfillment as we try to merge ourselves with the beloved, are so wrenching and frustrating and also absurd—an aspect of the erotic that Aristophanes' speech plays up—because erotic longing is meant to carry us on to merge with something far larger and more constant and more worthy than a mere *person.*

> Here is the life, Socrates, my friend, said the Manitinean visitor, that a human being should live—studying the beautiful itself. Should you ever see it, it will not seem to you to be on the level of gold, clothing, and beautiful boys and youths, who so astound you now when you look at them that you and many others are eager to gaze upon your darlings and be together with them all the time. . . . What do we think it would be like, she said, if someone should happen to see the beautiful itself, pure, clear, unmixed, and not contaminated with human flesh and color and a lot of other mortal silliness, but rather if he were able to look upon the divine, uniform beautiful itself? Do you think, she continued, it would be a worthless life for a human being to look at that, to study it in the required way, and to be together with it? Aren't you aware, she said, that only there with it, when a person sees the beautiful in the only way it can be seen, will he ever be able to give birth, not to imitations of beauty, since he would not be reaching out toward an imitation, but to true virtue, because he would be taking hold of what is true? By giving birth to true virtue and nourishing it, he would be able to become a friend of the gods, and if any human being could become immortal, he would. (211d–212a)

A running joke throughout the Platonic dialogues is Socrates' erotic alertness. In the *Lysis* he remarks that "though in most matters I am a poor useless creature, yet by some means or other I have received from heaven the gift of being able to detect at a glance both a lover and a beloved" (204b–c). In the *Charmides,* having just returned from the battle of Potidaea—one of the forerunners of the Peloponnesian War, fought in 432—the first thing he wants to know is who are the new boys on the scene, and if any of them are remarkable "for their wisdom or beauty or both." When Charmides sits down right next to Socrates and he gets a peek of the "inwards of his garment," he "takes flame and is overcome" (155d–e). Xenophon's *Symposium,* which is far more ribald than any of Plato's Socratic dialogues, exaggerates Socrates' libidinous fascinations even more, with Xenophon having Socrates announce that the personal virtue of his own that he values the most is his pimping. Plato, albeit more decorously, makes much the same point: "The only thing I say I know," Socrates says in the *Symposium,* "is the art of love *(ta erôtika)*" (177d).

Plato goes on to convert Socrates' wry statement into as ardent a statement of the ecstatic nature of knowledge as has ever been put forth. (Xenophon doesn't leave Socrates' scandalous announcement as it stands, either.) Socrates' Diotimaic speech merges the erotic ascent with the cognitive ascent, and merges both with the achievement of the extraordinary that makes a life worth living. "Do you think it would be a worthless life for a human being to look at that, to study it in the required way, and to be together with it?" Socrates has Diotima rhetorically asking about the vision of true beauty—inseparable from the True-the Beautiful-the Good—to which we will be led by our longing for beauty, when these erotic desires are properly educated by philosophy. Diotima had spoken "like a sophist," when she endorsed the fame and acclaim, the *kleos,* that people desperately pursue in their desire to defeat death; but now, in her own ascent, she is speaking like a philosopher. Our achieving the extraordinary in our lives may be the only solace we have in the face of death, but the only worthy way to achieve this extraordinary life is to reason our way to it.

It is at precisely this point of sublimity—Socrates having concluded his vision of cognitive bliss and erotic bliss made one—that Plato brings Alcibiades into the action. His presence is announced first by some undignified ruckus outside—those inside hear someone pounding on the door—and when Alcibiades makes his grand entrance he is so deep in his cups that he needs to be supported by a flute-girl and some other attendants. (This flute-girl had been sent outside to entertain the women when the men decided to make this symposium a temperate one.) Ivy and violets and ribbons are wreathed into his hair—a form of dressing up that mocks the garlands that have been genuinely won that day. (Perhaps this calls attention to another mockery he will allegedly perform, the parody of the Eleusinian mysteries.) He immediately takes command of the room in his outrageously adorable manner: "Are you laughing at me for being drunk?" he announces as he staggers in, letting us know the affectionate indulgence he provokes, the knowing laughter of the men in the room. "You may laugh," he lisps, "but I nevertheless know quite well that what I say is true," and we are meant, as Nussbaum stresses in her reading of the dialogue, to take this pronouncement seriously. Alcibiades is going to tell us something important, even if in a drunkenly disordered fashion. Informed that all of the men have been delivering praises to Erōs, Alcibiades agrees that he will do so as well, though first he drinks a huge vat of wine and gets others to start drinking as well, the disorder he requires as his backdrop already making itself felt.

He goes on to give a memorable account of Erōs. But he doesn't praise the god in the way that all the others have done, trying to capture the essence of the ideal love, concentrating on the transcendent divinity of Erōs in terms appropriately impersonal. Even Aristophanes had tried to do justice to the god in mythical terms. Alcibiades changes the terms of the assignment. It is a man he describes and not a god. It is Socrates. Or rather what Alcibiades describes is what it is like for him, for Alcibiades, to love this man—more unique, he declares, than Achilles himself (221c). It is the phenomenology of loving a particular individual, and an extraordinary one at that, which Alcibiades recounts. And now Alcibiades, in inverting everything Socrates has just said, ends up overtly declaring that the greatest philosopher in their presence, namely Socrates, does indeed exceed Achilles. But he exceeds Achilles in his personal uniqueness, in the achievement of being more *of himself*, of coming into possession of his own individuality. Socrates has just been preaching a sort of stripping away of all that is personal. The extraordinary individual not only is in love with the True-the Beautiful-the Good, but, to the extent that his love takes him over, he becomes one with it, which is why Socrates describes such love as a sort of immortality, everything individual, and therefore mortal, dropping away. But Alcibiades stresses the personal. He describes what is so extraordinary about Socrates in terms that are implicitly Nietzschean, an exaltation of the extraordinary individual precisely because he is irreducibly and irreplicably individual.[*]

Alcibiades resists Socrates' ascent into the impersonal. He *insists* on personal terms. Alcibiades' account of what it has been like for him to love Socrates is thoroughly human, filled with both the transformative sense of wonder that we feel when we are deeply in love, but also the absurdity of that state, the indignities that we are made to suffer when it is an embodied person we love, rather than a god or a mathematical theorem or the abstraction of Beauty itself, embedded in the structure of *to on*.

Alcibiades has loved Socrates in the only way that he knows how. That's what's so wrong with his love. It never transforms him. He could love the most extraordinary man in Athens and never be changed in the loving. He's Alcibiades when he falls in love with Socrates, and he's *the same* Alcibiades when the experience is over, and that's because Alcibiades is too intent on being Alcibiades. He never loses sight of what it is to be Alcibiades. He never *wants* to. And in that lack of wanting is his doom.

Having Alcibiades—drunk, no less—crash the party at a late hour not only makes the *Symposium* that much more dramatic but turns it

---

[*]Alexander Nehamas endorses this Nietzschean reading of Socrates: "Those who practice the individualist art of living need to be unforgettable. Like great artists, they must avoid imitation, backward and forward. They must not imitate others: if they do, they are no longer original but derivative and forgettable, leaving the field to those they imitate. They must not be imitated by too many others: if they are their own work will cease to be remembered as such and will appear as the normal way of doing things, as a fact of nature rather than as an individual accomplishment. . . . Nietzsche in particular was tyrannized by this problem. This aestheticist genre of the art of living forbids the direct imitation of models. Why is it, then, that Montaigne, Nietzsche, and Foucault all have a model? And why is their model always Socrates? . . . These philosophers care more about the fact that Socrates made something new out of himself, that he constituted himself as an unprecedented type of person, than about the particular type of person he became. What they take from him is not the specific mode of life, the particular self, he fashioned for himself but his self-fashioning in general" (*The Art of Living*, p. 10). Nehamas concedes that the only Socrates we know is essentially the literary creation of Plato. Does he then believe that Plato, by creating his Socrates, is endorsing the aestheticist conception of philosophy as the creation of an inimitable self for oneself? Nehamas's Nietzschean conception of Socrates is far more in line with Alcibiades' conception of Socrates, which is all about the inimitable uniqueness of this man, a conception which Plato seems to reject in the *Symposium*, not only within the content of Socrates' speech, but also by giving the contrasting view to mad, bad, and dangerous-to-know Alcibiades.

into a morality play. This is one of the aims that Plato accomplishes by bringing Alcibiades onstage so late in the evening. Alcibiades is the very inverse of all that Socrates has just been saying. Alcibiades, reveling in his excesses, projecting the edgy glamour of the dissolute outlaw, of living beyond the bounds, is also, in his own way, battling against the constraints of being merely human. It's a way of pretending we're not quite mortal, this abandonment and recklessness, exulting in doing what no ordinary mortals can get away with; and it proved to be the way Alcibiades not only loved but also lived and also died.

And by presenting Alcibiades as the very inverse of Socrates' teaching, Plato exonerates Socrates of the crimes of Alcibiades. He is underscoring the point that Alcibiades isn't the conclusion of Socrates' reasoning, but is rather its very negation. The *polis* might very well have had the ruinously willful defiance of Alcibiades in mind in charging Socrates with the corruption of the young. Alcibiades had defied the politicizing of *aretē,* and Socrates, too, defied the politicizing of *aretē.* For both of them it is the individual who must achieve exceptionality; the *polis* can't do it for him. It's true that Socrates' defiance is made in the name of moral philosophy, whereas Alcibiades' defiance is made in the name of Alcibiades. Small difference, the Athenians might be heard protesting in bringing their charges against Socrates; you opened the door to such rebellion. *Huge* difference, Plato is insisting in the *Symposium.* Alcibiades may have dramatically rejected the Athenian answer to what makes a life worth living—the politicized *aretē* that channels the energies of the extraordinary person so that it serves the extraordinary *polis*—but he rejected the Socratic answer just as dramatically. In fact, his rejection of the Socratic point of view, as Plato develops it in the *Symposium,* is even more emphatically rebellious.

The Socrates of the *Symposium* is asking a person to travel as far away from the personal point of view as is possible, to a place in which not only the particular beauty of a beloved will pale to insignificance but a person's very own identity will seem of little concern to the person himself. Reason itself can bring one to a state of ecstasy, literally standing outside oneself, that Bacchanalian frenzy of philosophy mentioned in chapter α. This rationalist ecstasy is pointedly referred to in Alcibiades' speech, describing what the arguments of philosophy can do when they "seize the soul of a not untalented youth." Socrates is here grouped with the other not untalented youths, for, from the point of view of Plato's mature philosophy perhaps that is what he was, though a youth who had the seminal intuition that was needed to extract *aretē* from its social and political embedding. It's politics that have to be shaped in the light of moral values, and not the other way round. The state toward which philosophy's arguments can bring us—a state of ecstatic estrangement from the self, slipping the bonds of one's own particular identity—is the state that Plato holds out as paradigmatically philosophical, the attitude toward which we are sojourning in making progress in philosophy. One's connection with one's own self has been attenuated to a degree that one can contemplate even one's own personal demise with equanimity. To live and to think and to love as a philosopher is to live *sub specie aeternitatis*—that is, under the form of eternity, which was the way that Spinoza, who philosophized much in the spirit of Plato, was to put it some millennia later.

Socrates' speech in the *Symposium* is urging us on in the direction of an impersonal vision, promising us that in losing our personal attachments, even our attachment to our own self, we will achieve a knowledge that will make us over in its light—the perfect proportions of the True-the Beautiful-the Good assimilated into our knowing minds, knowledge become *aretē*, which is what wisdom is. The implication is that it was toward this vision that a young man's love for Socrates was meant to take him. For here is where it took Plato, who, in loving Socrates, managed to carry philosophy, Socrates' own love, far past the point that Socrates had reached. To love Socrates is to have been impregnated with his intuitions—the *Symposium* is filled with references to impregnation—in the manner in which Plato was. The very proof of this impregnation—the living child, as it were—is the ideas that Socrates is given to speak in the *Symposium*. Because Plato loved Socrates, because he—and *not* Alcibiades—was Socrates' ideal *erōmenos,* just as Socrates was Plato's ideal *erastês,* Plato has given birth to a unity of metaphysics and epistemology and aesthetics and ethics that lay implicit in Socrates' intuitions.

In contrast, Alcibiades' love for Socrates was sterile. Nothing creative or beautiful ever came of it. His career, in which his connection to his own self—his impulses and his ambitions—is promoted before all else, the exaltation of individuality, could not have been more distant from what it was to have been truly a student and lover of Socrates, to have taken him into one's self, been penetrated by what he had to offer.

By bringing Alcibiades so vividly to life before us, reminding us of how abundantly he failed to transform himself in the light of Socratic love, Plato doesn't just provide a motivation for the difficult path of educating our erotic strivings (you don't want to end up like Alcibiades, do you?). He also provides an exoneration for Socrates: he no more corrupted Alcibiades than Athens did. Both Athens and Socrates failed to win Alcibiades over to a love greater than self-love. If anything, Athens was even more to blame, for allowing itself to be so manipulated and seduced by Alcibiades. Socrates, in contrast, refused to be seduced. "There is no reconciling you and me," Plato has Alcibiades declare to Socrates (213c), and though Alcibiades tosses off the remark in a flirty way, Plato doesn't want the irreconcilability to be lost on us. The story of Socrates and Alcibiades is a story of a mutually failed seduction.

It is rather Plato who was successfully seduced by Socrates. It is Plato who demonstrates the effect that loving Socrates was meant to have on the young men who counted themselves the lovers of Socrates. Plato's demonstration consists in the life he is living, the works he is writing, including the *Symposium*.

For here is the paradox that Plato's conception of love presents to us: In recounting this vision of a love so transcendent and impersonal, that "vision of the mind [that] begins to see keenly when that of the eyes starts to lose its edge" (219a), Plato still does not lose sight of Socrates. The very dialogues he creates to speak his vision of depersonalized philosophy keep the person of Socrates a constant before him, and so before us.

ed not having majored in philosophy. After hearing that
nt, I thought you wouldn't object to my sending a few of
ries your way.

 by name, of course, together with all your official titles.
now if you're game. You mentioned that you were eager
 our society, and believe me, honeypie, nothing gets you
ers more quickly than the sort of questions that I get sent.

ember you very well. Please feel free to send me any que-
 you believe I might be helpful. I have no official titles
f philosopher. As it means "lover of wisdom," it seems
at anyone could ever desire. The only hesitation is in the
t.

ale graduate student, and though I'm sexually adventur-
ut.

 professors has proposed that he be my "professor with
ou get my drift. We've both got partners, but our rela-
open, so there's no question of cheating. There's also no
at we'd get emotionally involved, since our affections are
gaged. Frankly, I see this as a good educational move on
 man is one of the best minds in my field, and I'd learn
 extra face time with him. Conversations with him are
ating (yes, pun intended). He's also got powerful connec-
mises he'd help me professionally. The job market in my
ed, and though I'm at a top-notch department, we grad
 all the help we can get. I trust my professor to keep his
l indications are that he's a man of honor, and he's been
 front with me. Do you think I should take him up on his
mutually enjoy and use each other?

,
Higher Dreams

casting couch has moved to academia. What theatrical
 good career move" you are calling "a good educational
ell you that your crystal ball can't be entirely trusted when
at neither of you will become emotionally involved. One
now these things will play out. But, hey, if you fell for each

5

xxxPLATO

denly regr
announce
the racier
    I'll cite
Just let m
to learn ab
under the
    xxxMar

Dear Marg
    Yes, I re
ries for wh
beyond tha
all the title
deserving
    xxxPlato

Dear Marg
    I'm a fe
ous, I'm no
    One of
benefits,"
tionships a
possibility
otherwise
my part. T
a lot from
always stin
tions and p
field has ta
students n
word, since
completely
offer that w
    Yours tr
    Pursuin

Dear PhD:
    Wow, th
people call
move." I ca
it forecasts
never know

Margo Howard is a journalist and t
columnist Esther (Eppie) Lederer,
ders. Margo eventually went into t
her advice column, Dear Prudence
syndicated in more than two hun
National Public Radio. She then w
umn for Women on the Web (wow

Ann Landers was renowned for
names of such eminent friends a
the theologian and university pre
and U.S. Supreme Court Justice W
consulted in offering her guidanc
her columns. Her daughter Margo
strates below:

Dear Plato,
    Well, I told you'd be hearing
don't remember me, I was at Mart
ties, the redhead to your immedia
gallantly retrieved when it someho
was whispering in your ear to tell y
were. Marty and Anne tell me the
the three of us go back to ancient ti
the ego department, cupcake (or sl
you that they just can't stop singir
there hasn't been another like you

    You were very gracious in agr
sultants for my advice column. Y
tions have a philosophical angle,
be. You sure got the attention of t
philosophy is really all about eros.

other and ditched your partners, think of the professional possibilities! You do seem clear-eyed, however, about mutually using each other . . . you and this "man of honor." What you are suggesting is commerce, my dear. There is a name for people who trade sex for money or entrée. If you are comfortable with that, fine with me.

But this being a full-service advice column I decided to consult one of the world's leading experts in moral philosophy. Plato is probably the most quoted thinker in the world, and now I'm joining the crowd. Here's what the philosopher has to say:

"I commend PhD for the high value she places on wisdom and knowledge. For wisdom's sake, there is no disgrace in being servant and slave to a lover, no reproach for a person willing to give honorable service in the passion to become wise (*Euthydemus* 282b).

(So says Plato, leading me to parenthetically remark that now I know why they call them "philosophers," which literally means "lovers of wisdom." These philosophers really have the hots for wisdom. But before you think Plato has given you the green light for knowledgeable nooky, PhD, he's got some serious reservations he wants to share with you.)

"What PhD must ask herself is whether the arrangement that she is contemplating with such cool and disinterested calculation is truly one that will end in her having acquired the knowledge for which she longs. There would be no passion in this relationship; those are the terms on which this affair would be conducted. Nothing transformative would occur, with each self remaining firmly in possession of itself. Again, those are the terms on which this affair would be conducted.

"But PhD should consider whether these very terms preclude the possibility of attaining the good she desires. Wisdom is an extraordinary state. It requires experience sufficiently out of the ordinary to break the hold over us of our habitual ways of seeing and being, which are, in truth, ways of *not* fully seeing and *not* fully being.

"There are those who have been granted genius—artistic, intellectual, spiritual*—and the extraordinary makes itself available to them in the sphere of their genius. But for those unvisited by genius's spirit, there is only eros to break the heavy sleep of ordinary life and lead the way to the extraordinary. Eros wrenches the soul from its lazy reliance on conventions that substitute for sight. Gripped by the intensity of erotic longing, a person begins to know the world and its beauty. She knows the world is beautiful because it contains the one she loves. Neither human judiciousness nor divine madness can provide a human being with any greater good than that (*Phaedrus* 256b).

"There is risk in approaching eros on these terms, but that is because

---

*Cf. *Phaedrus* 244a–245a, 265a–265b.

any transformative experience carries risk. But so, too, is there risk in forcing eros to surrender to our ordinary calculations: it is the risk of never surrendering oneself to eros. These are the risks that PhD, yearning for knowledge, should consider."

So there you have it, PhD. I guess I'm a bit more old-school than Plato when it comes to the kind of arrangement you're talking about here, but even he thinks that you've got more to lose than gain by your higher-education shenanigans. And since you seem like the kind of kid who tallies up her wins and losses pretty crassly—whoops, I mean closely—you'd probably best pay attention.

Margo, philosophically

Dear Margo,

I am an academic in a highly demanding and theoretical field. I'm gay (male) and married. My husband complains that I'm too emotionally distant. He says that he feels I married him not to indulge in passion but to escape it. The truth is that he's right. I have more important things to think about than personal relationships, including my relationship with him. I want that part of my life settled so that I can think about other things. He wants me to think more about him and so is always trying to unsettle things. Who's right?

Yours truly,

Maybe I'm Not Domesticated

Dear MIND:

I'm afraid your husband is the one who is right. I am not sure why you married in the first place. It sounds like cruel and unusual punishment to marry to *escape* passion, and all that goes with it, unless your partner is of like mind—which he clearly was not. Do your spouse a favor and become single again as you think about those more important things. In fact, you might do all gay-mankind a favor by forswearing marriage and devoting yourself, unhindered, to your highly demanding and theoretical field.

But I decided to run this one by my new consultant, Plato, just because I was pretty certain he would disagree with me, but once again he surprised me:

"Margo, there is little I have to add to your advice. I can't but be sympathetic to MIND's desire to think about his theoretical field. And yet, since love of his husband seems not an option for him, but instead he married so that no personal love would distract him, then I think he must end the relationship. Perhaps if his husband had the same attitude it would be permissible, though even here I must wonder. For activities that would otherwise be perfectly shameful are not so when through them the god of love is moving. The friendship of a lover can bring divine things, but a relationship with one who does not love is diluted by instrumental thinking by means of which one person tries to use the other for his ends. All it pays are cheap, human dividends (*Phaedrus* 256e). As your passion lies elsewhere, follow it there passionately."

Okay, MIND. It sounds like Plato is ordering you to put your passion where your mind is.

Margo, mindfully

Margo,

I wonder if you've ever gotten a complaint like this one. I am engaged to a wonderful man, who's highly successful in his field, which happens to be the same one that I'm in. My problem—if you can call it that—is that my fiancé thinks too highly of me! Somehow he's gotten an inflated view of how talented and brilliant I am, and no matter how banal a suggestion I make is he infuses it with profound insights. Most of the time these insights are really his own, loosely inspired by some half-baked thing I've said. Sometimes he takes "my" opinions so seriously that he uses them to challenge his own views, and ends up proclaiming that only "I" could have seen through his fallacies!

All of this makes me nervous, most of all because I think that eventually, after his infatuation wears off (which it's bound to do . . . right?), he's going to see me for what I am and feel he's been deceived—which, of course he has been, even if the deception is really self-deception. On the other hand, I have to admit that it feels great to be so valued, and I've gained a lot more confidence from hearing myself praised to the skies by someone I so respect. It feels so good, in fact, that I never correct him and just accept the credit and compliments as if I deserved them.

What should I do? I love this guy to bits and don't want to lose him—not now and not in the future when his fog of love lifts.

Yours truly,

Teetering on Pedestal

Dear TOP,

The idealization of you by your fiancé has several components—not one of which has anything to do with you. Accomplished though he may be, he doesn't sound as if he has much self-confidence. An offshoot of this is that, to feel important, he must partner with the most brilliant and insightful woman . . . making you a reflection of his superior taste.

Now that he has built you up with his compliments, and because you wisely see reality, I think a heart-to-heart is in order about the whole issue. If you initiate it now, you will greatly diminish the chance for him to arrive at these conclusions himself, later on, to your detriment. What I recommend is essentially to set him straight. Admit that it was wonderfully flattering to have him consider you his intellectual and solution-oriented superior, but such is not the case, and you do not want this to go on any longer. Point out that he became a success before you arrived on the scene, and that you really are equals—complementary and inspiring to each other, no doubt, but equals.

I thought it would be interesting to check what a philosopher had to say about your predicament and went right to the top, TOP, and consulted Plato. And I have to admit, his take is quite a bit different from mine. Here's what he has to say:

"The dynamic that you describe, TOP, seems to me altogether characteristic of love. To be in love with a person is to see reflected in that person all the values we most idealize.* This is how it comes to be the case that, in the lover's special sight, the one who is loved is bathed in a radiance of significance that sets him or her apart from all other things that exist. Of course, such an incomparable radiance is a delusion, but it is a delusion inseparable from love and love's desire to want all good to come to the one we love.†

"Your lover's perception of an intelligence in you that you yourself don't recognize is an indication of both his love for you and his love for intelligence, and I think you will agree that both of these are good. But there is even more good to be derived from his loving delusion, since it is, by your own description, mutually beneficial. He is being led on to ever better ideas by perceiving them as coming to him through the filter of you; and you, TOP, are coming more into your own as your confidence grows. This, too, is characteristic of lovers at their best. When they gain their wings, they shall do so together because of their love (*Phaedrus* 256e).

"So my advice to you, TOP, is don't even think about setting him straight and getting him to see you as you really are, since that would be inconsistent with his continuing to love you. Love has no truck with straight seeing, and for this reason is sometimes spoken of as a divine intoxication. Rather, continue to flourish in his tipsy sight so that the unrealistic way in which he sees you corresponds, over time, more and more to the way that you really are."

So there you have it, TOP, two very different responses to your quandary, one from the romantic Plato and the other from the pragmatic Margo.

Margo, realistically

---

* Cf. *Phaedrus* 252c–253c.
† Cf. ibid., 245c: "We must prove the opposite, that this sort of madness is given by the gods for the greatest possible good fortune."

Dear Margo,

I'm twenty-eight years old and engaged to be married to an unusual man who's ten years older than me. He's the minister of our church, and his preaching is so powerful that we are now bursting at the rafters and are raising money for a gigantic new facility. My family belongs to his congregation, and they think the world of him. In fact, ever since he and I have become engaged, my family has regarded me with new-found respect. Simply being his choice has cast me in an entirely different light in their eyes.

Recently, and bit by bit, my fiancé has been taking me into his confidence, and this involves his telling me about the special plans he believes God has for him. He told me that he's known about his "divine destiny" since he was sixteen, and every step of his life since then, including his engagement to me, has served to convince him that he has a unique providential role to play. I am the only person to whom he has made these revelations, and he has told me that every piece of information he has shared with me binds us more closely together.

On the one hand, I do believe that there have been prophets in times past, so why shouldn't there be people who are in special communion with God now? On the other hand, I can't help but wonder whether my fiancé is a little bit crazy. The question that's tormenting me, Margo, is how can I tell the difference? I think my doubts may even be the strongest evidence I have against his "divine destiny," since if he were truly being led by the hand of God, why would God lead him to a girl who has the doubts I do? Then again, maybe God is testing me to see if I'm worthy to marry such a man.

So my question is, first of all, how can I tell the difference between a man of God and a lunatic? And secondly, even if my future husband is somewhat delusional, does it really matter? Is there such a thing as a good kind of lunacy that might help a person like him in his special calling?

Yours truly,
Doubting Thomasina

Dear Doubting,

I suspect people who believe in God also regard their representatives and interpreters on earth as special, and perhaps one step closer to the deity than they are themselves. For this reason I am not surprised that your family sees you in a new light. A halo effect, as it were. (Sorry.) Regarding your fiancé believing God has "special plans for him," one may assume that many people choosing the clerical life have this feeling. In Catholicism this is referred to as "being called." I think, for believers, one need not be a prophet to feel in communion with God.

Just as there may be a fine line between mad and inspired, there may be just as narrow a difference between grandiose and thinking big. Your intended sounds as though he has Billy Graham–like charisma, hence the growing flock. Should you see him veering off into Reverend Schuller's Crystal Cathedral territory, then it might be time for you to step in and remind him of his main mission.

As for your question about divine destiny and God, my understanding of these things is that God does not pick partners for people—one assumes because he is too busy.

But since this seems to me a question bordering on the trickily philosophical, I turned to my expert consultant on all things philosophical, Plato. He agrees with me that you don't need to sound the loon alarm just yet, since your fiancé may be, as he puts it, "maddened and possessed in the right way" (*Phaedrus* 244e). I'll let him explain: "There is madness which is sickness, leading a person only to confusion and falsity, and then there is madness that is not sickness at all, since it leads to clarity and truth, even though the person struck by such madness can give no account of how he comes to know what he knows. This is why another name for such madness is inspiration, since it is as if the gods themselves had breathed into them, and it is why, in both your English and my Greek, the words "manic" and "mantic" are closely connected.*

"There are many gifts that come with a touch of madness, if we understand madness in this salubrious sense: that is, the state of knowing of a sudden and without knowing how we know—more possessed by, than possessing, the knowledge. Poets, when inspired, engender passages echoing with unworldly knowledge, so that trained hierophants can consume their lives in interpretation. And yet the poets, when we speak with them, are ordinary people of ordinary knowledge, so that there is no accounting for how they spoke the poetry that they did, and they themselves are at a loss both to say not only how they know but even what they know.† It is as if they are moved by a spirit greater than their own, their own small person taken over, *possessed,* by a genius altogether outside them. In all such cases, artistic or not, it is natural to put the experience, so profound and mysterious, in the language of the gods, or of God, as your fiancé does, Doubting, whether one deploys this lan-

---

* *Phaedrus* 244b–c.
† "I used to pick up what I thought were some of [the poets'] most perfect works and question them closely about the meaning of what they had written, in the hope of incidentally enlarging my own knowledge. Well, gentlemen, I hesitate to tell you the truth but it must be told. It is hardly an exaggeration to say that any of the bystanders could have explained those poems better than their actual authors." *Apology* 22b, translated by Hugh Tredennick in *The Collected Dialogues of Plato*, ed. Edith Hamilton and Huntington Cairns (Princeton, NJ: Princeton University Press, 1961). Also *Phaedrus* 244a–245c, and 265b, as well as *Ion,* especially 533d–534e.

guage metaphorically or not. From what you say, Doubting, it seems that your fiancé deploys the language of God non-metaphorically, but that in itself is not a sign of the bad madness. For whether these unaccountable insights come from sources divine or not is not for us to say.

"How then can a person tell which is the good madness and which the bad? The person himself who is subject to this experience is in the least favorable position to judge the difference, since both good and bad madness will feel for him equally compelling. This is just as true in romantic as in religious experience, where extraordinary experience explodes into a life and leaves ordinary life shattered. It is for others to judge the difference, based on whether what the person says seems to bespeak only confusion and falsity, on the one hand, or something more worthwhile, on the other, even if wonderous strange. If, Doubting, you, like your parents and the other congregants, feel that there is wisdom and beauty in your fiancé's madness, then think of him as no more frighteningly mad than a greatly inspired poet. But if your suspicions about your fiancé continue to nag at you, then it does not matter whether he is maddened in the right way or the wrong. Whatever his madness, it is not the right madness for *you*, since clearly the madness of eros has not entered into your apprehension of him."

I think Plato makes an excellent point, Doubting. The line between madness and inspiration, or between, as Plato puts it, bad madness and good madness, may be blurred; but if you find yourself irresolvably uneasy over where on the continuum your fiancé falls, then perhaps this just isn't the madman for you.

Margo, maddeningly

Dear Margo,

I am a twenty-six-year-old woman, married with three kids. Life is good for the most part, and I am happy as a stay-at-home mom. My husband and I have a great relationship; we are the best of friends, and when we're doing things together as a couple or a family we have the best time. Our sex life is good, and we're very happy, for the most part. Here's the problem: I don't find him attractive. I think I married him because we were such great friends. He finds me attractive and extremely sexy for a mom of three. I don't want to end the relationship, and I definitely don't want to cheat. But I find myself flirting and becoming attracted to very handsome men. Should I stay in a marriage in which I'm not attracted to my partner, or should I try to find happiness with a man I am attracted to? I don't want to lose my husband as a great friend. I do love him.

Yours truly,
Eye Rampantly Roving

Dear ERR,

I have redacted the city you live in to avert a stampede of women moving there in hopes of finding your husband. To be in a marriage with a great friend—enjoying a good relationship, great times, a good sex life, and general happiness—is pretty much all there is.

Take it from me, the handsome thing wears thin. (Plus, low lights—or no lights—often set the scene in the bedroom.) If it takes a therapist to get your head back on straight, go!

And what's more, this time I've got Plato, philosopher among philosophers, firmly on my side, even though he had a lot to say about how important beauty is.

"Of all the perfections—wisdom and courage and virtue and temperance—beauty is the only one that makes itself known to the sight, which gives us our keenest sensations coming to us through the body. Wisdom doesn't make itself known in this way because the sort of clear image of itself that would be required for sight would provoke terrible, loving desires, as would be the case with any other of those perfections we long for. It is beauty alone which has that fate, and thus is the most evident and the loveliest" (*Phaedrus* 250d).

So that's a big *yes* from Plato as regards the importance of beauty in making us fall for others. But just in case you're thinking that Plato, being one of those frivolous types who thinks that only looks matter, is giving you permission, ERR, to hustle yourself along in the direction where your roving eye is leading, Plato also has this to say:

"The beauty of souls is more valuable than that of the body so that if someone who has a becoming soul is not physically attractive, still the person who appreciates this beauty will be content to love him, to take care of him, and with him to search out and give birth to the higher forms of beauty" (*Symposium* 210c).

So listen to Margo and Plato, ERR, two wise old birds, and give up the flirting with handsome men. Come back to reality, which for you, lucky girl, is a pretty sweet deal.

Margo, pleadingly

Dear Margo,

I've been in an intense relationship with a guy for a little more than a year now. For a good chunk of this time I couldn't believe how great things were going, especially since, to be honest, he's the sort of man I would have thought was leagues out of my league. But lately he's been making demands on me in the bedroom (and everywhere else, including the kitchen sink) that I don't feel comfortable with. He says that if this is going to be a long-term deal then we have to keep working on it to keep it interesting, but in my book the stuff he wants me to do is just gross, not to speak of slightly frightening. He's taken to taunting me for being sexually vanilla, which is the least offensive way he puts it, and has even hinted that if I don't play along with his idea of fun then he's going to lose interest in me, which translates to my losing him. Since everything else about him is great—he's handsome, successful, and a lot of fun, at least when he's not berating me for being uptight—should I just give in and let him do what he wants (maybe getting good and drunk first)?

Yours truly,
Battered Over New Demands And Getting Exhausted

Dear BONDAGE,

There is an old rule subscribed to by mentally healthy people: Anything is permissible in the bedroom, including the kitchen sink, if both parties wish to do it. This man's ideas of "interesting" could escalate into God knows what (or whom) so I would tell him you disagree, philosophically, with his unilateral ideas of good sex, as well as with his lack of regard for your wishes. I would, indeed, lose him, and the sooner the better.

I was certain that Plato, whom faithful readers of this column know I've been regularly consulting on matters philosophical, would agree with me on this one. Not only does he agree, but, as usual, he's got some interesting theories to back up his advice. Here's what he has to say:

"Eros can bring out the very best in people and the very worst, and what is being brought out in your partner, BONDAGE, is the very worst of the worst. What has reared up in all its ugly violence is the essence of the tyrant, a lawless person who recognizes no reality beyond the peremptory urges of his own desires.

"To be sure, within each person, even the best, there are various impulses battling it out with one another, like a chariot that has two horses (*Phaedrus* 253d). One horse, ill formed and ill groomed, with eyes bloodshot and unfocused, is animated by the spirit of insolence and wanton hubris.* It wants to go where it wants to go and can barely be restrained with whip or goad. The other horse, well shaped and holding itself with discipline and dignity, is nobly guided by nothing harsher than a word or command. The character of a person is manifested by how it manages the two horses of its chariot. No situation

---

* Whereas we mean by "hubris" an excessive vanity or deluded arrogance, the ancient Greeks used the word to cover all cases of the unrestrained assertion of individual will without any regard for the wishes of others, or of the law, or of the gods. Depending on the context, then, *hubris* can be translated as wantonness, outrageousness, lawlessness, and violence against an individual, including rape. The most severe punishments in Athenian law were delivered against infractions deemed hubristic.

places a greater strain on character than eros, which engorges the urges of the wanton horse against which the good horse and the charioteer must exert themselves. The urges are not in themselves a cause for shame, so long as the chariot as a whole behaves in accordance with the dignity due both itself and the beloved, pulling hard on the reins of the wanton as he charges forth with bloodshot eyes. The wanton horse will barely have recovered its breath as the pain from the bit and its fall diminishes, when it will angrily rail against its charioteer and yokemate, heaping abuse upon them as unmanly cowards for backing out of the arrangement they had agreed upon (*Phaedrus* 254c) just as your partner, taking his passions from the unruly horse, rails against you.

"A tyrant is a person who lets the wanton horse have his way with him, and thus it is that the tyrant must have his way with all others.[*] Your partner, BONDAGE, is just such a tyrant. So I do not think, BONDAGE, that you are altogether correct in saying that 'everything else about him is great.' All tyrants are dangerous, and none more than the tyrant who is beloved by his subject."

So now you've heard it from both of us, BONDAGE. Lose the tyrant, find the therapist.

Margo, firmly

Dear Margo,

I've never been the type to fall head-over-heels in love, but all that changed a year ago. I met a girl and immediately found her irresistible. I couldn't tell what it was about her, but she seemed to conjure up the deepest feelings in me, a kind of tenderness I hadn't felt in a long time. It was only about six months into the affair that I realized that there's something about her—the way she moves, the angle at which she holds her head—that reminds me of a lost love of mine from long ago, someone I've never really gotten over. All the mysterious tenderness I felt when I first laid eyes on her is channeled from a different time in my life, stirred up by a different person. Is there anything wrong with this? Am I being unfaithful by being with one person only because she reminds me so irresistibly of someone else?

Yours truly,

Tethered In Memory Eternal

---

[*] See *Republic* 573d, where Plato draws a link between the tyrant and Erōs. The *Republic*, unlike the *Symposium* and the *Phaedrus*, shows Plato in a far less accommodating mood toward Erōs, and he is willing to condemn the whole business as the "indwelling tyrant Erōs." Plato's uncharacteristic tenderness toward Erōs, most especially in the *Phaedrus*, led Martha Nussbaum to speculate that he was himself in love when he wrote that dialogue. And if he was, then it's obvious, she informs us, with whom: Dion, the uncle of the tyrant of Syracuse. See her "'This Story Isn't True': Madness, Reason, and Recantation in the *Phaedrus*," in *The Fragility of Goodness*, pp. 200–233, especially pp. 228–231.

Dear TIME,

I do not find the dynamic you describe so unusual. If whatever qualities the lost love possessed reappeared in a new woman, it would be
perfectly logical for you to be attracted, again, to those qualities. This
situation would only be worrisome—and unfair to the new woman—if
it were just superficial likenesses that reeled you in, such as the angle
at which she holds her head. If you are pretending you have recaptured
and re-created the lost love, denying this girl her own individuality, that
I would consider problematic. If she is your "type," however, and *also*
has gestures or movements that are reminiscent of someone you loved,
that is all to the good (and is part of attraction). Just so long as you are
smitten with this woman's actual self, everything is fine. The trick for
you is not to feel like Pygmalion and imagine you've gotten the former
love back.

But since you've expressed worries about the whole ethics of your
situation, I decided to consult my expert moral philosopher, Plato.
Like me, he agrees that there's nothing unusual in the dynamic you
describe. In fact, he went much further, and asserted that "all falling in
love is due to reminding."[*]

"That is why a person whom you barely know—though you long to
know her as you long to know nothing else in all the world—can arouse
such a profound response in a lover. There is an elusive sense of the
familiar, escaping like a word on the tip of the tongue. There is a deep
and terrible ache to recover something of infinite value that was lost.
This sense and this ache are conveyers of a truth, telling us that this
person is a reminding of a love you had known before."

I have to admit this was a new one to me, and I asked Plato a pretty
obvious question. If falling in love is always a reminding of a former
person we once loved, how did we fall in love with the first person to
begin with? I should have guessed that Plato wouldn't get caught in
any logical traps I could lay, even though the way he sprang it certainly
surprised me. Here's what he said: "The reminding isn't of a person at
all; or if it is of a person, as in the case of TIME, then that formerly loved
person was herself a reminding of something that wasn't a person. A
beloved person is a signifier who bears intimations of all that stirs us in
existence. We are reminded of the nature of beauty and of all the other
mysteries of existence that come to us but obscurely and by which we
long to be overtaken and overwhelmed."

I had to push Plato: What if the trait that's doing the reminding is
only something haphazard and inconsequential about the person, like

---

[*] Cf. *Phaedrus* 245c–250d, which ends with the words: "Let that be our tribute to memory,
then, for the sake of which, and because of a longing for those earlier times, such a lengthy
statement has been made."

some particular body part—his upper lip, her inner thigh—or a sexy foreign accent? Doesn't the beloved person have a right to be loved as the person she is rather than for the random tilt of her head?

"There will always be the elements of the haphazard in eros. It is for this reason that eros is often portrayed as an irresponsible child let loose among us with bow and arrow. Our desires for another aren't the conclusions of reasoned arguments, though often enough we wish they were, with premises entailing the beloved's worthy attributes. Were it otherwise, we would all fall in love with the same estimable subjects. When we submit to eros we submit to the irrational and the random, allowing some perceived characteristic, whether insignificant or essential, apparent or real, to shatter us into bliss. All too often, the bliss proves ephemeral, cut short by further familiarity with the beloved. But if it does not, then who would be so foolish as to raise an objection?

"As for the rights of the beloved, she has her rights to a reciprocal bliss, which might reciprocally retreat in the face of further familiarity with TIME."

Well, as they say: TIME will tell.

Margo, temporizing

Dear Margo,

I've always been a girl who likes bad boys, and now I've found a bad, bad boy who breaks my heart on a regular basis. He cheats constantly—with other girls, married women, guys, for all I know with fresh fruit. He never lies about it, never apologizes. He expects to get away with everything, and he does. Did I mention he's gorgeous, exciting, charming, exciting, rich, reckless, powerful, charismatic, a leader, *exciting*? In addition to the breaking-the-heart part, he's talked me into some pretty risky adventures. So far nothing disastrous has happened but I'm afraid that given his tendency to push the boundaries and my tendency to turn to putty in his hands, I may end up doing something I'll regret. I know you're going to tell me to delete him quicker than spam from my in-box. But when he goes away the color drains out of everything and I feel like I'm sleepwalking through a beige-colored desert, and when he comes back—and he always does, eventually—all the colors come rushing back in. Should I resign myself to sleepwalking? Or should I just hang on for dear life on a joyride like no other, no matter where it takes me?

Signed,

I Don't Know How (Or If) to Love Him

Dear Don't Know How (Or If),

Women have gotten mixed up with VBBs (Very Bad Boys) against their better judgment and advice from others since boys started behaving badly. It is human nature to be attracted to the verboten, the bad, the dangerous. There's an element of gambling, as well, and the idea—as unrealistic as beating the odds in Vegas—that *you* will be the one special enough to get him to change his wanderin' ways. These entanglements are about living on the edge and satisfying one's need for excitement. It is my unpleasant duty to tell you, from everything and everyone I know, that staying away from these "colorful" lovers finally comes from having one humiliating experience too many. These relationships are basically S&M—his S and your M. The lucky girls are the ones who finally gain the strength and maturity to say "No more. I am better than this, and I deserve someone who values *me*—not the game."

I was curious whether Plato had something philosophical to add to this. He told me that he had given the matter a great deal of thought, having witnessed the devastation wrought by a certain VBB upon one of his friends at close hand. "Strange to me was it that even the best person whom I have ever known, a person whose inner moral voice spoke so clearly that he passed through his life without ever committing a great wrong, even such a one as he fell victim to a VBB." Plato pointed out that one of the deadliest aspects of VBBs—the part that earns them their V—is a sweetness mixed into the badness.

"The contrast between the two, the sweetness and the badness, wrenches the heart of the lover as such sweetness on its own would not, and the lover shudders all the more at dread of the beloved's recklessness, for the sake of the sweetness that is there, and the shudder only makes more violent the shuddering that announces love (*Phaedrus* 251). I do not think, but for that sweetness, the friend of whom I spoke would have become impassioned as he did and he would have recognized that such a one, entirely wanting in the desire to become better than what he knows himself to be, was not worthy of his love. She who signs herself "I Don't Know How (Or If) to Love Him" repeated the word "exciting" three times. A VBB (and let us remember that there are also, though perhaps they are rarer, VBGs) creates around himself or herself a separate world in which all that happens is exciting, for exciting it *must* be. Excitement is the air they breathe, and they cannot exist without it. And when they pull others into their world, then these others leave the world of common air and now they breathe the rare air of excitement, which they are not accustomed to, and in their confused state they are more apt to think that the excitement they breathe is the excitement of love. She asks whether she should continue to love her VBB, but I do not think she really loves him, just as he, and this for a certainty, does not love her. For I think even the best man of his day of whom I just wrote did not love that boy as he thought he did. Perhaps if your questioner thinks more on the true nature of the excitement she feels, she will be able to see the wisdom of the course of action that you and I both urge on her, and then she will find the strength to break the spell that her VBB casts upon her. Last, let her think on this, that though love is a profound disturbance, not all profound disturbances are love."

To which I say: Amen.

Margo, profoundly

Dear Margo,

I'm a twenty-two-year-old male, and I have zero desire for sex. I really don't want to have it, ever. However, I am attracted to the opposite sex and not against the idea of having a romantic relationship, with the extent of "sexual" contact being hugs and short kisses. I'd be a perfect candidate for the priesthood, if only I were Catholic, or even religious. What bothers me about my condition is that I'm afraid I'm going to be lonely my whole life, since whoever I'm involved with is almost certainly going to want sexual intimacy sooner or later. So far, that's been my experience. Do you think it's possible to have intimacy without sexual intimacy? I guess the kind of relationships I want would be called Platonic. Do you think I'd be more likely to find a soul mate in philosophy departments?

Yours truly,

Thanks But No Thanks

Dear TBNT,

I do believe what you desire is possible, because—and I hope you are sitting down—you are not the only person with the feelings you describe. Asexuality is more common than many people think. Should you be interested, a good psychotherapist might be able to help you learn the genesis of your aversion to the physical aspects of intimacy. This understanding, in turn, if it did not change your mind, would at least make you more comfortable with something that is obviously troublesome, but part of your emotional makeup. To answer your direct question, I do believe there can be intimacy without sex. I have some friendships that would fall into this category. (A meeting of the minds, as it were, without any joining of other things.) People do not talk about "Platonic relationships" for nothing. As for scouting around philosophy departments, I think not. The philosophers of my acquaintance, and one in France I have read about, are sexually exuberant people, which is not to say that there is not a chaste philosopher out there somewhere. If you're up for being a pioneer, the recent success of all the LGBT groups suggests that you might try to organize a group for asexual people, perhaps with a snappy name—something like CHASTE: Choosing Asexuality Together Electively.

Of course, I couldn't pass up this opportunity to go to the source himself and ask the eponymous Plato what he makes of your question, wondering most of all what he thinks of the term "Platonic relationship" that we all find so useful in getting out of awkward situations. He confessed to being confused about what we call Platonic love, especially since he's been busy on the Internet (and you might want to follow his lead) and acquired familiarity with such terms as asexual, aromantic, heteroromantic, homoromantic, biromantic, and quite a few others he was eager to explain to me. Ironically, the only category he confessed confusion about was the one derived from his own name. "There seems to be a lot of ambiguity surrounding 'Platonic love.' Must it be asexual? Is it romantic or aromantic? And if it is, by definition, aromantic, what distinguishes it from any sound friendship? Platonic love seems to me a confused category."

Is Plato's confessing confusion over Platonic love something like Charles Darwin scratching his head over Darwinism? No, he told me, it's just our own confused use of the term that has him muddled. He knows what he had in mind: "The kind of love whose praises I have sung focuses more on knowledge than on carnal pleasures, although it does not necessarily preclude those carnal pleasures.* But whatever the other pleasures, the particular quality of its rare intimacy and pleasure is one of knowledge, of which there are two kinds, which can exist in separation from one another or in unity."

Do tell, I urged him.

"There is first of all the intimacy of coming to know the very person that you love, eager to possess every detail of her being, jealous for the hours of the days he lived before you knew him, the felt quality of her experiences that you can never recover for yourself and so mourn, exercising all your faculties to find ways to make up the lack in knowledge. And for this reason it can truly be said that all who truly love are savants on the subjects of their darlings, no detail too small or insignificant, as much in love with knowledge as any philosopher just so long as the knowledge is of the one whom they love. And that is one of the ways in which the intimacy of eros is the intimacy of knowledge.

"But there is another way, too, even more rare, which consists in gaining a knowledge which is not focused on the beloved but is an intimacy nevertheless. It is the rapture of together gaining knowledge. It is the intimacy of making progress together in an understanding which is soul-shaking and therefore soul-making, enclosed with one another in an *épiphanie à deux*, ideas from the one flowing into the other, the hard-

---

* In the *Phaedrus*, he is tolerant of those who can't resist the carnal: "In the end they lack their wings, but they emerge from the body with the impulse to grow wings, so they carry away no small prize from their erotic madness" (256d).

ened dryness of their twoness moistened into one as they are thinking with one mind and seeing with one vision. This is an intimacy which our clumsy bodies can only try to imitate as if in a comedy of crude pantomime."

Well, I have to admit that that's not exactly what I had in mind when I used to say, "Let's just keep it Platonic, shall we," but there we have it, straight from the philosopher's mouth. And I think, No Thanks, Plato has answered your question as to whether there can be intimacy between two persons even in the absence of anything more, ahem, seminal than ideas being exchanged between them.

Margo, platonically

ζ
_____

SOCRATES MUST DIE

*One of the numerous poison vials found in the
vicinity of the ancient Athenian prison.
Photo from the display in the
Agora Museum, Athens*

A MOST PECULIAR MAN

For if he [the philosopher] holds aloof, it's not for reputation's sake.
The fact is that it's his body that's in the state, here on a visit, while his
thought, disdaining all such things as worthless, takes wings.

—*Theaetetus* 173e

And all the people said, "What a shame that he's dead,
But wasn't he a most peculiar man?"

—Simon and Garfunkel

On one of the rare occasions that he found himself in the country-side, he couldn't stop exclaiming on the prettiness that he saw around him. Look at the spreading branches of the sycamore tree! Look at the splashing of the sparkling stream! It made the boy walking the path beside him burst into laughter to hear him exclaim over these commonplaces, natural beauty being such a novelty to him. He hardly ever wandered far outside the city's walls, since what he wanted to learn the charming scenery couldn't teach him (*Phaedrus* 230d).

Even during the hardest years of the war that brought so much filth and disease into the crowded city packed inside its walls, he did not let the conditions undermine his passion and his pleasure.* Chances were

---

* Pericles' tactic had been to engage the Spartans, who were superior fighters on land, only on sea. Toward that end, he built up the city walls of Athens and extended the long walls down to the port of Piraeus, guarded by the navy, so that goods could be brought in. He had all the Athenians, including those living in the outlying *khora*, move into the *astu*. The Spartans could come and burn everything outside the city walls, which is what they periodically did, but, according to Pericles' plan, the Athenians would not be drawn into a land fight. Pericles knew that the Spartans would not stay away from their own city for long, since they were always wary of a slave uprising. In between the attacks the Athenians could go out and replant their fields. Sparta's secret ally was the plague, which most scholars believe came by way of boat, perhaps carried by rats—or the flees on the rats—and entered the city by way of Piraeus. The plague broke out three times during the course of the Peloponnesian War, spreading quickly through the overcrowded city. Thucydides, one of the few to contract the plague and live to tell of it, gives a lurid description of plague-gripped Athens. See Appendix B.

good on any given day that you would find him in the agora, barefoot and eager, striking up an inquiry into the true nature of some virtue or other, having cornered someone in the long columned southern stoa where the merchants set up their stalls, or at the palestra of Taureas (*Charmides* 153da) or the gymnasium at the Lyceum (*Euthyphro* 2a, *Lysis* 203a, *Euthydemus* 271a, *Symposium* 223d),* where the boys gathered to exercise and he was always on the lookout for new talent.

He showed up every day conscientiously, even after a night of hard drinking (*Symposium* 223d), for the work for which he declined to receive remuneration (*Apology* 19d–e). Perhaps his wife's reputation as a shrew had something to do his exalted decision to work without compensation; that might make any wife complain, especially as there were three sons. His wife's name, Xanthippe, has come to mean a nagging ill-tempered woman. In *The Taming of the Shrew*, Shakespeare has Petruchio describe Katherina as a "Xanthippe or a worse" (act 1, scene 2). Their youngest was still a toddler when Xanthippe became his widow (*Phaedo* 60a). His net worth, including his house, was five *minae* (Xenophon, *Oeconomicus* 2.3.4–5), the equivalent of what a sophist might charge for a single course (Xenophon, *Apology* 209b), and less than a skilled laborer could earn in a year and a half. He saw no dishonor in his impecuniousness, and carried himself with the imperturbability of a man of independent means who lives exactly as he pleases. It was hard to know whether to laugh or be awed at his pose.

He was often the butt of the jokes of the most popular comic playwright in town, his activities lampooned as "hairsplitting twaddle" (Aristophanes, *The Frogs* 1495). He laughed at the poet's stanzas with the same ironic distance with which he regarded all things that mattered not at all, though some of his friends, including Plato, will blame Aristophanes for contributing to his defamation (*Apology* 18d, 19c).†

He looked on his job as a high calling, to which he was singularly suited, and not even when his working conditions had deteriorated to the point of threatening his life would he consider quitting his post (*Apology*, passim).

He did not leave the city when, in its defeat, a garrison of Spartans

---

*The Lyceum was a place for exercise and conversation right outside the city walls, on the east side, and was the future site of Aristotle's school.

† In the *Symposium*, however, Socrates and Aristophanes are represented as being on good terms.

camped out on the Acropolis, lending ominous support to the oligarchs who were set up after the victors sailed in, effectively dismantling the democracy for which the *polis* had been famous. It wasn't only in the other *poleis* that many had looked on the political experiments within Athens as bizarre.* Within Athens, too, there had always been those who held firm to the innate *aretē* of the aristocracy and despised the granting of citizenship to the unpropertied hoi polloi. They resented that any *thēte*'s vote counted as much as any aristocrat's, and that that same *thēte*, coming straight to the Pnyx from his field or shop, could get up and address the *ekklêsia* as if his opinion mattered just as much as a worthier man's. Pericles had long been dead, carried off by the plague in the early years of the decades-long war.†

The Thirty, who ruled for barely a year, were scions of the city's oldest families, and in their brief and violent reign they allowed their hostility to the government that they had always regarded as rabble-rule to roil into vengeful lawlessness intensified by greed for power and property. Voting was restricted to the few and then to the even fewer, a designated Three Thousand who were allowed the privilege of bearing arms and of trial by jury.‡ Those who didn't make the cut went to live in the narrow streets and alleyways surrounding the harbor of Piraeus and could be summarily brought up on any trumped-up charges, a system of informers reticulating to entrap them. The dead and the exiled had their property confiscated, so that greed soon rivaled politics as a motivation. Metics, the foreign workers who had never been allowed citizenship in the city, were particularly vulnerable, especially if they were rich.

---

* "Only in Athens, where the traditional forces of family, blood, and religion were at a single blow deprived of political significance, did a genuine political democracy grow up out of the older organization. In the few other democratic states in Greece, the democracies had almost all come into being as the result of outside and usually Athenian pressure and consequently did not count, psychologically, in the eyes of their Greek contemporaries. . . . This is why Alcibiades attacks democracy in his speech at Sparta as 'acknowledged lunacy.' It was a thing virtually unique and Athenian in fifth-century Greece." David Grene, *Greek Political Theory*, pp. 35–36.

† Pericles died in 429. The Second Peloponnesian War, which was the main one, lasted from 431 to 404.

‡ Actually, whether there was a formal list drawn up is still debated. Like the Five Thousand, who were said to have been the government after the fall of the the Four Hundred in 411, another brief period of an earlier oligarchy, there may have been no actual list. Rather, in both cases of temporary oligarchy, the larger number might well have pretended to a broader base of oligarchic support than there is reason to credit. Dramatic rhetoric about striking names from a scroll comes decades afterward.

The Thirty had been "elected" by the *ekklêsia,* though the vote had been primed by Pausanius, king of the victorious Spartans, and the citizens had little choice, with the city half-starved by blockade. They were mandated to produce a constitution that would restore "the ancestral laws," but the Thirty showed no inclination to produce a constitution, ruling instead by non-legitimized fiat. From September 404 to May 403, fifteen hundred Athenians were killed, which exceeded those killed in the last decade of the Peloponnesian War. Three hundred "whip-bearing servants" carried out the orders and induced a reign of terror. Besides the many who were dispensed with hemlock, thousands more had been sent into exile or had voluntarily fled, some of them organizing to retake their city and restore its democracy.

But not he. He continued on in the same way as he always had, trekking to his favorite haunts to pursue the impractical inquiries which almost always ended far short of resolution. Is virtue a matter of knowledge or of something else? If it isn't a matter of knowledge, then how can it be trusted as reliable? But if it is a matter of knowledge, then can it not be taught? But then why do virtuous men so often have vicious sons, and virtuous teachers vicious students? It was as if he were oblivious to the great political events transpiring around him.

But then, among his various oddities was an oblivion that could settle over him, a fit of distraction that could lift him out of his immediate circumstances (*Symposium* 174d). Once he became intellectually engaged—and he was almost always intellectually engaged—his attention narrowly focused in. If circumstances weren't relevant to his peculiar preoccupations, he paid them little notice, relegating them to the wide margins he kept for subjects worthy only of his mischievous sense of humor. The particularities of the political upheavals racking his city belonged to those margins, and he expressed the seriousness with which he regarded local politics by subjecting even the most deadly, if ephemerally, powerful men of his city to his quips.

He was an inveterate quipster.

Some might interpret his staying put during the year of mounting terror as de facto sympathy for the Spartan-sponsored Thirty. Any citizen who had failed to join the democrats in exile by 403 could later be said to have "remained in the city" and that alone could raise suspicions of having been one of the Three Thousand, the few allowed to retain citizenship. The list of the Three Thousand was never made public, so we still do not know his status. "Remained in the city" became, in the years after the Thirty had been dispatched—most to their graves—code in the courts for having been one of their fellow travelers. And even though that particular suggestion isn't lodged against him in any extant account, still it's natural to conclude that he was suspected of Spartan leanings—especially since, years before the reign of the Thirty, rumors about his split loyalties had already been bruited, quite publicly in the plays of Aristophanes.

Among the aristocratic young men who flitted in and out of his influence were some with decidedly anti-democratic leanings. The famous comic playwright had hinted broadly that he was implicated in subversive politics, making up two words to send home his message, one of which meant "to be mad for Sparta," and the other of which was based on his name.* The play was performed in 414, during the period when Alcibiades was making himself the toast of Sparta, Athens having suffered disaster at Sicily. Of course, Sparta was just a passing phase on the part of the irrepressible Alcibiades, who would be held to no accounting but that of his own transgressive nature. But others who had hung on the man's words had proved to be more lastingly mad for Sparta, including the most notorious of the Thirty, Critias. So what should we say then of the man whose name had been linked with them? Was he not also mad for Sparta?

No, he wasn't. His cynicism about the Athenian democracy doesn't mean he was keen on the oligarchy. To see subversive politics in his stance is to miss the point of what he was about.† To see subversive politics hovering around him is to join the crowd of Athenians who never grasped the nature of his questions. He was about something far more subversive to Athenian values than mere party politics.

It is true that he had not joined the democrats in exile, and that he had not gone out of his way to protest the abuses of the Thirty, though

---

*Aristophanes had coined *lakōnomanein*, meaning "to be mad for Sparta," Laconia, or Lacedaemonia, being the region of the Peloponnesian peninsula of which Sparta was the administrative capital. His other neologism was *sōkratein*, "to socratize." The verse reads: "All men had gone Sparta-mad; they went / Long-haired, half-starved, unwashed, Socratified / With scytales in their hands" (*The Birds* 1281–83).

†This is, sadly, true of I. F. Stone's book *The Trial of Socrates* (New York: Anchor Books, 1989). Perhaps it is no surprise that Stone, whose lifelong orientation was political, would interpret Socrates' death in exclusively political terms. The conclusion he draws is that Socrates was anti-democratic, rejecting the civic goddess of persuasion, Peitho, as understood by the nature of Athenian democracy.

he had balked when they ordered him to take part in one of their errands of evil. This was their way, to try to implicate as many as they could in their dirty deeds (*Apology* 32a–d). Those who share in guilt are unlikely to call out the guilt. So he kept himself unstained, refusing to fetch an innocent man, Leon of Salamis, for summary execution. Instead he took himself home, an act of passive resistance that, though mortally dangerous under the circumstances, was perhaps not sufficiently resistant—not sufficiently *political*—to convince those disposed to hold him in suspicion.

His pronounced apolitical stance can arouse, still, incredulity, suspicion that some deeper hidden political allegiance must be lurking below. How can a person of gravitas hold himself aloof from contemporary politics? And in his city in particular, which looked on participation in public policy as a measure of the superiority that each citizen could achieve, he seemed to take a perverse pride in not even knowing how, when he was elected to the Council of 500, to vote (*Gorgias* 473e). To be politically indifferent is to remove oneself from one's own time. But that is precisely what he intended to do: to step out of his time. In so intensely a political culture as his, which could barely conceive of virtue except in terms of politics, his stance bordered on incoherent.

If conventional Athenian politics struck him as being beside the point of the elusive notion of *aretē* that he was endlessly trying to chase down, then the politics of the oligarchy were beneath his contempt. He never identified with the Thirty, any more than he had identified with the democrats, and, in fact, a good deal less.

As for the Thirty, they showed themselves increasingly ill disposed toward him. They framed a law expressly forbidding the art of disputation, which was, more than likely, specifically aimed at him (Xenophon, *Memorabilia* 1.2.31). The art of disputation is hardly compatible with the mood of intimidation their unconstitutional government required.

And then Critias had long had it in for him, the result of an insult that must have festered for years, when, in his irrepressible frank speaking, he had remarked on the unseemly erotic attentions that Critias was showing the beautiful young Euthydemus. He had compared the future oligarch's behavior to a pig rubbing himself against the stones (ibid.). Critias, a ragingly proud man, with a noble ancestry that reached back to Solon the Lawgiver and a cultivated mind that made the man consider himself a philosopher and a poet of the first rank, was unlikely to forget such insolence.

And who were this man's forebears that he took such liberties with the high and mighty? His father, Sophroniscus, had been a stonemason. His mother, Phaenarete, had been a midwife. He sometimes, at least according to Plato, likened his own profession to hers, remarking that he, too, helped people give birth, only to ideas rather than children, aborting those that weren't worth rearing. He used the comparison to remark the lack of definite conclusions to which he ever came (*Theaetetus* 149a and 210d). Like his mother, too old herself to give birth, he helped others bring forth living ideas, but never bore his own. Plato at least puts those words into his mouth.

He loved to use homely analogies in his philosophical arguments, taking examples from the work of carpenters and shoemakers and others who worked with their hands. It was a way to show the continuity of philosophy with everyday affairs, as was his method of plying his profession in the streets of Athens and among all varieties of people.

But if by his language and rhetorical style he claimed solidarity with the people, his philosophical populism did not make him a democrat in his political sympathies, at least not Athenian democracy. He scoffed at the Athenians' delusion of their participatory exceptionalism. He scoffed at how the artful speakers could sway the unsophisticated in the crowd.

They do their praising so splendidly that they cast a spell over our souls, attributing to each individual man, with the most varied and beautiful verbal embellishments, both praise he merits and praise he does not, extolling the city in every way, and praising the war-dead, all our ancestors before us, and us ourselves, the living. The result is, Menexenus, that I am put into an exalted frame of mind when I am praised by them. Each time, as I listen and fall under their spell, I become a different man—I'm convinced that I have become taller and nobler and better looking all of a sudden. It often happens, too, that all of a sudden I inspire greater awe in the friends from other cities who tag along and listen with me every year. For they are affected in their view of me and the rest of the city just as I am: won over by the speaker, they think the city more wonderful than they thought it before. And this high-and-mighty feeling remains with me more than three days. The speaker's words and the sound of his voice sink into my ears with so much resonance that it is only with difficulty that on the third or fourth day I recover myself and realize where I am. Until then I could imagine that I dwell in the Islands of the Blessed. That's how clever our orators are. (*Menexenus* 234e–235c)

Rhetoric is dangerous when the average citizen is empowered to make decisions, because the average citizen is in no position to withstand the rhetorician's savvy manipulations. The way to get ahead in Athens' democracy, he complained, was to flatter and ingratiate, to bypass the intellect and go for the emotions. The crowds wanted good theater in the *ekklêsia* and good theater demands large emotions, and large emotions overpower reason.

Still, he couldn't help but cherish the spirit of free speech for which his *polis* was celebrated. He must have cherished it, since he so lavishly availed himself of it. And what hope could there be for arriving at some consensus about the most important questions a person could ask himself, what hope for blasting through the biases that make each person see the world at a severe slant depending on his own individual positioning, if we don't bring lots of slants into the dialogue, so that, colliding with each other, something straighter can emerge?

Herodotus, in running down a list of Greek *poleis,* citing what it was that defined the distinctive exceptionality of each *polis,* commends the citizens of Athens for their wonderful discourse. The democracy had formalized the privilege of speaking one's mind, legally guaranteeing *isegoria,* the right of all citizens to speak their piece in the *ekklêsia.*

But beyond the formal legal right, there was a spirit pervading the city's culture at large, known as *parrhêsia,* which means "frank speaking." Athenians prided themselves on *parrhêsia,* and he, who was its very avatar, must have been highly prized by the friends of democracy.

Or so one might have thought.

In any case, the oligarchy brought an end to all such logorrheic liberty. And so, as sarcastic as he had often been about his city's democracy, he despised the oligarchs. He defied them on the one occasion when they tried to implicate him in their wrongdoing, and he defied them, daily and uninterruptedly, by continuing his art of disputation, thumbing his nose at the law which might well have been framed with him specifically in mind, especially since, in the time of the Thirty, there were no more sophists visiting the city, and it hardly needed saying that the privilege of *parrhêsia* had been revoked.

Only he still partook of the privilege. Liberty of expression was one aspect of the democratic spirit that he was not prepared to forgo. As the deaths mounted he quipped, using his favored form of homely analogies, that just as a herdsman who randomly thins his flock and worsens their overall condition must regard himself as a poor herdsman, so a ruler who randomly thins his populace and worsens their condition must regard himself a poor ruler. The wisecrack earned him a summons before two of the tyrants, Charicles and Critias, who reminded him that his diurnal activities put him in violation of the law against disputation and then forbade him to engage in all conversations with young men. He immediately began to have his fun with the two, disputing with them over the meaning of the prohibition against disputation. Charicles, falling into the verbal trap, answered the increasingly ridiculous questions, until Critias, who had been silently listening, put an end to the farce, threatening him with a none-too-veiled threat that Charicles then emphasized: "And, if you continue," Charicles put in, "you'll find the herdsman has thinned the herd by one more" (Xenophon, *Memorabilia* 2, 32–138).

Still, can't we ask whether his simply remaining in the city under such circumstances in itself constituted tacit collaboration? Was it possible to avoid the moral contagion spreading like pestilence, infecting anyone who didn't actively, politically, resist? Can't he, at the very least, be charged with moral cowardice, this man who made himself so conspicuous with his incessant moral interrogating, the smug suggestion that he alone knew something of which everybody else was ignorant?

But then he *did* regard himself as actively resisting. His refusal to take any political stand was itself a stand against the politicizing of virtue. His refusal to partake in any political actions was the only action compatible with the inquiries he conducted daily, trying to pry the notion of *aretē* away from its encasement in the politics of Athenian exceptionalism. This prying away was the very point of his inquiries.

Politics was a dirty business, as far as he was concerned, which was so radical a position to take in his *polis* as to be almost unintelligible. The *polis* was the source of normativity. Certainly this was true in Athens, where the political reforms of the evolving democracy had destroyed the old tribal ties, so that they didn't interfere with loyalty to the city. This particular *polis*'s radical experiments with self-government, its faith in the capacity of the average Athenian to participate directly in every decision, was a testament to its extraordinary nature, which had seemed to be abundantly confirmed in the days of its ascendance. Given the presuppositions that grounded his society, his asking for a definition of *aretē* independent of civic duties was like asking someone to translate a sentence into a language for which no vocabulary or grammar has been provided. It was like asking someone to declare the winner of a game when nobody has specified its rules.

And yet these were precisely the questions he was asking: Tell me what is essential about a worthwhile life, no matter whether the life is lived in Athens or elsewhere, whether one is born into high circumstances or into low, whether one is a free citizen or a slave. Virtue must be something over which people can take responsibility for themselves, not something that makes them hostage to the gods, which is to say hostage to what lies beyond their control. We are hostage in all other things, but surely not in whether we live virtuously. That is an aim people must achieve for themselves, the only aim worth the achieving. But if it is an aim we must achieve for ourselves, then the collective achievement of a *polis* can do it for us no better than the gods can.

With his distinctive use of the word, *aretē* begins its slide into what we now would translate as "virtue." The word is loosened from its conceptual entanglement with *kleos*, the social aspect of it shed. For him, there is no contradiction in saying that a person has achieved *aretē* even when the majority of his peers condemn him. But the achievement still specifies the life worth living. The subtle changes he renders to the word *aretē* tell of the revolution he is after with his questions. We read Plato's dialogues in English, with *aretē* straightforwardly translated as virtue, and in that straightforward translation the normative shift that he was after is contained.

But to the majority of those whom he questioned, his questions were barely intelligible. What made his ceaseless importuning all the more incoherent is that he himself protested that he didn't know the answers to his questions, that in truth he knew nothing except that he knew nothing. But if he knew nothing, then on the basis of what, pray tell, did he reject the answers of his community? Why did he persist, day in and day out, in setting himself up as a one-man display of normative confoundment?

How could one make sense of his interrogations when the constituent norms and values of one's society rendered them incoherent? To appreciate the difficulties of the Athenians in the face of his questions, think of those in our own day who are at a loss to say how there can be virtue independent of the word of God. Even if you don't share that reaction, as long as you are able to *imagine* yourself into it, you should likewise be able to imagine yourself into the reaction of the citizens of

Athens when they were confronted with this man's inquisitions. And to try to imagine one's way back to their perplexity helps one gain a clear vision of the progress that's been made.

The ideology of Athenian hegemony intensified the politicizing of *aretē*. To live a life worth living was to do one's civic duty, which included duty to the city's gods. The religious questions, too, were bound up with the political, melded together in the tacit normative assumptions constituting the lives deemed best. But even were religious and civic duties not bound up with one another, would it be any wonder that he would be charged with impiety? Impiety signals the deepest of normative disturbances, and that is precisely what he aimed to provoke with his normative impudence.

And when he pushed his city too far—or rather, when his city had been pushed too far by historical circumstances to tolerate his irreverance any longer—he was brought up on charges of impiety, as well as on the charges of corrupting the young, since it seemed his shenanigans had only served to undermine the constraints of politicized *aretē*, which reined in the more dangerous antinomian strains of the Ethos of the Extraordinary. And, his investigations ending in the inconclusivity of aporia, he had offered nothing to replace the norms he'd undermined. Wasn't Alcibiades proof of the normative vacuum he had created? And what about Critias? Both of these characters had presented a spectacle of vicious lawlessness not seen for a long time in the internal affairs of Athens.

It is much like those who argue today that our own recently experienced atrocities—perpetrated by such villains as Hitler, Stalin, and Pol Pot—were the results of questions that had removed the restraints of religion, which alone can keep violence and savagery at bay. So, too, might an Athenian have believed that nothing could replace the moral constraints imposed by the institutions of the *polis*.

Soon the tyrants who had disputed with him about the art of disputation were themselves dead or exiled. Those among the Thirty and their immediate henchmen who weren't killed in the fighting to recapture the city—unlike Critias and Charmides, who died in battle in May 403—fled to nearby Eleusis, a *deme* of Athens that the Tyrants had secured as a retreat should things not go their way, taking the precaution of first slaughtering all its inhabitants on the false charge that they were democratic subversives. And when the *polis* suspected that these oligarchs were planning, once again, to march on Athens, they themselves marched on

Eleusis, and preemptively killed the last of the Thirty. Still, Athens was filled with those who had been complicit, either actively or passively.

With the restoration of the democracy, something quite extraordinary occurred. The customary bloodbath never happened. In all the other *poleis* that had undergone revolutions and civil wars, a vicious round of retribution and counterretribution had been the pattern, but it was not so in Athens. The cycle was forestalled by a declaration of a general amnesty, granted to all but a notorious few at the top. "Those of Piraeus" and "those who remained in the city" shed their labels and came together to restore the city, though it would never again be the empire it had been.

The amnesty was an act of political brilliance, fostering a renewed sense of solidarity, easing the way toward an ameliorating fiction that all the Athenians, with the exception of the Thirty, had been victims, that none had been collaborators. It was a collective act of willful forgetting. In fact, the citizens swore an oath, *me mnesikakein,* which means "not to remember past wrongs."

Of course, no oath or legislation can blot out memories, especially not those of atrocities committed by neighbors who had born false witness out of cowardice or worse. But the amnesty proved to be surprisingly successful in stabilizing the city, knitting its social fabric together again—so successful, in fact, that a Harvard law professor recently published a paper offering it as a case study for "the design of modern transitional justice institutions."[*] Her verdict: "The Athenian experience suggests that the current preoccupation with uncovering the truth may be misguided." The Athenians managed the delicate balancing act of both forgetting and remembering. No crimes committed at the behest of the Thirty could be prosecuted. Nobody could bring charges against someone for having been a sympathizer. In this sense, there was willful forgetting. But they allowed for remembering by allowing lawsuits to cite past behavior under the Thirty as evidence for good character or bad. (The Athenian courts, lacking both professional lawyers as well as professional judges, were more freewheeling than the judicial systems to which we are accustomed.) In this way, the city both refused to get bogged down in endless prosecution, but ensured that there wasn't

---

[*] Adriaan Lanni, "Transitional Justice in Ancient Athens: A Case Study," *University of Pennsylvania Journal of International Law* 32, no. 2 (2010): 551–594.

total impunity either. "The Athenian case suggests that at least in some situations pursuing a true account of who bears responsibility for atrocities may not be necessary, or even desirable, if the primary aim is to ensure an enduring, peaceful reconciliation."

And an enduring, peaceful reconciliation is exactly what they accomplished. The Spartan garrison withdrew, and the stability of the reconstituted democracy persisted right up until Athens, together with the other *poleis,* fell before the imperial conquest of Philip II of Macedon. Despite the horrors, despite the widespread complicity during the reign of terror, the *polis* managed to put itself to rights again, its citizens participating together in their reestablished institutions, exercising moderation and a judicious tolerance that again put them on the other side of the ordinary.

Nothing like this had ever been seen in the ancient world, a bloody civil war settled with such wisdom and prudence. Once again, in their righting of what had gone so wrong, the Athenians revealed themselves as beyond their fellow Greeks, which meant beyond all mortals, as they lost no time in telling themselves. "This, too, is worthy of our remembrance that, although our forefathers performed many glorious deeds in war, not the least of its glory our city has won through these treaties of reconciliation. For whereas many cities might be found which have waged war gloriously, in dealing with civil discord there is none which could be shown to have taken wiser measures than ours. Furthermore, the great majority of all those achievements that have been accomplished by fighting may be attributed to Fortune; but for the moderation we showed towards one another no one could find any other cause than our good judgement. Consequently it is not fitting that we should prove false to this glorious reputation."* Yes, others, such as Sparta, might have fought nobly and heroically, but Athens had topped such nobility and heroism, or so the Athenians took to telling themselves. Athens had done something, once again, that had never before been known. Athens' response to their defeat and the horrors that had followed was a demonstration of rationality, generosity, and overall largeness of mind. The speechifiers of the restored democracy—whether

---

* Isocrates, *"Against Callimachus,"* 31, http://www.perseus.tufts.edu/hopper/text?doc=Perseus %3Atext%3A1999.01.0144%3Aspeech%3D18%3Asection%3D31, and 32). http://www.perseus .tufts.edu/hopper/text?doc=Perseus%3Atext%3A1999.01.0144%3Aspeech%3D18%3A section%3D32.

speaking in the Assemby or before juries—all participated in elaborating a story in which the amnesty was not so much a compromise in the face of terrible realities as an opportunity for demonstrating a new kind of extraordinariness. It was a cause for celebration, and the Athenians were once again *kleos*-worthy. "All the world thought our city exceptionally wise," the rhetorician Aeschines writes. Nobody could do defeat the way the Athenians could.

Athenian exceptionalism had taken a hit since the glory days of Pericles. Not only had they been vanquished and occupied by the Spartans, their protective walls torn down, the bulk of their navy relinquished; but they themselves had sunk, under the pressures of war, to a level of irrationalism and cruel depravity whose forgetting they might very well have wished to legislate at the *ekklêsia,* along with the amnesty's *me mnesikakein.*

They had converted a league that had valiantly repulsed the Persian invasion of Europe into a ruthless imperialism over their former Greek allies, and they had enslaved and exterminated thousands of their fellow Greeks. There had been atrocities committed in the prosecution of the war, atrocities that went beyond the realpolitik toward other *poleis.* There had been coldly calculated cruelty, as in the destruction of Melos, a *polis* that had resisted joining the Delian League and devastatingly lost its argument with Athens.[*] There had been even worse, as at the massacre at Mycalessus, committed by Thracians whom the Athenians had contracted as mercenaries for the Sicilian expedition and then dismissed because they hadn't arrived in time. The atrocity at Mycalessus, which Thucydides relates, had happened when the

---

[*] In chapter 17 of the *History of the Peloponnesian War,* Thucydides presents the dialogue—whether personally recalled or reconstructed from others' accounts—between the Athenian envoys and the Melian magistrates. The slaughter that follows the exchange between the adversaries is all the more chilling for the dialogue that went before. The Athenians, with full civility, laid out their realpolitik, explaining that real negotiations can only take place between parties equal in power; otherwise, the stronger will do as they wish—in other words, Thrasymachus applied at the political level. The Melians bring up fairness and the favor of the gods, to which the Athenians respond: "When you speak of the favor of the gods, we may as fairly hope for that as yourselves; neither our pretensions nor our conduct being in any way contrary to what men believe of the gods, or practice among themselves. Of the gods we believe, and of men we know, that by a necessary law of their nature they rule wherever they can. And it is not as if we were the first to make this law, or to act upon it when made: we found it existing before us, and shall leave it to exist for ever after us; all we do is to make use of it, knowing that you and everybody else, having the same power as we have, would do the same as we do. Thus, as far as the gods are concerned, we have no fear and no reason to fear that we shall be at a disadvantage."

Thracians were being escorted back to Thrace by the Athenian general Diitrephes, so the Athenians could well feel it had happened on their watch. Thucydides had dwelled on the horror of what had occurred at Mycalessus, a small *polis* that had not taken either side in the war and so had taken no pains to protect itself, never expecting anyone to take notice of them. Thucydides—whose own covert attitude either for or against his former *polis* of Athens is still debated—tells of the pity and terror of Mycalessus, of the slaughter of young boys beginning their schoolday, and breaks out of his strictly enforced impartiality to state: "This is what happened to Mycalessus, a thing which is as much worth our tears as anything that occurred in this war, considering the small size of the town."*

If a hazard of the Ethos of the Extraordinary is individual hubris, Athenian exceptionalism had bred collective political hubris. And political hubris, no less than individual hubris, had spawned tragedy. The Athenians had rejected a peace treaty Sparta had offered in 410, which in hindsight would have been to their advantage. Now in their defeat they were beholden—passively, cravenly beholden—to victorious Sparta for not doing to them what they had done to other *poleis*— sacking the city, slaughtering its males, ravishing and enslaving its females. They had ingloriously survived on the sufferance of Sparta. And then there had been the period of the Thirty, best forgotten.

Pericles had been able to compare his contemporaries to Homeric heroes, in fact telling them that they had surpassed the heroes of Homeric epic: "We don't need a Homer to sing our praises," he had said, using a trope of the genre of the funeral oration *(epitaphios logos)*, that the deeds of the Athenians of the present surpass the deeds of mytho-historical heroes. Now, after what they had seen and what they had done, comparison to their former selves, much less to their legendary ancestors, was humbling to the point of shame. But with their amnesty, so different from anything accomplished before, together with the fiction of victimization it helped to create, they recovered the redemptive sense of themselves as daring innovators, brilliant pragmatists exquisitely adaptable to new circumstances. Their protective walls may have been razed by their conquerors, but there was still the protection of a fiction formalized in the phrase *me mnesikakein*.

*Thucydides, *History of the Peloponnesian War*, chapter 17, vii, 30.

Only he—tenacious, teasing, taunting—did not participate in the sustained delusion of continuing supremacy. That was the point of his maddening inquisitions that never seemed to go anywhere but were always driving home the same point: you Athenians live off the myth that you are living lives worth the telling and so worth the living. What a whopper, more incredible even than the bizarre stories that the manic poets tell of the gods. Being an Athenian doesn't make you extraordinary in any of the ways that matter. It never did, not even in the days of your self-proclaimed glory, your collective hubris fed by your most famous statesmen—much less now. Don't be so hasty in donning yourselves with laurels.

And who was he to talk, to taunt? Just how much good had he done his young men anyway—or more to the point, how much good had his young men done the city? The two worst exemplars of the opposing political sides had emerged from his sphere of influence, Alcibiades the lawless democrat and Critias the lawless oligarch. On the surface they might seem different, but what the two had in common was a raging individuality that burst out into fearsome hubris—in other words, precisely the dangers that the politicizing of *aretē* was meant to prevent. His questioning had only served to remove the restraints on the sort of ambition that was always a danger in a society that valorized *kleos*-measured *aretē*.

The fragile fiction of their non-complicity, on which they were trying to resurrect their sense of exceptionalism, had a great deal of pragmatic wisdom to it. They were pulling together to reassemble a ruptured social unity. This was not the time for rethinking first principles.

And that was precisely what he was asking of them, this insufferable man, constantly pelting them with questions whose point it was almost impossible to discern, rejecting everything they tried to say in response, they who were being spoken about throughout the world (at least so they liked to tell themselves) for the godlike wisdom of their reconciliation, their reputation still glorious.

"Forgive me, you most excellent fellow," he had answered the boy who had laughed at him for being so enthusiastic about a sycamore perched by a babbling stream. "I'm a friend of learning. The countryside and the trees don't want to teach me anything, but the people in town do."

Did he then hope that people could teach him what he wanted to know? He made a great show of expecting the people he constantly questioned to provide him enlightenment, but the show was a sham. He was insincere in the enthusiasm with which he greeted their initial complacent responses, and he was insincere in the disappointment he expressed when their answers were invalidated in a sad little pop of self-contradiction. He was convinced, even before he heard the specifics of their answers, that the people he questioned didn't know what they were talking about, and his questioning was designed to convince them of the same.

Why was he so sure that the answers he would hear would be inadequate? Had he arrived at answers that conflicted with everyone else's? But he was adamant in denying he had the answers. He denied it up until the very end, when his very life was at stake.

He didn't supply answers at the end of the elenctic exercises he was always conducting, after he had destroyed the increasingly unsteady responses of his interlocutors. What better circumstances could there have been to reveal—ta-da!—his own discoveries to the questions he foisted on others. So was it true that he lacked the answers? If he had the answers, then why didn't he come out and share them with his fellow citizens, yield up his superiority for the good of the *polis*? And if he didn't have them, then on what basis could he be so certain that his neighbors lacked them? Why always the insincerity masking the smug certainty that he would get no satisfaction from his fellow Athenians?

It had been one thing to tolerate his impudence when they had been riding high. They could afford to tolerate a genuine Athenian eccentric like him in the days when, as Pericles had made so clear for them, their worthiness was so manifest as to need no publicizing Homer. But now, clinging to a sense of their collective exceptionalism as best they could, his incessant challenges had simply become too much. They were no longer to be tolerated.

They would not be tolerated.

And so at the first opportunity, with the Spartan forces withdrawn and the democratic government once again on a stable footing, an indictment was entered at the stoa of the archon basileus by one Meletus, son of Meletus.

Meletus was a young and obscure poet. The accused, who spent his days wandering the streets of Athens in search of conversation and got himself invited to the best parties in town, had never heard of his accuser, neither by reputation nor by personal dealings. "I do not really know him myself, Euthyphro. He is apparently young and unknown," he says as he waits to go before the archon basileus, continuing with his brand of irony: "[I]t is no small thing for a young man to have knowledge of such an important subject. He says he knows how our young men are corrupted and who corrupts them. He is likely to be wise, and when he sees my ignorance corrupting his contemporaries, he proceeds to accuse me to the city as to their mother" (*Euthyphro* 2c). Always with the sarcasm.

Two others, more prominent in the *polis,* stood as Meletus' supporting speakers, or *synegoroi.*

There was a rich man, Anytus. He had both a tannery and a problematic son, who had displayed an enthusiasm for Bacchus and a taste for vice. The accused had warned Anytus that his scion was in need of guidance higher than what an inherited tannery could offer him (Xenophon, *Apology* 31.1–4), which no doubt offended the father. Like many of the *polis*'s citizens, Anytus' political allegiances were complicated. He had supported the regime of the Thirty until it deemed him either too unreliable or too rich and banished him from the city, helping itself to his property. He became a general for the exiled democrats and then a leader in the restored democracy.

The other supporter of the accusation was Lycon, an orator of the *polis.* Lycon also had a son, Autolycus, who had associated with the accused. Sadly, Autolycus had been executed by the Thirty.[*] Lycon, like Anytus, had a complicated history of allegiances; he had been accused of having betrayed the city of Naupactus to the Spartans during the Peloponnesian War.[†]

But it was Meletus who entered the formal indictment, which brought the famous eccentric of Athens, now a man of seventy, to trial:

> This indictment *(graphê)* is brought on oath by Meletus, son of Meletus of Pithus, against Socrates, son of Sophroniscus, of Alopece. Socrates is

---

[*] In Xenophon's *Symposium,* Autolycus is a young boy of astonishing beauty, who sits modestly on the floor near where his father reclines on a couch. "Noting the scene presented, the first idea to strike the mind of any one must certainly have been that beauty has by nature something regal in it; and the more so, if it chance to be combined (as now in the person of Autolycus) with modesty and self-respect."

[†] Debra Nails, "The Trial and Death of Socrates," in *A Companion to Socrates,* ed. Sara Ahbel-Rappe and Rachana Kamtekar (Oxford: Blackwell, 2006).

guilty of not believing in the gods the city believes in, and of introducing other divinities *(daimōnia)* and he is guilty of corrupting the young.
The penalty assessed is death.[*]

CONTRETEMPS

Meletus made his formal indictment against Socrates in the spring of
399. The archon basileus, an official who had jurisdiction over cases
of both homicide and impiety, ruled that the case had sufficient merit
to go to court. The trial took place a month or two later, in Thargelion
(May/June). It was held outdoors to accommodate the full crowd, not
only the 501 jurors,[†] but the large crowd of onlookers. Great swaths of
material were laid out as awnings to protect them from the blazing sun.
The trial lasted for the better part of the day, the three accusers together
having three hours to make their case, and the accused accorded three
hours in defense. The time was measured out by a water clock. The day
before the trial, the Athenians had sent a ship, dedicated to Apollo, to
the island of Delos. This was a yearly event, in commemoration of the
legendary victory of Theseus over the Minotaur, which the Athenians
celebrated as part of their history. To preserve the ritual purity, Athenian
law prohibited any executions from taking place until the ship returned
(*Phaedo* 58a–b). The duration of the voyage varied with conditions, but
in this year it took thirty-one days for the ship to return (Xenophon,
Memorabilia 4.8.2), which means that Socrates lived thirty days beyond
his trial, into the month of Skirophorion (June–July).

Of the twenty-six dialogues of Plato, the internal dramas of seven of
them are set during the spring and summer of 399: *Theaetetus, Euthyphro, Sophist, Statesman, Apology, Crito,* and *Phaedo.* No matter how
scholars argue the chronology of the dialogues, it seems safe to say
that these seven dialogues were written over the course of Plato's long
life. Not only was Plato's ongoing philosophical imagination organized
around the figure of Socrates, but it continuously reverted to those
months in the spring and summer of 399. Even in the *Sophist* and the

---

[*]This indictment comes to us by way of Diogenes Laertius (third century C.E.), who got it
from Favorinus (second century C.E.), who said he saw it in the public archive, the Metroön.
[†]The number became 501 sometime around the time of Socrates' trial. The odd number
ensured that there were no ties. The large number discouraged bribery.

Statesman, late dialogues in which the figure of Socrates is withdrawn from the philosophical center into the margins, still the temporal settings give singular priority to the drama of Socrates' death.

Other dialogues as well, set before those few months of 399, make allusions to the Socratic death drama. For example, an irascible Anytus makes a late entrance in the *Meno,* pulled into the conversation Socrates is conducting with the eponymous visitor from Thessaly on the subject of whether *aretē* can be taught. Scholars agree that the character is meant to be the same Anytus who will play a part in Socrates' downfall. When Socrates asks Anytus who are those who can teach virtue to the rich visitor Meno, Anytus impatiently answers, much like Cheryl the media escort of chapter β, that any decent Athenian citizen whom Meno happens to meet can teach him *aretē* (92e). This is telling, as is Anytus' threatening outburst, after he has heard Socrates argue that *aretē* cannot be taught, given the failure of such men as Themistocles and Pericles to raise exemplary sons. "I think, Socrates, that you easily speak ill of people. I would advise you, if you will listen to me, to be careful" (94e). Socrates remarks, "I think, Meno, that Anytus is angry and I am not at all surprised. He thinks, to begin with, that I am slandering those men, and then he believes himself to be one of them. If he ever realizes what slander is, he will cease from anger, but he does not know it now" (95a). Knowing what lay ahead in Socrates' future, we read this passage and hear its ominous undertone.[*]

There are large historical events transpiring as Plato writes. The Ionian *poleis,* which had served as the casus belli of the last century's Persian Wars, were absorbed once again into Persia. Philip of Macedon was making steady encroachments into Greece. None of this is paid any mind in Plato's writing. Instead, time is altogether stalled in the last quarter of the fifth century. It is life as it was then, during Socrates' heyday, that Plato re-creates the bulk of the dialogues set either before Plato was alive or when he was a boy.

The earliest dialogue, according to its internal dramatic chronology, is the *Parmenides,* though it was probably written relatively late in Plato's life. Its internal date is the summer of 450, and Socrates is a young

---

[*] And surely, it's telling, too, that having argued the inabilities of such men as Pericles to transmit a knowledge of *aretē* even to sons, and so, by implication, to the citizens of Athens, Socrates proceeds, in that same dialogue, to demonstrate that, with the right methods, knowledge of mathematics can be drawn forth from a slave.

man. The last of the dialogues, according to its internal chronology, is the *Phaedo,* extended to a bare few moments after Socrates' death.

But it is most particularly those seven dialogues crowded into the spring and summer of 399 that reveal how the Socratic death drama functioned in the ongoing philosophical project of Plato's life. Plato presents Socrates as always maintaining a certain distance from the personal crisis in which he finds himself. He is not going to let a contretemps like being brought up on a capital offense interfere with his pursuing the philosophical subjects that interest him.* The positioning of the personal drama—being accused, convicted, imprisoned, and executed—as a mere backdrop to discussions of timeless questions is meant itself to convey a moral lesson. To reflect on the conditions that make a life worth living is to remove oneself from the circumstances of that life as much as possible. It is to see that life in the context of a perspective that does not take the contingencies of that particular life that you happen to be living overly seriously. To philosophize is to prepare to die. Or, to truly take your life with the seriousness that philosophy demands, you can't take your life all that seriously. This is to give a new philosophical spin, more dizzyingly paradoxical, to the old Greek idea that the hero is, like Achilles, prepared to shorten his life in order for that life to be something extraordinary, *aretē* achieved.

The first of the dialogues that is set against the death drama of Socrates is the *Theaetatus,* a dialogue which is often grouped (for those who group) with Plato's vigorously doctrinal Middle Period, even though it ends, as the earlier dialogues do, in aporia. It deals with the nature of knowledge and how it differs from belief even when that belief happens to be true. The *Theaetetus* is set on the day that Socrates is to answer the summons of the archon basileus. This might be an anxious time for an ordinary person, but not for Socrates, at least as Plato tells it. It turns out to be an excellent day for Socrates because he

---

*Xenophon also presents Socrates as being indifferent to the mortal danger he was facing. "And now I will mention further certain things which I have heard from Hermogenes, the son of Hipponicus, concerning him. He said that even after Meletus had drawn up the indictment, he himself used to hear Socrates conversing and discussing everything rather than the suit impending, and had ventured to suggest that he ought to be considering the line of his defense, to which, in the first instance, the master answered: 'Do I not seem to you to have been practicing that my whole life long?' And upon his asking 'How?' added in explanation that he had passed his days in nothing else save in distinguishing between what is just and unjust, and in doing what is right and abstaining from what is wrong, which conduct, he added, 'I hold to be the finest possible practice for my defense'" (*Memorabilia* VIII, 8–9).

meets an excellent boy, Theaetetus,* the prized student of the mathematician Theodorus. Theodorus tells Socrates that he's not ashamed to praise Theaetetus because nobody will suspect he's in love with him, since he's not handsome "but—forgive my saying so—he resembles you in being snub-nosed and having prominent eyes" (143e). Socrates employs his maieutic methods on Theaetetus, and neither is altogether satisfied with the results. (The term "maieutic" derives from the Greek *maieutikos*, pertaining to midwifery. Maieutics is the pedagogical method that tries to draw a conclusion out of a mind where it latently resides. As Leibniz remarks of the method, it consists in drawing valid inferences.)

The conclusion that Socrates draws out of Theaetetus—that knowledge is true belief fortified with an account of why it is true, does not seem altogether adequate to them. (Future epistemologists, seeing in Plato's *Theaetetus* the core analysis of knowledge as "justified true belief," would be far more appreciative of what that dialogue accomplished than its own author might have been.) But by the end, Socrates is ready to see the young mathematician as beautiful. Here are the closing words of the dialogue, and, they, together with the framing drama of the dialogue, which reports that Theaetetus, now a respected mathematician, has been fatally wounded in battle and is being carried back to his city to die, provide a certain pathos to the epistemological work that is done in the *Theaetetus:*†

> And so, Theaetetus, if ever in the future you should attempt to conceive or should succeed in conceiving other theories, they will be better ones as the result of this inquiry. And if you remain barren, your companions will find you gentler and less tiresome; you will be modest and not think you know what you don't know. This is all my art can achieve—nothing more. I do not know any of the things that other men know—the great and inspired men of today and yesterday. But this art

---

* He is indeed excellent. He'll be among the mathematicians whom Euclid cites in his *Elements*. Other sources on whom Euclid drew are Leon and Theudius, both of whom were fourth-century mathematicians who spent time in the Academy; and also Eudoxus, with whom Plato also had significant contact. See Burnyeat, "Plato on Why Mathematics Is Good for the Soul."

† The framing drama is set in 391 (see Nails, *The People of Plato*, pp. 274–278), eight years after the internal drama. The battle in which Theaetetus is fatally wounded would have been fought during the Corinthian War. Nails calculates that he would have been twenty-four, remarking, "Theaetetus is thus no exception to the rule that mathematicians do most of their creative work while very young" (p. 277).

of midwifery my mother and I had allotted to us by God; she to deliver women, I to deliver men that are young and generous of spirit; all that have any beauty. And now I must go to the King's Porch to meet the indictment that Meletus has brought against me; but let us meet here again in the morning, Theodorus. (209b–210d)

The *Euthyphro*, which is one of Plato's earlier dialogues and deals with the relationship between theism and morality—an issue still fraught for us today—takes place later that same day, the date of the preliminary hearing, and it transpires while Socrates is on the portico of the Royal Stoa, awaiting his turn to appear before the archon basileus. Unwilling to squander any opportunity for meaningful discussion, he falls into conversation with a diviner-priest named Euthyphro, a priceless character whose sacerdotal vanity cannot be pierced. A self-declared expert on all things holy, Euthyphro has come to the Royal Stoa to indict his own father on a charge of homicide for having accidentally killed a hireling, who had himself killed another worker in a fit of anger. Socrates is amazed to hear that Euthyphro is so secure in his moral certitude as to charge his own father. (The ancient Athenian codes of family loyalty make Euthyphro's actions seem all the more questionable.) Euthyphro responds with the telltale conviction of the self-righteous. Socrates immediately launches in, having his fun, declaring that Euthyphro alone can save him, in this his moment of need, by instructing him on the nature of piety or holiness so that he can present himself as chastened to Meletus—though "Meletus, I perceive, along presumably with everybody else, appears to overlook you." With an interlocutor as deaf to sarcasm as to philosophical subtlety, Plato's Socrates proceeds to formulate a line of reasoning that will prove to be of fundamental importance in the history of secularism, one that will be adapted by freethinkers from Spinoza to Bertrand Russell to the so-called new atheists of today, persuasively arguing that a belief in the gods—or God—cannot provide the philosophical grounding for morality.

Plato begins the inquisition innocently enough, with Socrates asking Euthyphro, "Is what is holy holy because the gods approve it, or do they approve it because it's holy?" (10a). Plato uses this question to pry apart the notion of an action's being divinely ordained from its having moral worth. The argument is formulated in terms of "the gods," but is, without loss of force, susceptible to the substituting of "God" for "the gods." Plato's argument, in a nutshell, is this: If God approves of an action, either he approves of it arbitrarily, for no reason at all, so that it is only his approving it that confers on it moral value; or else there is a reason for his approving it, so that it is not simply an arbitrary whim on God's part but rather he has a reason for his approval, that reason being the independent moral worth of that of which he approves. If the former is the case, then how does this arbitrary whim, even if it is a *divine* arbitrary whim, confer moral value? How can something be good just because someone up there *feels* like calling it good, when, if he were of a different disposition or in a different mood, he could just as easily call the opposite act good? But if the latter is the case, then there is a *reason* for the divine normative attitude, and that reason is the reason both for God's approval and for the moral worth of that of which he approves. This makes God's approval, normatively speaking, redundant—he is, as we say today, a rubber stamp. In neither case—whether the approval is arbitrary or whether it is not—does the supernatural approval make any difference to whether an act is genuinely right or wrong.

What is still referred to as "the Euthyphro Dilemma" or "the Euthyphro Argument" remains one of the most frequently utilized arguments against the claim that morality can be grounded only in theology, that it is only the belief in God that stands between us and the moral abyss of nihilism.* Dostoevsky may have declared that "without God all is permissible," but Plato's preemptive riposte, sent out to us across the millennia, is that any act morally impermissible with God is morally impermissible without him, making clear how little the addition of God helps to clarify the ethical situation.

The argument Plato has Socrates make in the *Euthyphro* is one of the most important in the history of moral philosophy. When it is joined with another of Plato's claims, namely that a person's action is virtuous only if he can supply a reason for its being so, the Euthyphro Argument

---

*The word "nihilism" has an interesting history. It was coined by Heinrich Jacobi, who formulated it in the context of his attack on the Enlightenment, and most specifically his attack on all philosophers for attempting to ground moral truths on reason alone, without regard to theism. Jacobi put Spinoza (dead for a hundred years by this time) front and center in his attack on the Enlightenment. Indeed Spinoza's magnum opus, *The Ethics*, was one of the first attempts, after the long period in which Christian thought dominated Europe, to return to the project of placing ethics on a firm secular foundation. See my "Literary Spinoza," in *The Oxford Handbook of Spinoza*, ed. Michael della Rocca (New York: Oxford University Press, forthcoming).

demonstrates the need for moral philosophy. We humans must reason our way to morality or we will not get there at all. Relying on fiats, even should they emanate from on high, will not allow us to achieve an understanding of virtue. Any progress in our moral understanding—progress that, in time, would take us some distance away from the slave-abusing, captive-slaughtering, philosopher-executing, misogynistic Athens that held itself up as the very standard of *aretē*—has been made on the basis of an argument that Plato put into the mouth of a man awaiting a hearing on charges of impiety and corruption of the young.

This moment in Socrates' life, as Plato has rendered it, is sufficiently important to step away from it, and reflect. It has a bearing on the question that is always hovering over this book, as it traces the sources of philosophy as we know it, and that is the question of philosophy's progress. If one evaluates what the ancient Greek philosophers did solely in terms of Thales and Co., then of course one will conclude something like "Philosophy used to be a field that had content, but then 'natural philosophy' became physics, and physics has only continued to make inroads." But this is to focus on only one type of question the ancient Greek philosophers posed to self-critical reason, the protoscientific questions that awaited the mature sciences. It is to ignore such questions as those that Plato has Socrates raising with Euthyphro on the portico of the archon basileus. It is to ignore Plato's argument that, since religious authority can't answer these questions, we had better get to work on formulating the reasons that make right actions right and wrong actions wrong. It is to ignore the work that has since been done, not only on the normative questions of ethics but on the normative questions of epistemology, the work that is necessary to speak about rationality at all. It is to ignore the conclusions to which philosophy-jeerers freely help themselves, most certainly when they speak in the name of rationality.

When philosophy-jeerers are scientific, then their jeering frequently takes on religion as well as philosophy. Typically, they do not differentiate between philosophy and theology. Anything that isn't science is philosophy/theology. Lawrence Krauss, whom I keep mentioning only because he conveniently articulated a viewpoint that many scientists share, lumps philosophers and theologians together. Such jeerers should pause and reflect on this moment of the *Euthyphro*. Plato is arguing that ethical questions can only be answered by way of human reason, with no religious input, and that is a conclusion that radically separates philosophy from theology. It is ironic when freethinkers like Krauss lump philosophy and theology together. The Enlightenment came about when philosophers like Baruch Spinoza went back to the work of grounding ethics on purely secular reasoning, a project which had been interrupted by the centuries of theological ideology. Spinoza made his goal clear when he titled his magnum opus *The Ethics.* This was to take up again work which had been initiated by Plato's Socrates, there on the portico of the archon basileus.

Plato certainly did not do all the work that is required to answer the questions he is raising. He is only arguing, at least here in the *Euthyphro,* that it is a job for human reason, and not for the gods. Aristotle, who grandly advanced moral philosophy, didn't do all the work, either. They no more arrived at the definitive answers to their questions than did Thales and Co. in raising questions of physics and cosmology. Progress in one subject is as difficult as progress in the other, though for quite different reasons. Progress in philosophy is difficult for the same reason that its progress, once made, becomes invisible. What is in need of changing for progress to be made are convictions constitutive of points of view. It is both hard to discover these convictions— assimilated into our points of view as they are—and, once the changes are assimilated, it is hard to see that anything has changed. Since progress in philosophy is as difficult as progress in the sciences, it is unreasonable to expect one man—or one generation, or one millennium—to do all that needs doing.

And of course, despite the Euthyphro Argument, religious authorities managed to monopolize the moral discussion through the millennia, temporarily interrupting the kind of work that Plato was arguing had to be done for us to begin to acquire the knowledge we need. This is not to say that Plato is not himself respectful of religion, most especially of its power to keep the non-philosophical masses on the straight and narrow. One need only read his chilling Book X of the *Laws,* in which he comes dangerously close to making freethinking illegal, to know how essential for social stability he judged religion to be. The non-philosophical masses cannot be expected to grasp the required subtle reasoning, and for them there can only be religion—or so he seems to conclude in the very last years of his life.

Though Plato was optimistic that moral actions grounded on knowledge could be achieved, he was not optimistic about how *many* can achieve it. It's no wonder, since ultimately his notion of moral excellence—of the *aretē* worth the achieving—demands that the excellence of the cosmos be assimilated into oneself, become a part of one's own moral and intellectual constitution. To live the life worth the living one must be able to grasp and internalize the goodness that makes the cosmos worth the existing. One must integrate the beautiful proportionality of the character of the physical universe into one's own moral character, and then, and only then, will one see oneself in relation to all else—and all others—in the right perspective, the distortions of the cave corrected. This is not a dispassionate process. Plato always stressed how much love is involved in the process. But it's love of an impersonal kind, not love for persons, that reforms one's moral being. Plato would have approved this paragraph from Spinoza's *Ethics:* "Therefore, without intelligence, there is not rational life: and things are only good, in so far as they aid man in his enjoyment of the intellectual life, which is defined by intelligence. Contrariwise, whatsoever things hinder man's perfecting of his reason, and capability to enjoy the rational life, are alone called evil" (Appendix, Part IV, v). Morality necessarily crosses, for these philosophers, through the headiest of intellectual terrain, and that is a path that—I don't think they're particularly happy about this— few can follow. I think that Plato—and Spinoza—would have wished it otherwise, but there it is. Reality doesn't conform itself to our wishes. But given his view of how the moral perspective is to be achieved, it's no wonder that Plato wasn't hopeful that universal progress could be made in its achievement.

Plato stood at the beginning of the self-critical process of our reasoning out our moral convictions. He overestimated the unilateral role of moral reason, underestimated the role of the moral emotions—our sense of fairness, our capacity for empathy. He could not have known how moral arguments and moral emotions, together with social movements and political agitation, could join together in complicated ways that have slowly and jerkily brought us to where we are today, which is hardly at the finish of the process. There, where he stood, which was at the beginning, he could not possibly have foreseen how it would play out in the moral lives of successive generations, nor how it's still being played out. . . .

·   ·   ·

So that is how Plato presents Socrates amusing himself before going in to meet his accuser. Plato sets two of his late dialogues, the *Sophist* and the *Statesman,* the very next day. Socrates had parted from Theodorus and Theaetetus the day before, as presented in the *Theaetetus,* with the promise that they would meet on the morrow to continue their exploration of the nature of knowledge (210d).

The *Sophist* opens with Theodorus, Theaetetus, and Socrates showing up for that appointment (216a). It's the morning after what would have been, for anyone else, a disheartening appearance before the archon basileus, who judged Meletus' case to be of sufficient merit to go to trial. But no mention is made of the event. It is we, the readers, who supply the missing information by knowing the order of the internal dramas of the *Theaetetus,* the *Euthyphro,* and the *Sophist.*

Even so indefatigable an arguer as Socrates might be feeling his age after what must have been a dispiriting demonstration of insincere moralizing on the part of Meletus (who will be shown up, in the *Apology,* as just such a moralizer), on top of the theologically confused moralizing of Euthyphro. That's quite a lot of bad moralizing for someone like Socrates to have to ingest, and perhaps it has left him feeling philosophically dyspeptic. Or perhaps there is some other explanation for the odd thing that Plato proceeds to do in the *Sophist,* and to repeat in the *Statesman,* which is to have his Socrates bow out of a philosophical discussion in favor of a younger man, the mathematically gifted Theaetetus, who had so impressed him the day before, according to the internal chronology of the dialogues. And there is someone else who gets thrown into the thick of the philosophical action as well. Theodorus has brought along a guest, referred to as the Stranger, a native of Elea and a follower of Parmenides, the great metaphysician whom Plato, in the *Parmenides,* had discoursing with a far younger Socrates, whom he had philosophically bested. But here Plato goes even further in relegating Socrates to the philosophical sidelines: "Then you may choose any of the company you will," Socrates says to the Stranger. "They will all follow you and respond amenably. But if you take my advice, you will choose one of the younger men—Theaetetus here or any other you may prefer." So it is the Stranger and Theaetetus who together worry the concepts of being and non-being, the one and the many, ultimately chasing down the problem of how it is that we can intelligibly speak of that which is not. Plato's intuitions tell him this is a technical problem; and he was right. Gottlob Frege was quoted back in chapter α as a logician with strong Platonist tendencies. The developments in mathematical logic that he made have significantly sharpened the issues—and disagreements—that still linger over the logical structure of propositions containing non-referential terms, such as "the present Czar of Russia," "the highest prime number," or "the current liberal members of the Republican Party."

It has been a busy couple of days for Plato's Socrates, including meeting a wonderful young mathematician and formulating the basic questions of epistemology with him, trying to get a diviner-priest to see how moral theory cannot be grounded in the received arbitrariness of divine choices; a preliminary showdown with those accusing him of impiety and corruption of the young; followed by a long session of heavy metaphysics and philosophy of language. But Plato is still not ready to have his Socrates kick back and relax. There is still the *Statesman*.

To the *Sophist*'s gathering of Socrates, Theodorus, Theaetetus, and the Eleatic Stranger, Plato adds one more to join in the task of dialectically defining the statesman. This is a bland young man with the startling name of Socrates, referred to throughout the dialogue as the Younger Socrates. (He has been present, albeit silently, during the discussions of both the *Theaetetus* and the *Sophist*.)

The methodological problem that concerns both the *Sophist* and the *Statesman* is the problem of identifying the joint at which to cut between concepts. So though the *Statesman* is ultimately concerned with the kind of person who is the best ruler, and the kind of knowledge he is required to have, it's also self-consciously devoted to methodology. Overseen by the Stranger, the two youngsters, Theaetetus and the Younger Socrates, carry on the dialectical discussion, Socrates designating them worthy replacements with the odd remark that Theaetetus physically resembles Socrates and the Younger Socrates has his name (257d). The temporal setting of the dialogue manages to keep the death drama of Socrates in the frame, even while Socrates as a character is unceremoniously retired, the flimsiest of similarities to his person— his ugliness, his appellation—offered as justifications for having others take his place. Is Plato, who is experimenting with different philosophical techniques, perhaps suggesting that some are suitable for one sort of problem, others for other problems, but none is indispensable? This is an interpretation for Socrates' marginalization that some have suggested. Or is it rather another kind of indispensability that Plato is denying here? No one person is indispensable to philosophy. The process has been initiated, and it will overtake and improve upon any one philosopher, no matter how extraordinary. Process over persons. Love of philosophy should never be narrowed down into love of a particular philosopher, for then it will deteriorate into the hermeneutics of dogma, become yet another, albeit recondite, way for our thinking to be rendered thoughtless, and we'll find ourselves back inside the cave, chained together with our learned colleagues, staring at PowerPoint projections instead of those cast by the puppets of Plato's cave. A technological advance, but we'll still be in the dark. It's still a danger among professional philosophers that some who devote their lives to excavating the system of some chosen dense figure—a Kant, a Wittgenstein, a Heidegger—can tolerate no suggestion that the chosen one didn't achieve, all by himself, the completion of philosophy. Plato is perhaps here warning us against this tendency, by unceremoniously retiring the thinker who has become so identified with the Platonic dialogue that it's now all but impossible to disentangle Plato's Socrates from the historical man. This time it's Plato, and not the Athenians, who is decreeing that Socrates must die.

The three remaining dialogues of the seven crowded into the summer of 399 are the *Apology*, which gives us Socrates' defense at his trial; the *Crito*, which shows us Socrates in prison, a day or two before the end, visited in the middle of the night by his childhood friend, a distraught, insomniac Crito, who has arranged for Socrates' escape and has only to convince Socrates to flout his city's laws and save his skin, which he declines to do, on the basis, of course, of a philosophical argument; and the *Phaedo*, which gives us a daylong conversation on the last day of Socrates' life, as well as an account of his dying.

The *Phaedo* presents Socrates in conversation with his friends, mostly young men, though the faithful heartbroken Crito is present, as is Apollodorus, another interesting character. Apollodorus had been a successful businessman but left his affairs to follow Socrates in the last few years of the philosopher's life. Xenophon describes him as one of those who never left Socrates' side, and he seems to have had a reputation as an eccentric. He is the narrator of the *Symposium*, telling his secondhand account of the long-ago party—he mentions that at the time of the related events he would have been a mere boy—to an unnamed friend, who at one point makes reference to Apollodorus' nickname, "Maniac." Apollodorus is obviously an emotional fellow, as he demonstrates in the *Phaedo,* unable to stop crying through the whole dialogue (117d).

As the devotees arrive before daybreak at Socrates' prison cell, his wailing wife, their child on her knee, is dismissed from the room. "Socrates looked at Crito. Crito, he said, someone had better take her home" (60a). As the wife of a man soon to die it was appropriate for her, according to custom, to lead the mourners, but Socrates is not going to depart at this date from his *atopia,* his strangeness, meaning that what is appropriate for others does not suit him. (But did Socrates really send Xanthippe and his son off so coolly, or is this Plato's notion of how a philosopher regards the conventional sentimentality of family life?) It is only right that Socrates not be deprived of his pleasure before he drains his cup of hemlock—this pleasure being, of course, philosophical discussion, as Xanthippe, as a knowing wife, had already foreseen, exclaiming as his friends come trooping in that this is the last opportunity they'll have for discussing philosophy.

The philosophical theme, on this occasion, is the timely question of the soul's immortality. Are there any grounds to believe that a person can survive his death? "I suppose for one who is soon to leave this world there is no more suitable occupation than inquiring into our views about the future life, and trying to imagine what it is like. What else can one do in the time before sunset?" Plato has his Socrates asking with a poet's double entendre.

The *Apology* and the *Crito* are both chronologically sorted as belonging near the beginning of Plato's writing career, by those who countenance such sorting at all. In fact, the *Apology* is often regarded as the first of Plato's dialogues, and, some claim, the most historically faithful among them. Many had attended Socrates' trial that day in 399 and would have known for themselves what Socrates had said in his ineffectual defense, so perhaps Plato didn't stray too far from the historical record.*

---

*Though perhaps he did. Xenophon's account of Socrates' trial differs in many important respects from Plato's, and there were apparently other accounts of Socrates' trial in circulation. Since part of the rationale for the apologetic Socratic literature was a concern for the long view, rather than a journalistic record of what had happened, the historical accuracy of Plato's *Apology* is a matter of continuing debate.

The *Phaedo,* in contrast, was written later, and the opinions and arguments that Plato has Socrates offering on the subject of the soul's immortality depend not only on attitudes more temperamentally characteristic, perhaps, of Plato than of Socrates—for example, a marked antipathy for the bodily—but also on the metaphysical and epistemological ideas that Plato most probably explored on his own, such as the Theory of Forms. So, for example, there is an argument that appeals to the Theory of Forms and to Plato's theory of knowledge as recollection *(anamnesis)* of those Forms: Since none of our embodied experience in this world could provide us with the knowledge of absolutes that our knowledge of the Forms yields us; since everything we experience in this life is judged as falling short of the perfection we nevertheless must know, if only to judge all things of this world as falling short of perfection; it follows that we must have become acquainted with these exemplars, of which particulars are "only imperfect copies" (75b), in a disembodied previous existence (75c). So if there was personal existence *before* our being birthed into this embodied life, establishing the metaphysical possibility of disembodied existence, then why not personal existence *after* our departing from this embodied life? There is still controversy over how committed to the thesis of immortality Plato was, even though it became such a cornerstone of Christian Platonism. This is the dialogue where he most rigorously explores reasons to accept the proposition, and he doesn't end the dialogue with a conclusive endorsement of it. But then there is still controversy over whether Plato endorses any substantive propositions, or even endorses the endorsing of propositions.*

Also relevant in demarcating the historical Socrates from the Socrates of Plato's *Phaedo* is the strong strain of Pythagoreanism that runs throughout the dialogue. The cult that surrounded the seer Pythagoras

---

* Interestingly, Ruby Blondell, who argues with great force and persuasiveness that it is a "basic methodological mistake" to assume or even infer "the equivalence of any of Plato's characters with the voice of the author," and that we therefore must be extremely wary of attributing any doctrines to Plato, holds it to be hard not to believe that Plato himself believed in the immortality of the soul "in some shape or form." Blondell, *Play of Character in Plato's Dialogues,* p. 18. I confess I don't find it hard to question whether Plato believed in the soul's immortality. The notion of our personal identity inhering in "something" that could survive the body's death was a doctrine of the Pythagoreans, and Plato gives it a sympathetic hearing in the *Phaedo,* as well as using it as an element in many of his "myths." But there is a contrasting view of personal identity—and the possibilities (or not) for personal immortality—to be read out of Plato, one that places him far closer to his Greek ethos than to the Christianity to come, namely that it is our attainment of the extraordinary—for Plato, attained by way of reason—in this incurably mortal life that allows us whatever participation in immortality is possible. This idea will be taken up in the remainder of the chapter.

(who died in 495 B.C.E.) combined mathematics and otherworldly mysticism, including a belief in the transmigration of souls. After Socrates' execution, Plato removed himself from Athens for around ten years—perhaps out of aversion to his city or out of a feeling of danger—and spent time in the Pythagorean communities of southern Italy. Pythagoreanism looked on this earthly life as an opportunity to purify the soul so that it might be liberated from the incessantly revolving wheel of birth and rebirth, with mathematics—which still, to many, offers intimations of eternity—the wormhole to escape. The Pythagoreanism to which Plato was exposed, in the years after Socrates' death, separated him further from Socrates. The Pythagorean intuition that the form for rendering reality intelligible is supplied by mathematical ratios influenced him profoundly, ultimately yielding him his conception of the Sublime Braid and the means to make good on Socrates' search for the kind of knowledge that is also virtue.

There are references to Pythagoreanism throughout the *Phaedo*. Two of the most active participants, the lovers Simmias and Cebes, have ties with the Pythagorean community. And the slant toward Pythagoreanism is announced in the framing story:

It is a few weeks or months after Socrates' execution. Phaedo, one of Socrates' young men, is recounting the story of Socrates' death to Echecrates of Phlius, who is a Pythagorean. Phlius, which lay between Athens and Elis, was one of the refuges to which the mathematico-mystical Pythagoreans fled after the destruction of their original settlement in Croton, where they had been politically active. It was their political activism that had stirred up trouble, leaving the nonagenarian Pythagoras and many of his followers dead.

We aren't told the background story of the eponymous character in *Phaedo*. According to Diogenes, who got it from Hieronymus Cardianus, Phaedo had been a former aristocrat from Elis, brought to Athens as a captive and sold into the most degrading of all slavery, that of a catamite. According to the same source, it was Socrates who appealed to Crito, a man of means, to buy the boy's freedom. Christian writers, relying on Diogenes, drew moral lessons from Phaedo's reform, but, it goes without saying, there is debate over whether Diogenes' gossip about Phaedo can be trusted.* During the dialogue, Socrates absent-

---

* In "Phaedo, Socrates, and the Chronology of the Spartan War with Elis," E. I. McQueen and Christopher J. Rowe establish that the story is at least—as had been disputed—possible, since there was a Spartan-Elis war, in which Athens became involved. The defeat of Elis could then

mindedly strokes Phaedo's hair (89b), presumably worn long in the Spartan style.

Phaedo gives Echecrates the Pythagorean a complete account of Socrates' last hours. It begins with Phaedo, together with a handful of other devotees, entering the prison cell to see Socrates "just released from his chains" (59e), preparing the way for the metaphorical depiction of death as a freeing of the soul from the "shackles of the body" (67d), a very Pythagorean conception. The extreme denigration of the body, its description as "a contamination" (66b) for which the soul's purification is necessary, is an aspect of Pythagoreanism that Plato displays in the *Phaedo,* which is brimming with proto-Christian body-disgust. The best that this life can provide effects a separation from the body. This is what philosophy, strengthening our affinities to the abstract and impersonal, is (64a). Whereas in other dialogues the abstract can be interpreted as embedded in the structure of this world, an immanence rather than a transcendence—this is certainly true of the *Timaeus,* and can be read into other dialogues, including the *Republic*—the *Phaedo* seems to remove the abstract to another place, beyond space and time, to which our better selves, with their affinity for the abstract, may just possibly retreat after the final separation from our bodies is effected. This is a notion of immortality which got absorbed into Christian visions of heaven.

Phaedo, the former slave and catamite, is an appropriate narrator for a dialogue that conceives of death as both manumission and purification. He plays the role of Mary Magdalene to a Socratic Jesus. Plato makes a point of having Phaedo mention that Plato wasn't present at this last conversation (59b), which I've always read as a distancing move on the part of Plato from the position that is being explored, together with Socrates' argumentative tentativeness in actually affirming immortality.[*] In other words, I don't think that Plato was committed to the soul's immortality in any robust heaven-is-there-to-receive-us kind of way. It was another proposition that Plato put on the table to explore, and, as usual, he sets the scene carefully. Socrates, his cheer-

---

have been the occasion of Phaedo's being made a captive and sold to an Athenian brothel. Diogenes also reports that Phaedo founded a Socratic school of philosophy in his native Elis. He cites eight titles written by Phaedo, though none are extant.

[*] Plato uses this vision of a soul able to survive the death of its body in various of his myths, including the myths that end the *Republic* and the *Gorgias.*

fulness and objectivity intact,* is facing a once-in-a-lifetime experience, and in a mood to give the Pythagorean proposition of survival after death his best shot, though either way—survival or not—he's reconciled to his end. But the Christian Platonists, starting from the fourth century C.E., latched on to the *Phaedo* as giving Plato's honest-to-God point of view, and through them the doctrine of the immortality of the soul has had a long afterlife.

In other dialogues, most especially the *Timaeus,* a more attenuated kind of immortality is proposed. To the extent that we take into ourselves, in knowledge and in love, the True-the Beautiful-the Good, to that extent we achieve a kind of immortality. This is an immortality as impersonal as true knowledge. In fact, it's nothing more nor less than wisdom, that state of being that fuses together knowledge of, and love for, *to on.* It is an impersonal form of immortality in the sense that it offers no promise that a something uniquely personal—one's own self, bearer of one's attitudes and memories—will survive the body's death. No disembodied soul is appealed to in the *Timaeus* as it is in the *Phaedo.*† Instead the kind of immortality that we can achieve doesn't negate our mortality. We are immortal only to the extent that we lose ourselves in the knowledge of reality, letting its sublimeness overtake us. We are immortal only to the extent that we allow our own selves to be rationalized by the sublime ontological rationality, ordering our own processes of thinking, desiring, and acting in accordance with the perfect proportions realized in the cosmos. We are then, *while in this life,* living *sub specie aeternitatus,* as Spinoza was to put it, expanding our finitude to encapture as much of infinity as we are able. Or as Plato put it in the *Timaeus:*

> So if a man has become absorbed in his appetites or his ambitions and takes great pains to further them, all his thoughts are bound to become merely mortal. And so far as it is at all possible for a man to become thoroughly mortal, he cannot help but fully succeed in this, seeing that he has cultivated his mortality all along. On the other hand, if a

---

*A beloved friend of mine—the mathematician Bob Osserman—recently died. As his family and friends gathered around him, he said, "Well, this is the funniest thing I ever did!" Bob died in much the same spirit as Plato's Socrates, with just the same sort of cheerful dispassion.

† Consistent with its vision of an "attenuated immortality," the *Timaeus* suggests a non-dualist view of the human soul. Not only is our thinking a matter of the "marrow" inside our heads, but even our moral characters can be located there. (*Timaeus* 86c–d. See chapter 1, below.)

man has seriously devoted himself to the love of learning and to true wisdom, if he has exercised these aspects of himself above all, then there is absolutely no way that his thoughts can fail to be immortal, he can in no way fail to achieve this: constantly caring for his divine part as he does, keeping well-ordered the guiding spirit that lives within him, he must indeed be supremely happy. And there is but one way to care for anything, and that is to provide for it the nourishment and the motions that are proper to it. And the motions that have an affinity to the divine part within us are the thought and revolutions of the universe. (90a–b)

This notion of an immortality achieved within our (incurably) mortal lives is quite different from the possibility explored in the *Phaedo*. It is less Christian and more Greek. Just as the pre-philosophical Ethos of the Extraordinary had it, it is what we make of our lives in the short time that we're given that alone can expand our lives—not outward into everlasting time but still into something extraordinary and "godlike," and this is all the immortality that we mortals can know. Only it's not in the "auditory renown" of widespread *kleos* that we can achieve this form of immortality. It's by having, while we still live, our lives infused with infinity, our finitude "infinitized" by the vastness of beauty outside ourselves, allowing our love for it to overtake and dim even our love for ourselves. Outsized egos—even when attached to outsized intellects—are inconsistent with the life worth living as Plato envisions it.

Plato states in the *Timaeus* that it is only the very few who can achieve this kind of life. And so, to the extent that the best life is conceived in these terms, the many are excluded. "And of true belief, it must be said all men have a share, but of understanding, only the gods and a small group of people do." Understanding consists in seeing "the best reason," which is, for him, mathematical in essence. Using the mathematics of his day, he makes a stab in the *Timaeus* at offering such best reasons, though fully cognizant that better mathematics, and so better reasons, most probably lie in the future. Plato confides that he'll be happy to be trumped by the superior reasons offered by thinkers of the future mathematics—why else did he gather the best mathematicians of his day to his Academy?—describing their "victory" as that of his friends and not his enemies (*Timaeus* 54a). And again we remember how Socrates is set aside in the *Sophist* and the *Statesman* so that younger thinkers may carry the process forward.

But then is the good life to be attained only by mathematical physicists? (I can imagine certain philosophy-jeerers perceptibly warming to Plato.) Is that "small group of people" so very exclusive? Not quite. Any of us who allow our circumscribed earthly existence to be opened up to the vast and beautiful reaches of *all that is not ourselves*—another word for which is "reality"—is among the small group of people Plato has in mind. The means for opening to the infinite doesn't only come in the form of pursuing—or appreciating—mathematical physics, not even for Plato. He mentions, for example, music as holding the power to strengthen our affinity with the True-the Beautiful-the Good and so ethically reforming us. "And harmony, whose movements are akin to the orbits within our souls, is a gift of the Muses, if our dealings with them are guided by understanding, not for irrational pleasure, for which people nowadays seem to make use of it, but to serve as an ally in the fight to bring order in any orbit in our souls that has become unharmonized, and make it concordant with itself. Rhythm, too, has likewise been given to us by Muses for the same purpose, to assist us. For with most of us our condition is such that we have lost all sense of measure, and are lacking in grace" (*Timaeus* 47d–e).

Beautiful language, too, he says in the *Timaeus,* can carry us out of ourselves, providing the arrangements of words that, echoing with the harmonies of music, echo with the harmonies of the infinite (47c–d). This homage, brief as it is, to the transcendent powers of musical language offers the way for poetry to find its way back into Plato's city of reason, a re-entry for which Plato confessed himself hopeful (*Republic* 607d). It's not irrelevant that Percy Bysshe Shelley produced a stunning translation of the *Symposium,* nor that his friend, John Keats, would compose the immortal lines identifying beauty and truth, nor that many poets should have felt themselves directly addressed by Plato. Poetry that brings us in contact with the vastness beyond us—cracking us open to it and letting infinity seep in to expand our finitude in knowledge and in love—receives the Platonic seal of approval. Like mathematics and music and cosmology and philosophy, poetry, too, can "infinitize" us, granting us what immortality there is to be had in this mortal life. And all those who vibrate in harmony to language that itself vibrates to the harmonies of the infinite are entitled to inclusion among the "small group of people."

We can read into this passage an explanation for why Plato wrote as he did, lavishing his own soaring literary talents on the philosophical writings he left for us, however deep his misgivings concerning the linguistic enchantments that can override language's all-important role, which is to state the truth. Beauty, for Plato, always has a leading cognitive role in guiding us to the truth and in allowing the truth to work us over; and philosophy, in trying to impose the beauty of the infinite on our being, should therefore strive to be as beautiful as it can. And so it was that Plato wrote the works of art he did, allowing the poet in him to escape and soar. (But to a human-oriented art, no matter how great, that wrings the pity and the terror from our incurable finitude, Plato can't be reconciled.)

But what of Socrates' own possibilities for having been "infinitized"? Did he have what, according to Plato, it took? There's no indication that he was on fire with the beauty of mathematics or music or cosmology. And he certainly wasn't a poet. At one point in the *Phaedo,* Plato has Socrates explain—lots of people are asking, reports Cebes—why he's suddenly writing poetry in jail, setting the fables of Aesop to verse. Socrates explains how he's had, throughout his life, a recurring dream in which he was enjoined to "practice and cultivate the arts." He'd assumed that the dream was just encouraging him to keep on doing what he was doing, namely practicing the art of philosophy, but now that his life is about to end he's been worried that maybe the dream was exhorting him to "practice this popular art . . . and compose poetry. I thought it safer not to leave here until I had satisfied my conscience by writing poems in obedience to the dream," (60e–61b).

Now who knows whether Socrates actually undertook what sounds like a pretty lame attempt at becoming a poet in the last month of his life? I'm the mother of a professional poet, and I know what goes into the making of such a creature. You might as well try to become a mathematician or a cosmologist in the last thirty days of your life. And Plato, in whom the wings of a poet guide the flight of the philosopher, would have known this as well as anybody.

So Socrates was no poet, either. And yet, he did undeniably belong, in Plato's eyes, to "the small group of people" whose lives throw them open to the infinite, with thoughts that, no matter how tethered to a mortal being they are, can't "fail to be immortal, he can in no way fail to achieve this: constantly caring for his divine part as he does, keeping well-ordered the guiding spirit that lives within him, he must indeed be supremely happy." That was the picture that Socrates presented, not only to Plato but also to many of his contemporaries. He presented a picture of a man who, though constantly confessing his own ignorance, yet seemed to have come by a mysterious knowledge of how to live. The way in which Socrates combined his godlike certainty with his human confusion was a paradox that held great power. Socrates stood before Plato, wreathed in implications.

And nothing more powerfully convinced Plato of the implicative power of Socrates than the uncompromising stance he took in the summer of 399 on behalf of the philosophical project. It was that stance that had made the Athenian jury draw the conclusion: *Socrates must die.*

Socrates had always been a performer, and he might well have put on the performance of a lifetime on that day. We'd be naive to think that Plato, in the *Apology*, recorded that performance as a journalist might. Still, even if what we're getting from the *Apology* is what Socrates' stance had meant to Plato, then that's more than enough. Socrates' performance that day convinced Plato that Socrates would continue to perform for as long as there are people who care about philosophy. As far as Plato was concerned, Socrates had been infinitized.

WHAT WOULD ACHILLES DO?

We see him in all his maddening glory, making a fool of the obscure young poet Meletus, whom he effortlessly reduces to absurdity, using the dialectical tactic he's perfected throughout his life.

Yes, Meletus confirms, Socrates is guilty of atheism, that is, of believing in no gods; and yes, Meletus also confirms, Socrates is guilty of introducing new gods, unrecognized by the state. So he believes in no gods while at the same time he believes in the gods. Socrates likens the blatant inconsistency to playing like a child (27a).

And yes, Meletus responds, everybody in Athens, every last voting citizen, has a salubrious influence on the youth of the city, and, yes, it is only Socrates who harms them, a situation so preposterous that Socrates mocks it as "a singular dispensation of fortune for our young people" (25b).

Meletus is dispatched so easily by Socrates because, says Socrates, he is not sincere in the matters he professes to care about; he demonstrates this insincerity by his failure to think through the implications of his statements. "You see, Meletus, that you are tongue-tied and cannot answer. Do you not feel that this is discreditable, and a sufficient proof in itself of what I said, that you have no interest in the subject?"

In other words, Socrates is accusing Meletus, unknown poet, of indeed being an artist: a bullshit-artist.

The term "bullshit" has been usefully incorporated into polite philosophical parlance through the work of the contemporary American philosopher Harry Frankfurt, who published a little philosophical tract called *On Bullshit* in 2005. The work had first been published as an article in a scholarly journal, *Raritan,* in which form I had been assigning it for years to my students in Introduction to Philosophy, hoping to disabuse them at the onset of the expectations with which many entered the class. The essay was then republished in a collection of Frankfurt's called *The Importance of What We Care About: Philosophical Essays.* Finally, it was republished as its own little book, in which, its third incarnation, it achieved a surprising state of bestsellerdom, the somewhat bemused philosopher even making an appearance on *The Daily Show with Jon Stewart.*

Frankfurt's book opens with this observation: "One of the most salient features of our culture is that there is so much bullshit." Frankfurt went on to offer a theory of bullshit, homing in on its essential features with the analytic precision for which Anglo-American philosophy is justifiably celebrated.

Bullshit must be pried apart from related concepts, such as humbug and, most importantly, lying. Both liars and bullshitters have a problematic relationship to the truth. Both liars and bullshitters misrepresent their relationship to the truth, but there are essential differences between them. A liar may—because he himself is confused about the truth—end up saying something true, but his intention is to say something that is *not* true. His intention is to induce a false belief in the person to whom he lies. A liar, therefore, is someone who is keeping track of the truth, or in any case trying to, for the purpose of deceiving.

A bullshitter, too, may end up saying something that is true. But unlike the liar, the bullshitter is not trying to keep track of what is true. The truth-conditions of his statements, their correspondence to the facts they purport to convey, are irrelevant to his motives for saying what he does. His motive for saying what he does is not to induce a false belief, as the liar's is. He is not intent on deceiving with regard to the content of his assertion. His motive is to deceive with regard to his own bullshitting self, passing himself off as someone who cares about the truth when he does not.

Upon further clarification of the concept of bullshit, Frankfurt ventures the normative conclusion that bullshitting is more pernicious than lying. Acts of lying are, in the typical (non-pathological) liar, localized events; whereas a tendency to bullshit affects a person globally. Frankfurt closes his essay with this judgment:

> Both in lying and in telling the truth people are guided by their beliefs concerning the way things are. These guide them as they endeavor either to describe the world correctly or to describe it deceitfully. For this reason, telling lies does not tend to unfit a person for telling the truth in the same way that bullshitting tends to. . . . The bullshitter ignores these demands altogether. He does not reject the authority of the truth, as the liar does, and oppose himself to it. He pays no attention to it at all. By virtue of this, bullshit is a greater enemy of the truth than lies are.

Surveying the history of Western philosophy, the impression is that not all philosophers have shared Frankfurt's moral repugnance to bullshit. But it is undeniable that many have, and Socrates was one. In fact, it is surprising that it has taken philosophers so long to get around to analyzing the concept, given that the offended reaction to bullshit helped to fertilize the original grounds for the field. Socrates, as he comes across in the *Apology*, would have loved Frankfurt's essay, right down to the normative conclusion that bullshit is a greater offense to truth than lies are.

He would also have assented to the essay's opening statement. One can well imagine Socrates turning to the assembled crowd on that summer's day in 399 and declaring, "One of the most salient features of our culture is that there is so much bullshit." That would have been exactly in the spirit of Socrates' performance at his trial, as he went out of his way to offend Athenian normative sensibilities, a large part of which was bound up in its sense of exceptionalism. It had been Pericles' genius to strengthen and spread the Athenian sense of themselves as collectively extraordinary. Socrates seemed intent on getting in a last-ditch effort at undermining such a sense.

His jurors were citizens chosen by lots who were drawn from every segment of Athenian society. It's probable that most of them would have been farmers or shopkeepers—in any case not aristocrats, since the aristocracy composed the smallest proportion of the citizenry. But that doesn't mean that the jurors, as well as the many who gathered to watch, wouldn't be offended by the suggestion that they didn't partake of a certain portion of superiority just in virtue of being Athenian citizens. In countries in which nationalist exceptionalism still flourishes today it is by no means the most privileged classes who feel the most strongly that their sense of superior worth derives from their citizenship.

Socrates was able to trip up a Meletus who could not articulate his sense that Socrates was, somehow or other, impious. But certainly Meletus was justified—Socrates' own performance in the *Apology* supplies plenty of justification—that Socrates was attacking the Athenian normative framework—its structure of values—at a fundamental level. It's not only Socrates' performance at his trial, but his performance throughout his whole life that shows his dissent from the Athenian value system. Is it any wonder that Meletus feels the normative affront as impiety? "Recall how closely a Greek community's sense of its own identity and stability is bound up with its religious observances and the myths that support them. If Socrates rejects the city's religion, he attacks the city. Conversely, if he says the city has got its public and private life all wrong, he attacks its religion; for its life and its religion are inseparable."*

And then enormous changes had taken place in Athens since the days of Pericles, which would put the assumptions that went into its sense of identity under intense threat. Athens' claim to extraordinariness, especially now that its empire was no more, its great wealth squandered in hubristic overreaching, lay in its claim to cleverness, its rationalism, as demonstrated by the amnesty, as well as its continued celebrity in the rhetorical arts. But here was one of Athens' most famous talkers, an endlessly inventive channeler of disputatious energy, who was intent on turning his talent for talk *against* the city's sense of itself. Throughout his speech he is careful to couch his claims in terms of the rhetoric of the Athenian law courts and the civic values that prevailed

*M. F. Burnyeat, "The Impiety of Socrates," in *The Trial and Execution of Socrates*, ed. Thomas C. Brickhouse and Nicholas D. Smith (New York: Oxford University Press), p. 138.

there, while at the same time subverting them and reversing them, playing fast and loose with conventions that were meant to pin everything down.

Socrates is not overly concerned in his trial to go after Meletus, an insignificant person trying to jump the queue in order to become a footnote to history. Socrates wields his dialectic prowess to swat his prime accuser away as if he were an annoying gnat.

Well, actually, it's he himself whom he compares to an insect, telling the jurors, before they have yet put to a vote the question of his innocence: "[I]f you kill me you will not easily find another like me. I was attached to this city by the god—though it seems a ridiculous thing to say—as upon a great and noble horse which was somewhat sluggish because of its size and needed to be stirred by a kind of gadfly. . . . I never cease to arouse each and every one of you, to persuade and reproach you all day long and everywhere I find myself in your company" (30e–31a). It hardly seems a self-aggrandizing description that Socrates is offering here, a bothersome bug buzzing incessantly around the city. But the real insult is delivered to the *polis* of Athens, described as a great lazy horse, dozing in its complacent moral obtuseness. Socrates knows full well how profoundly disturbing he is being, knows that, in challenging the normative preconceptions of his city, he is outraging those who are about to vote on whether he lives or dies.

He could not be more explicit in denying the fundamental assumptions of those normative fortifications, the politicizing of *aretē*. *Aretē*, as Socrates conceives it, requires individuals' thrashing through the moral complexities as honestly as they know how, taking responsibility for their own beliefs and actions, not picking them up ready-made in some public place, like pottery you can buy in a stall of the agora. This kind of thrashing has been what he's been trying his whole life to provoke, which means smashing the assumptions that seem to settle such matters, making his fellow citizens think anew about what would make their lives worth living.

He actually, outrageously, states that engagement in Athenian politics—no matter whether under the democrats or the oligarchs—is *inconsistent* with virtue: "The true champion of justice," he tells them, "if he intends to survive even for a short time, must necessarily confine himself to private life and leave politics alone" (31e). And again: "Do you suppose I should have lived as long as I have if I had moved in the sphere of public life, and conducting myself in that sphere like an honorable man, had always upheld the cause of right, and conscientiously set this end above all other things? Not by a very long way, gentlemen; neither would any other man" (32e–33a).*

Eschew the public life of the Athenian *polis in order* to be just? Socrates is choosing his words in the same spirit with which the late Christopher Hitchens chose the title of his book *God Is Not Great: Why Religion Poisons Everything*. They are the words of a professional provocateur. And provoke he does. Several times, they try to shout him down, to judge by Socrates' repeated admonishments to the crowd to let him speak his piece: "As I said before, gentlemen, please do not interrupt" (21a); "Do not create a disturbance, gentlemen, but abide by my request not to cry out at what I say but to listen, for I think it will be to your advantage to listen, and I am about to say other things at which you will perhaps cry out. By no means do this" (30c).

Many who have looked back on that day in 399, trying to deconstruct its drama, have thought first and foremost of politics. Of course, it *had* to be politics, because what else could have been so important as to demand the life of the man? The serious designs are always the political designs, and that's where the savvy minds know to look. Socrates is anti-democratic, and that's why he must die. Socrates was soft on Sparta, and that's why he must die. This is political payback for Alcibiades or Critias or both, and that's why he must die. The irony of these political readings is excruciating. In prioritizing the political, such interpreters make themselves part of the Athenian crowd that Socrates is trying so desperately for one last time to arouse from out of their normative stupor.

And he keeps hammering it home. He doesn't hold back from quoting the Delphic oracle, who had declared that no man was wiser than Socrates, reporting that he, too, had been dubious when he first heard this report, since he was so keenly aware of his ignorance. And so, in order to prove the oracle wrong, he had gone about questioning the citizens of Athens, starting with the most prominent citizens first, the politicians, the sophists, the poets. All of them satisfied the conditions for the concept that Harry Frankfurt would analyze several millennia

---

*The translations in this paragraph of the *Apology* are by Hugh Tredennick, from Princeton University Press, 1961.

later. Socrates concluded that the oracle had bestowed her kind words on him because he alone didn't pretend to a knowledge he didn't have. Or to put it in Frankfurtian terms: Socrates cared, as his fellow Athenians did not, about what the truth-conditions of his statements were, even if, for a great number of those statements, he couldn't determine whether those truth-conditions held or not.

And then he simply comes out and says it, at a point when he's been judged guilty and is supposed to be negotiating a penalty to counterpose against the death penalty that Meletus is demanding. He's expected to offer to go into exile or to pay a heavy fine. And this is the juncture in the proceedings when he makes the most provocative statement a man in his position could make: A life in which he would cease questioning the value of his society would be a life not worth living, implying that the people about to judge him are living exactly such lives. "If I say that it is impossible for me to keep quiet because that means disobeying the god [Apollo, the god of the Delphic oracle], you will not believe me and will think I am being ironical. On the other hand, if I say that it is the greatest good for a man to discuss *aretē* every day and those other things about which you hear me conversing and testing myself and others, for the unexamined life is not worth living for men, you will believe me even less" (37e–38a).

Not worth living? What harsher words could he have used to those who hold the fate of his life in their hands? What more horrifying counter-condemnation could he have voiced to his condemners? *Not worth living?* This is to speak the unspeakable right into the open, to rip the bandages off the shame-provoking wound, like the legendary open wound of Philoctetes, the Greek who was left behind on an island by his comrades-in-arms on their way to Troy because of a festering wound that reeked to high heaven.

But it's not Philoctetes who's called out by name during Socrates' defense, but rather Achilles (28c–d). This is the hero with whom Socrates draws a comparison. Achilles was given the chance to choose a long and ordinary life or a short and extraordinary one, and chose the latter. Achilles is the poster boy for the Ethos of the Extraordinary, the very apotheosis of the hero. And at least one person who hears Socrates that day fully accepts the comparison.

Socrates assumes the shape of a hero for Plato, even though he hadn't achieved the knowledge at which his examined life had been aimed. Plato doesn't think Socrates had achieved that knowledge, since for him the answers will require the metaphysical conception of the True-the Beautiful-the Good in terms of which human virtue can be formulated. Plato's approach to the question of *aretē* will require him to raise up the whole submerged continent of philosophy. Nevertheless, the way Socrates had lived—and certainly the choice he made on that day in 399—lifts Socrates, for Plato, into the sphere of the extraordinary.

After he has been voted guilty by his fellow Athenians, Socrates considers what penalty he ought to request. "I have deliberately not led a quiet life—*ouk hēsychian ēgon*" (36b), and by pronouncing *hēsychian* he is distancing himself from the Athenian elite, many of whom maintained a tradition of quietism either because they were afraid of being misunderstood or they had genuine oligarch sympathies.* "I have neglected what occupies most people: wealth, household affairs, the position of general or public orator or the other offices, the political clubs and factions that exist in the city. I thought myself too honest to survive if I occupied myself with those things." He cannot stop rubbing it in. All the political activity and intrigues by means of which people make themselves important in the city run counter to his scruples. And he has singled out two hallmarks of the elite, including their factions *(staseis)* and secret societies *(sunōmosia).*† ‡

And now he moves on to another outrageous flouting of Athenian sensibilities: "What do I deserve for being such a man? Some good, men of Athens, if I must truly make an assessment according to my deserts, and something suitable. What is suitable for a poor benefactor *(anēr penēs eurgetēs)* who needs leisure to exhort you?" By emphasizing his poverty and yet claiming himself as a benefactor, he is again disrupting Athenian expectations about the typical benefactor, a wealthy member of the elite. And then comes the coup de grâce: "Nothing is more suitable, gentlemen, than for such a man to be fed in the Pryta-

---

* See L. B. Carter, *The Quiet Athenian* (Oxford: Oxford University Press, 1986).

† For example, the citizens who conspired to mutilate the herms the night before the fleet sailed to Sicily belonged to a *sunōmosia*.

‡ There are other places in which Plato has Socrates flaunt his non-involvement in Athenian politics, for example, in the *Gorgias*, "I'm not one of the politicians. Last year I was elected to the Council by lot, and when our tribe was presiding and I had to call for a vote, I came in for a laugh. I didn't know how to do it. . . . I do know how to produce one witness to whatever I'm saying, and that's the man I'm having a discussion with. The majority I disregard" (473e–474a).

neum, much more suitable for him than for any one of you who has won a victory at Olympia with a pair or a team of horses. The Olympian victor makes you think yourself happy. I make you be happy. Besides he does not need food, but I do. So if I must make a just assessment of what I deserve, I assess it as this: free meals in the Prytaneum" (36e). As one classicist put it to me in discussing this passage, "Plato/Socrates is so shrewd in his manipulation of Athenian values. There is a subtle process of transvaluation at work in every sentence."

In the midst of that transvaluation, the most famous slogan of philosophy emerges: *the unexamined life is not worth living.* Some—like Cheryl, Plato's media escort at the Googleplex—might bristle at a sensed elitism. But Socrates' statement is only elitist if one assumes that only the few have it in them to examine their lives. The democracy that became again, in the eighteenth century, the most daring political experiment on earth embodies the hope that we many have it in us as well.* Instead of an indictment of his city's daring experiment in democracy, Socrates' statement can be read as putting forth the stringent condition that alone would allow democracy to flourish. Perhaps Plato did eventually come to a viewpoint that excluded the many from being capable of thinking through their lives in the light of the True-the Beautiful-the Good; but the entire life of Socrates, including that day of his trial as it's presented by Plato, is evidence of the fact that Socrates continued to cling to hopes for us many. He says that, if only he had more than the one day that Athens allows for such cases, he knows that he could have prevailed in convincing his jurors of his vision of the examined life.

The unexamined life is not worth living. What is supposed to follow from this? Is the examined life merely necessary for a life worth living, or is it sufficient as well? And is it even really necessary? Could you live a life worth living, the life of *aretē,* by fortunate accident—that is without living the examined life? Socrates is saying no. The examined life is, at the very least, a prerequisite for living a worthwhile life. *Aretē* can't happen by happy accident, no more than knowledge can. *Aretē,* like knowledge, requires an accounting, a *logos.* This is a theme that will resonate through all of Plato's dialogues. There is no such thing

---

*Of course, it would take several social and political revolutions to expand "the many" so as to include such groups as women and the impoverished. And even now this issue of *just how many?* isn't altogether closed, even in the United States.

as just happening to get it right—not when it comes to beliefs and not when it comes to actions. Having the good fortune to be born an Athenian, or any other such contingency—including one's religious affiliation, as Plato powerfully argues in the *Euthyphro*—does not add up to an accounting, a justification, a *logos*. And without an accounting there can no more be virtue than there can be knowledge. Indoctrination, even should the doctrines be happily right, can't produce the life that is most worth living.

At the end of the *Republic*, Plato presents yet another myth of his own devising, the Myth of Er. Plato has Socrates telling the myth to Glaucon, whose sophisticated game-theoretic reworking of politicized *aretē* the preceding dialogue has subverted with its image of the *kallipolis*. Now he tells a story of a warrior, Er, resuscitated after twelve days of seeming death, about to be burned on the funeral pyre, though his corpse remains "quite fresh." Er reports what he has experienced while apparently dead, rather a bit like those near-death experiences that have become familiar to us from our improved medical technologies able to revive comatose patients. Er leaves out any mention of a bright light or a black tunnel. Instead he tells of a "marvelous place, where there were two adjacent openings in the earth, and opposite and above them two others in the heavens, and between them judges sat. These, having rendered their judgment, ordered the just to go upwards into the heavens through the door on the right, with signs of the judgment attached to their chests, and the unjust to travel downward through the opening on the left, with signs of all their deeds on their backs. When Er himself came forward, they told him that he was to be a messenger for human beings about the things that were there, and that he was to listen and look at everything in the place" (614b–d).

The sojourns upward and downward are only the first phase. After the souls of the departed have spent their allotted thousand years in the places to which they'd been sent, they return and journey to the center of the universe. Plato pauses for a vision of this unearthly beautiful center of the universe. Here the mathematics of perfect proportion and harmony, which lies at the core of the True-the Beautiful-the Good, becomes audible in the music of the eight spheres, spinning one within the other and turned by the spindle of Necessity. "And up above and on each of the rims of the circles stood a Siren, who accompanied its revolution, uttering a single sound, one single note. And the concord of the eight notes produced a single harmony" (617b). In

1619, Johannes Kepler, who, like so many of the new physicists of the seventeenth century, found his inspiration in Plato, will publish his *Harmonices Mundi*. Kepler had discovered that the difference between the maximum and minimum angular speeds of a planet approximates a harmonic proportion, and with this discovery he tried to provide a physical reality to Plato's poetry of the "music of the spheres."

But eventually Plato gets back to the point of the myth. A prophet appears and has the assembly of souls choose their new lives, having first drawn lots to decide the order of the choosing. And the outcome is not as one would predict based simply on the experiences these souls have had in the thousand years they have spent above or below. The very first who gets to pick—who has the greatest variety of options— chooses, "out of his folly and his greed" (619c), the life of a tyrant.

> He was one of those who had come down from heaven, having lived his previous life under an orderly constitution, where he had participated in virtue through habit and without philosophy. Broadly speaking, indeed, most of those who were caught out in this way were souls who had come down from heaven and who were untrained in suffering. (619c)

An interchange of good lives and bad lives. Neither the lives they have lived nor the supernatural rewards or punishments they have endured edify them so that they're prepared to make good choices. Many of those who suffered in their earthly lives automatically choose lives that are diametrically opposed. Orpheus, hating women because of how he had died, chooses to be a swan so that he won't have to be born of a woman (620a). Agamemnon, reacting to his last life with a hatred of all men, chooses to be an eagle (620c). And Odysseus, who is the last to go, so recoils from his last life, filled with enough *kleos*-worthy adventures to fill yet another Homeric epic, searches around until he finds a perfectly ordinary life and says that even had he chosen first, this ordinariness is what he would have chosen. Plato is hardly endorsing ordinariness here; there is nothing ordinary about the kind of life that he has been at pains to depict throughout the *Republic* as the one that fully realizes *aretē*. In fact, just before introducing the closing Myth of Er, he describes a person who lives such a life—chosen for his mettle, intelligence, and love of truth, and then vigorously trained so as to be able to absorb into himself the True-the Beautiful-the Good—as

one "who makes himself as much like a god as a human can" (613a).
That's hardly ordinary. Odysseus, like all the others who have just been
named, is making an unreflective choice, a mechanical action of recoiling from the circumstances of his former life, and Er, watching the
souls making their choices, says it makes for "a sight worth seeing,
since it was pitiful, funny, and surprising to watch" (619e–620).

All of the circumstances that are determining the souls' choices are,
from the point of view of logos-supported *aretē,* mere accidents of happenstance, and the moral character of a person cannot be a matter of
mere accidents of happenstance. This would be a point developed with
great force by Immanuel Kant. There can be nothing accidental in a
person's being moral. Whether people are moral must be something
within their control—a matter of their will, is the way Kant put it—
and so a matter over which they have choice and can be held accountable. And to be held accountable they must be prepared to offer an
account for why they behaved as they did. That's the sense of free will
that matters. If I am a respectable specimen of humanity because I
happen to have enjoyed the good fortune of having been born into
a nice family, which inculcated the habits of behaving well, then,
although well-behaved, I am not a virtuous moral agent. Change my
circumstances—put me into a family of cheaters and exploiters—and
just watch what I'll become. That is how the returned souls in the Myth
of Er behave. They are morally hostage to their circumstances and not
moral agents. Not even the thousand years spent either in reward or
punishment could transform them into moral agents.

If there really is to be such a thing as moral accountability, believed
Plato and Kant, then a person's habits of character (our word "ethics"
derives from the Greek *ethōs,* which, as has already been mentioned,
means "habit") must be an intimate part of that person, arrived at *deliberately,* reflected on and chosen and becoming the habitual manifesta-
tion of moral character. One must be in a position to *own* one's ethics.
Kant argues that only a person making the moral choices for the right
moral reasons can be regarded as morally worthy, which, of course,
implies that there are right moral reasons, and that it lies within our
power both to comprehend their rightness and to act upon them
because of their rightness: to let these reasons be the determining factor in our behavior.

All of these claims can be read into Plato's Myth of Er. It is a plea for
thinking one's way out of the contingent circumstances of one's life;

even divinely imposed circumstances count in this myth as merely contingent and offer nothing by way of the knowledge we need to live as we should—which is a resounding rejoinder to those who still threaten us, even still, that without the fear of heaven and hell nobody would behave as they should. Plato's souls actually return from their thousand years of either above or below in no better position to choose the life they ought to live.

> Now it seems that it is here, Glaucon, that a human being faces the greatest danger of all. And because of this, each of us must neglect all other subjects and be most concerned to seek out and learn those that will enable him to distinguish the good life from the bad and always to make the best choice possible in every situation. He should think over all the things we have mentioned and how they jointly and severally determine what the virtuous life is like. That way he will know what the good and bad effects of beauty are when it is mixed with wealth, poverty, and a particular state of the soul. He will know the effects of high or low birth, private life or ruling office, physical strength or weakness, ease or difficulty in learning and all the things that are naturally part of the soul or are acquired, and he will know what they achieve when mixed with one another. And from all this he will be able, by considering the nature of the soul, to reason out which life is better and which worse and to choose accordingly, call a life worse if it leads the soul to become more unjust, better if it leads the soul to become more just, and ignoring everything else. We have seen that this is the best way to choose, whether in life or death. (618c–e)

Plato is unpacking for us, in his Myth of Er, all that he heard resonating that day in Socrates' rousing cry to the Athenians that the unexamined life is not worth living; just as Kant will later unpack for us the implications of Plato's Myth of Er, and as others will unpack for us the implications of Kant's categorical imperative, in a process that has been cumulative and ongoing, if also excruciatingly halting, with periods when the direction is altogether reversed.

Anyone who denies this ongoing process, arguing that such normative concerns as Socrates was urging can never make a difference and thus have produced no consequences in the way life has been lived over the centuries, has an affinity with the Athenians that summer day over two millennia ago, unwilling to give Socrates his due, shouting him down as he strived to be heard over their jeers.

PLATO ON CABLE NEWS

ROY McCOY: Okay, so they tell me you're a big deal in philosophy, Plato. I'm going to tell you up front—because that's the kind of guy I am, up-front—that I don't think much of philosophers.

PLATO: Many don't. The term attracts a wide range of reaction, from admiration to amusement to animadversion. Some people think philosophers are worthless, and others that they are worth everything in the world. Sometimes they take on the appearance of statesmen, and sometimes of sophists. Sometimes, too, they might give the impression that they're completely insane (*Sophist* 216c–d).

McCOY, *laughing:* Well, I'm going with that last one, just as long as you add that it's the kind of insane that makes right-thinking Americans the world over want to smack you upside the head.

PLATO: It comes sometimes to reactions even more violent. My friend, the best of men, was charged with the crime of doing nothing more than practicing philosophy as best as he knew how. He was found guilty and executed.

McCOY: Where was that, Texas?

PLATO: No. It was in Greece, though many years ago.

McCOY: I'm sorry to hear about your friend's ordeal, but I have to ask you a question: What was he doing to tick people off so much?

PLATO: It is a good question.

McCOY: Since you apparently don't know me, you don't know that I only ask the one kind of question, and that kind is good. So unless your friend was prosecuted under that military junta you Greeks had going on back there in the late sixties, early seventies—

PLATO: He was brought to trial by our democracy. And in fact it was by popular vote that he was condemned to die, even though the Delphic oracle had declared that there was no person who was wiser.

McCOY: Yeah, well, I have two Peabodys. Listen, not to disrespect the memory of your friend, but there's got to be more to the story than what you're saying. You're spinning, and I'm calling you on it. We don't call this the No Bull Bin for nothing. See, democracies don't go around bumping people off just for being annoying types who think they know better than everybody else, which, from what I can gather, is what you philosophers specialize in. In fact, I have to say, Plato, I haven't spoken to you two minutes and you're already beginning to irk me. But we live in a democracy that protects everyone's right to be a royal pain.

PLATO: Progress has been made.

McCOY: Debatable.

PLATO: I am impressed by the progress.

McCOY: Then maybe your initial expectations were too low.

PLATO: Maybe they were.

McCOY: You sure you're a philosopher? You seem a little too ready to change your mind. Or maybe you just don't have the courage of your convictions.

PLATO: I would prefer the courage of my questions.

McCOY: Okay, I'm getting the picture. The pixels are depixelating. Maximum questions, minimum convictions. I don't have to ask you on which side of the political divide you stand. I'm glad we got that one sniffed out early. Although really, what else should I have been expecting? Philosophy. It's one of those we-don't-have-anything-to-teach-so-we'll-just-lecture-you-on-and-on-about-our-own-moral-superiority subjects. Here's the setup in your so-called higher education. You've got your humanities on the one side, and your sciences on the other.

PLATO: I cannot see it that way. They must both be on the same side, or there is no knowledge at all.

McCOY: It's like I'm saying. Your humanities get off on saying there are no answers to anything, while your sciences get off on saying that they have the answers to everything. And you know what I say? I say a pox on both their houses.

PLATO: If it were truly as you describe, I would join you in your condemnation.

McCOY: Yeah, well, it is as I'm describing. The humanities say no one can know anything, the scientists say that they know everything. I've had some of the scientists here in the No Bull Bin when they were hawking their atheist books, and they made a poor showing of themselves. I haven't been able to get a single one of them to even explain to me how you get the tides to work without the Deity stepping in to keep it all going.

PLATO: The tides?

McCOY: Yeah, the tides. You must see plenty of tide action down your way in Greece. The tides come in and the tides go out, and the scientists can't explain it. There's never a miscommunication.

PLATO: Never a miscommunication among the scientists?

McCOY: No, among the tides. Not one scientist I've had on the show could explain to me how they seem to know all on their own when to go out and come back in. They do it day after day—

PLATO: In fact, twice a day, roughly every twelve hours—

McCOY: Nothing rough about it. It's like the Marines, a precision operation all the way. Still, at least they have something to show for themselves.

PLATO: The tides? The Marines?

McCOY: No, the scientists. Where do you even start, what with the smartphones and all the other whatsits and whichits that make it impossible for me to keep the attention of anybody on my staff for more than two minutes at a time? They're all ADD, but science provides the cure for that, too. There's no end to the benefits pouring out of science, whereas I don't see you philosophers putting any merchandise on the market.

PLATO: And when we speak of the *parerga*—

McCOY: *Parerga?* What's that? Some new smartphone?

PLATO: I meant only the by-products of science, among which let us not forget the computer—

McCOY: Yeah, I noticed you carry yours around like your blankie—

PLATO: —and the Internet, and Google's cloud with the vast amounts of information that it stores, even though in no particular place, and which we can summon, exactly what we need, to our own personal devices because of an algorithm, powerful and secret, but nevertheless the outpouring of human minds inflamed by the Promethean fire.

McCOY: Okay, let's not break out into strains of poetry here. We've established that science has got some useful stuff to show for itself.

PLATO: The technology is wondrous—

McCOY: Except when it breaks and you're on the phone for an hour and a half with someone whose first language is definitely not English—

PLATO: But it's not in such usefulness that science demonstrates its true worth, but rather in answering to the power in the soul that is of a nature to love truth and to do everything for its sake (*Republic* 527b–c; *Philebus* 58b–d).

McCOY, *laughing:* What, you want to praise useless knowledge?

PLATO: Yes.

McCOY: Is that the title of the book you want me to plug? *In Praise of Useless Knowledge?*

PLATO: No.

McCOY: Well, it's a good title, even if it's a stupid idea. Why would anyone want useless knowledge?

PLATO: I only meant that the most desirable knowledge isn't desirable because of its instrumental value in yielding such by-products as those of which we've just been speaking—

McCOY: Zero of which you've been able to show for yourself—

PLATO: But rather the most desirable knowledge is useful in the search for the beautiful and the good. Otherwise, if pursued for any other purpose, it is useless (*Republic* 351c).

McCOY, *laughing:* The beautiful and the good? How'd we suddenly get to the beautiful and the good? One minute we're speaking about the tides coming in and going out, the next you're off and running on about the beautiful and the good.

PLATO: Because the reason that will most please you, in your search for the best explanation for the tides, is the same reason that most pleases the world itself. You and the cosmos are as one.

McCOY: I'm as one with the cosmos?

PLATO: In your search to understand it.

McCOY, *laughing:* There are pinheads who accuse me of having an inflated ego, but I'm not about to equate myself with the whole frigging cosmos. I'm not that delusional. Are you?

PLATO: It was not the assertion of an equation between you and the cosmos, but rather an affinity.

McCOY: You mean that the cosmos likes me? Okay, I'll go with that. That's my take on the situation.

PLATO: Here is what I had in mind: You, in your search to understand the cosmos, are most satisfied with the most beautiful reason; and so, too, is the cosmos. You will find the most beautiful reason to be the one that most elegantly does justice to your notion of intelligibility, which form is given by none other than mathematics; and so, too, will the cosmos. In that sense, you and the cosmos have an affinity with each other, in your recognition of the beautiful. The force of the good has taken refuge in an alliance with the nature of the beautiful. For measure and proportion manifest themselves in all areas as beauty and virtue (*Philebus* 64e).

McCOY: Listen, I've already warned you once about that flowery stuff. We tell it like it is on this network. And I don't know where to even begin to attack all the nonsense you're trying to spin here. First off, the last place I'd look for a beautiful reason is in mathematics. I hate math, except when my accountant uses it to give me good news. In the second place, the cosmos isn't looking for its reasons in math either, because the cosmos isn't looking for any reasons at all. The cosmos doesn't do any looking. We're the ones who do the looking.

PLATO: Of course, the cosmos doesn't look for any reasons. The cosmos has already found the best reason, which is what has made it exactly what it is. The cosmos is full of reasons, which are the very ones you seek. It is in this sense that the saying "all things are full of gods" is entirely right and sufficient (*Laws* 991d).

McCOY: Full of gods! Oh boy. Most of you academics we get passing through here aren't willing to give me one God. You want to give me a whole cosmos full of them. Is it so hard for you big thinkers to get the number right? It's either one, or it's blasphemy.

PLATO: I meant no blasphemy. In fact, it seems to me that, in speaking now of the forms with which the world becomes intelligible, both to itself and to us, we speak of what things a person is to learn about reverence towards the gods and how he is to learn them. When you hear what it is, you will find it strange (*Laws* 989e).

McCOY, *drily:* No doubt.

PLATO: I say its name is astronomy (ibid.).

McCOY: Listen, if you somehow think that going out on a starlit night and staring up at the Big Dipper beats going down on your knees and praying to the Deity, then okay, what can I say? People find all sorts of ways of getting out of going to church. But let's just clear up this confusion about any so-called gods. It's *God,* singular, not plural. Just because something is good that doesn't mean that you improve the situation by bringing in multiples of the thing. You don't go making the situation any better by multiplying gods. That's kind of the biggest no-no there is. Have we got that squared away?

PLATO: That there is a unity that binds together all the diverse best reasons: who would want to dispute this? The person who learns in the right way always fixes his or her eye, as we say, on unity. Every diagram and complex system of numbers, and every structure of harmony and the uniform pattern of the processes of nature, are in the end, a single thing applying to all these phenomena. To one who studies these subjects in this way, there will be revealed a single natural bond that links them all. But anyone who is going to pursue these studies in any other way must "call on good fortune for help," as we also say. For without them, no one in cities will ever become happy. This is the right way, this is the upbringing, these are the studies. Whether they are difficult, whether they are easy, this is the way we must proceed (*Epinomis* 991e).[*]

McCOY: Sounds to me like you want to set up your own No Bull Bin, buddy. Only you wouldn't let in anybody but the math nerds. I'd like to see them running our cities when they can't even manage to put on two matching socks in the morning. What was that some sort of grand theory of everything that you were spinning there a moment ago.

PLATO: There is always the vision of such a unity guiding those who think rightly on these subjects. The beauty toward which we are led could only be one that would unify all.

McCOY: Some kind of $E=mc^2$ magic bullet, only stronger?

PLATO: And even deeper and even more beautiful, the expression of the elegance that the cosmos itself gives as its own reason for being whatever way it is. In this way, the world becomes intelligible to us when we grasp how the world is intelligible to itself.

McCOY: Yeah, only here's the difference, pal. The world can be intelligible to me but that doesn't mean it's intelligible to itself, because— and here's the Network News Alert—I'm an intelligent being and that's how I come to know so much, but the cosmos knows squat. There's intelligence behind it, all right, but not *in* it. Got that? *Behind,* not *in.* You seem to be a guy who puts a lot of stock in math stuff. Am I right?

PLATO: You are right.

---

[*] *Epinomis* is, as the name implies, an addition to the *Laws,* which is *Nomoi* in Greek. *Epinomis* features the same three old men reconvening at some unspecified time after their first conversation in the *Laws.* The words Plato speaks above, from "Every diagram and complex system of numbers" until the end of the paragraph, are verbatim. Translation by Richard D. McKirahan Jr. in Hackett 1997.

McCOY: Well, then, if the cosmos is so smart, how come it doesn't know any math? It doesn't even know how to count!

PLATO: It knows that the very thing that it teaches us and that we learn is number and how to count. If it did not know this, it would be the least intelligent thing of all. It would really not "know itself," as the proverb goes (*Laws* 988b).

McCOY: But that's exactly what I'm trying to tell you! It doesn't know itself! It's just dopey mindless matter. Just because it's huge doesn't mean it's smart. Just take a look at the federal government.

PLATO: Is it because these physical processes are unalterable that you deny them intelligible reasons? Do you see intelligence in humans—

McCOY: The few who aren't pinheads—

PLATO: —but not in the cosmos because humans are unpredictable, whereas the cosmos isn't? But it is their very intelligibility that renders the workings of the cosmos so unalterable. Not even adamant could ever be mightier or more unalterable (*Laws* 982c).

McCOY: You actually got an institution of higher learning to fund you for spinning this stuff?

PLATO: I founded the Academy.

McCOY: That explains a lot. Okay, let me see if I've got this straight. You're saying that the cosmos, by which you mean stuff like the tides coming in and going out in military formation, and the sun rising and setting so that you can set your clocks to it, and rain becoming snow if the temperature falls below freezing, and all the rest of it, you're say-ing that these things, these dumb physical things, actually know what they're doing? You're saying I could walk up to the tides and ask them "So how the heck are you doing that?" and they'd be able to tell me?

PLATO: Is that not what our men and women of science have always been doing, pressing the processes of nature so that they yield up the reasons why they are as they are?

McCOY: But they're just dumb physical things that are moving around! And I don't just mean the scientists here, but the things the scientists are studying, the tides and what all. And any scientists who question the tides and think they hear them answering back ought to be medicated.

PLATO: Their movements speak the language of mathematics.[*]

---

[*] It was a Galileo inspired by Plato (and, most particularly, by the *Timaeus*, which was the one dialogue of Plato's still read in the centuries of Scholasticism) who wrote: "Philosophy [nature] is written in that great book which ever is before our eyes—I mean the universe—but

MCCOY: No, pal, *we* speak the language of mathematics. Or at least some of us do. Others of us would rather go get a colonoscopy. The tides aren't saying a thing. They just keep going in and out, never a miscommunication, just like I keep trying to tell you.

PLATO: That is exactly right. Never a miscommunication because they speak the language of mathematics. Hence, they are unalterable. You yourself keep repeating the answer, as repetitive as the tides.

MCCOY: Are you trying to tell me that you can actually hear the tides talking to you? And you started out this conversation complaining that people call you insane?

PLATO: I did not complain of it. I only reported it.

MCCOY: Well, is it any wonder? I mean, the rest of us go to the shore and hear splish, splash, but you hear . . . Well, I don't know what the hell you're claiming to be hearing. The Pythagorean theorem or something or other. Splish, *a*, splash, *b*, splish, *equals*, splash *c*.

PLATO: $a^2 + b^2 = c^2$. In right-angled triangles the square on the side subtending the right angle is equal to the squares on the sides containing the right angle.*

MCCOY: Show-off.

PLATO: But the Pythagorean theorem is not the mathematics relevant to the tides, but rather the lunar gravity differential field at the earth's surface, the primary cause of the equipotential tidal bulges that account for the two daily high tides.

MCCOY: Okay, I told you before, pal, that this is the No Bull Bin, and that means speaking so that people can understand you, so just cut that out, otherwise I'm going to have to cut your mic.

PLATO: Still, in another sense you are very right about Pythagoras, friend of number, who guessed the secret long ago that the most difficult task of reconciling the realm of change with the realm of the eternal necessities could happen only by way of mathematics, running between the two realms, like winged Hermes. It is number and measure that provide the structure of all that comes to be; in this way

---

we cannot understand it if we do not first learn the language and grasp the symbols in which it is written. The book is written in mathematical language, and the symbols are triangles, circles and other geometrical figures, without whose help it is impossible to comprehend a single word of it; without which one wanders in vain through a dark labyrinth."

*This is how it is stated in Euclid's *Elements* (I, 47). As was mentioned earlier, Euclid collated and formalized a great deal of mathematics that had been done before, including by mathematicians at Plato's Academy.

the limitless possibilities are rendered intelligible by the exactitude of number (*Philebus* 25c–d, 30c; *Timaeus,* passim).

MCCOY: Did you say Pythagoras was a friend of yours? That guy and his stupid theorem made my life a living hell in seventh grade. And you can tell him that for me personally when you see him.

PLATO: It would dismay him to hear it.

MCCOY: Good. Let him be dismayed by all the trouble he's inflicted on defenseless children through the ages. Boy, talk about useless knowledge. You think that in all the years that I've been able to achieve so much—writing best sellers and winning awards for excellence in broadcast journalism and having the biggest audience on any cable network—you think I ever once had to think about the Pythagorean theorem? I should have known you'd be friends with a killjoy like that.

PLATO: He did not kill our joy but showed us the painless path to it, the path that avoids the pains that cannot be avoided in seeking other pleasures, for the pleasures of learning are unmixed with pain (*Philebus* 52b)—

MCCOY: Maybe for you, pal.

PLATO: It is the pleasure you yourself seek as you ask the tides to yield up their explanations, seeking the best of reasons—

MCCOY: Those, of course, ultimately residing in the inscrutable will of the Deity—

PLATO: Not inscrutable at all, but expressed in the most perfect of mathematical formulations, itself an expression of intelligence—

MCCOY: Okay, you got that right, if what you mean is intelligent design, that being the Deity's—

PLATO: —so that to see that most perfect reason realized within those processes is to see that reason is forever the ruler over the universe (*Philebus* 30d).

MCCOY: If what you're saying is that the Deity is forever the ruler over the universe, then okay, you finally got something right. Otherwise, you're batting zero. Those explanations you seem so enamored of are all brought to you courtesy of the Lord God Almighty. And if they're so beautiful that you're having trouble restraining your poetry—which I've already warned you about twice—then you've got Him, and only Him, with a capital H, to thank for that. And that's why when I look at the tides coming in and going out, never a miscommunication, I'm struck dumb with awe.

PLATO: You do well to be struck with awe—

McCOY: Struck dumb with awe.

PLATO: Anyone who is happy began by being struck with awe at this cosmos and conceived a passion for learning all that a mortal can, believing that this is how to live the best and most fortunate life and that when he dies he will go to places where virtue is at home (*Laws* 986c–d).

McCOY: Well, thank you, I guess. It's nice of you to say so. I certainly hope that when I die I'm going to the place where virtue is at home, because right now the only place where it's at home is right here in the No Bull Bin, where I'm the one in total control.

PLATO: Further, once he is really and truly initiated and has achieved perfect unity and a share of the wisdom of the cosmos, he continues for the rest of his days as an observer of the fairest things that sight can see (*Laws* 986d).

McCOY: It's true that I've been a keen observer all my days, which is how I got into the broadcasting business, but not all that I observe can exactly be described as "the fairest things that sight can see." I mean, I actually live in the real world.

PLATO: Then you cannot help being both virtuous and happy.

McCOY: I like to think so. If seeing the unvarnished truth has anything to do with being virtuous—

PLATO: Oh, it has everything to do with being virtuous.

McCOY: Then okay, I'd have to say I'm virtuous. By why "happy"? I mean, I'll admit it, I'm a happy guy. I'm very happy. I've got everything it takes to be happy. Money, fame, power. And as someone once wisely said, it's not enough to succeed—

PLATO: That is certainly most wisely said.

McCOY: —others must fail.

PLATO: But living in the real world as you do, you know that not one of these things just mentioned—money, fame, power, much less the downfall of others, whether friends or foes—has anything to do with your being happy.

McCOY: Well, I don't know what kind of real world you're talking about there, pal. Sounds to me like the kind of real world dreamed up by losers. I'm talking about the *real* real world. And in the *real* real world, as opposed to the losers' version, money, fame, and power ain't half bad.

PLATO: Perhaps, it is as you say that they ain't half bad. But that still means they fall far short of perfectly good.

McCOY: Again, I'm going to remind you that here in the *real* real world, there's nothing that's perfectly good, excluding the Deity, in the singular. Otherwise, it's all a mixed bag.

PLATO: Yes, quite right you are that almost everything in the world is mixed of many parts, so that there is precious little of which we can say: Yes, that is always worth choosing; to choose so-and-so will always result in the better life. And you are certainly right that money, fame, and power are of this mixed kind, for in many circumstances to choose them results in a worse life.

McCOY: Still, if you've got to settle for the mixed bag, what could be better than the bag stuffed with money, fame, and power?

PLATO: But perhaps we don't have to settle? Would that not be better?

McCOY: Better than money, fame, and power? Forget about it. I'll settle.

PLATO: Even if we could find a good that was far more reliable, so that choosing it would lead to a better life no matter what else happened?

McCOY: Something in that loser version of the real world you're peddling? I've got an even better idea for the title of your next book: *Knowledge You Can't Use and Goodness You Don't Want.*

PLATO: Everything wants goodness. Everything that has a notion of it hunts for it and desires to get hold of it and secure it for its very own, caring nothing for anything else except for what is connected with the acquisition of some good (*Philebus* 20d).*

McCOY: You're right. And nobody wants what you're offering. Ergo what you're offering isn't goodness.

PLATO: That was an argument.

McCOY: What, you think you're the only one who knows his way around a syllogism? I went to Catholic school. Those Jesuits knew a thing or two about tortured logic, God bless their souls.

PLATO: But how do you know you don't want the unmixed good that I have in mind if you haven't even let me tell you what I think it is? And I mean here by an unmixed good not only a good that always results in a better life, but also—and these two conditions are connected—one which leads to pleasure that has no admixture of pain in it.

McCOY: I don't need to hear what you have in mind in your losers'

---

*This is repeated in many dialogues, including the *Republic:* "Every soul pursues the good and does whatever it does for its sake" (505e).

real world. Whatever it is, I don't want it. I don't want anything to replace what I have.

PLATO: Your money, fame, and power.

McCOY: And all the *parerga*. See, you taught me something.

PLATO: But do you agree that sometimes money, fame, and power lead people to do what's wrong?

McCOY: Listen, I never said that these things would necessarily make you a good person in the way that the nuns back in school meant for us to be. I meant they're good things to have. They make a person happy. But that doesn't mean they necessarily make a person nun-approved good. Maybe the happy person is good, maybe he isn't. It depends on how he got that money, fame, and power, and what he does with it. Sometimes the virtuous are happy and sometimes they're not. Sometimes the sinners are happy and sometimes they're not. That's why we need heaven and hell to settle up the score.

PLATO: And what if I told you I think being virtuous and being happy can't be pried apart from each other? The person who does wrong is unhappy, and the person who does the most wrong is the most unhappy.

McCOY: Yeah, once he gets caught. Once he's Bernie Madoff sitting in his prison cell, his empire toppled around him and everybody hating on him and his family members changing their last names so that they won't have people hating on them, too.

PLATO: He is far happier then, sitting alone in his prison cell, than he was before when he went unpunished (*Gorgias,* passim).

McCOY: Okay, this is frankly getting ridiculous. Next thing, you're going to tell me that you'd rather be the guy ripped off by Bernie Madoff than Bernie Maddoff when he was at the top of his game.

PLATO: I would say that.

McCOY: Then you're an idiot. You'd rather be done wrong than be the perp doing the wrong and getting clean away with it?

PLATO: For my part, I wouldn't want either, but if it had to be one or the other, I would choose suffering wrong over doing what's wrong (*Gorgias* 469b–c).

McCOY: So you wouldn't want to be a tyrant if you could get away with it?

PLATO: Not for anything in the world. Would you?

McCOY: You better believe it. And so would anybody with an IQ above a chicken's and who isn't living in your delusional world.

PLATO: And yet I do not believe that you believe it. For you do not desire your own unhappiness.

McCOY: Look, this is just getting too ridiculous for words, not to speak of unchristian. Not even the nuns would have ever tried to sell us on the crazy idea that being good was going to necessarily make us happy. In fact, they were pretty big on talking up suffering as a kind of moral prosthetic. You a Christian, Plato?

PLATO: No.

McCOY: I thought not. You're lacking the whole Christian sense of things, including about suffering. See, here's the thing about moral struggles. The reason they're such a struggle is that what makes us good and what makes us happy aren't the same things. The things we want, which are just the things that I have—

PLATO: Money, fame, and power.

McCOY: —and their *parerga,* those are the things that make us happy, which is why I'm such a happy guy, and nothing you can say is going to convince me otherwise. What, you're going to tell me I'm not really happy, that it only seems to me like I'm happy? Well, you know what? I'll settle for the seeming happy, because I don't see any difference. What do I care if money, fame, and power are a mixed bag and not everybody who goes after them is necessarily going to be happy, much less virtuous? They're not a mixed bag for me. When you look at the statistics, and you're factoring in the guy who's throwing all the numbers off because his life is so freaking awesome, then that would be me. I'm probably second in power only to the occupant of the Oval Office in terms of my opinions. Do you know how great that feels? Can you begin to imagine what it feels like to have that much influence over what people think?

PLATO: I am almost afraid to imagine it, so pitiful does it seem to me.

McCOY: *Pitiful?* Is that what you just said, pal? Are you frigging kidding me? You think it's *pitiful* to have so much power over people's minds? That's got to rank up there as the dumbest thing I've ever heard. You keep outdoing yourself, Plato. Only a guy who goes in for showing off how smart he is would ever say anything so obviously stupid.

PLATO: I always regarded having influence over the opinions of others as one of the greatest misfortunes a person could suffer.

McCOY, *laughing:* Will you listen to this guy? Okay, this has got to be some kind of shtick to sell books. And by the way, if you want to have so little influence over people, why do you even write your books?

PLATO: I think a writer can only hope, at the very best, to provide reminders for those who already know (*Phaedrus* 278a).

McCOY: Wait a minute, here. You're a guy from the Academy who's actually confessing that he's never learned anything from a book? What kind of syllabuses do you give your classes?

PLATO: You understand that I'm speaking of knowledge and not information. Knowledge can't be passed from one person to another as information can.

McCOY: Knowledge, information. I don't see the difference. What is this, semantics? Aren't you ashamed, at your great age, to quibble over words? (*Gorgias* 489b). Anyway, I don't think you should lose any sleep worrying over any undue influence you may have. I don't think you're ever going to have that problem.

PLATO: Nevertheless, undue influence was always a concern of mine. My position as head of the Academy made me worry that some would simply take my position there as sufficient reason to agree with me, using me as an authority to ground their positions.

McCOY: Frankly, I don't see that happening. Nobody listening right now to this show—and the following is huge, the largest on cable—is going to be swayed by anything you say, including your claiming that it's a misfortune to have a huge following. You should only be so unfortunate!

PLATO: I hope not, for then any mistakes I made would be compounded many times over, blighting not only my own point of view but that of all the others over whom I have influence. And were my influence so great, it would then only increase the likelihood of my thinking erroneously, for it would stifle both my own and others' abilities to assess what I am saying. I would be like the person who makes a great voyage in order to see the world but travels with the windows of his carriage draped over with images of the view from his own bedroom window.

McCOY: Your carriage? We're getting really high-tech here.

PLATO: I meant only that the person of great influence lacks access to views that challenge his own.

McCOY: Oh, there isn't any shortage of views clamoring to challenge my own. That's what we call the viewpoints of the pinheads, and fortunately nothing forces me to pay any attention to them.

PLATO: Except your own self-interest.

McCOY, *laughing:* This just keeps getting better. I'm supposed to pay attention to the pinheads out of my own self-interest?

PLATO: Otherwise you must do all the hard work of challenging your own positions all by yourself. Isn't it better to get some help with so difficult a task? And wouldn't you call those who help you out your friends?

McCOY: Why should I challenge my own positions? That's the job of my enemies, who it's my job to vilify.

PLATO: I would have thought it the job of your most valued friends.

McCOY: I can't tell whether you're putting me on or not. Is this some kind of Ali G or Borat scam you're trying to pull here? Just answer me that. Are you putting me on? Have my stupid staff screwed up again and let in some Sacha Baron Cohen operative?

PLATO: I am sincere.

McCOY: So I'm actually supposed to believe that you think friends are the ones who try to refute you?

PLATO: Certainly, when what I say is wrong; and I can't be certain it's not wrong unless I hear the best of the refutations that can be offered. And I hope that I am a good enough friend to return the favor.

McCOY: But obviously you'd rather refute than be refuted.

PLATO: I wouldn't be any less pleased to be refuted than to refute. For I count being refuted a greater good, insofar as it is a greater good for oneself to be delivered from something bad than to deliver someone else from it (*Gorgias* 458b).

McCOY: Well, then let me become your BFF by ripping everything you've said to shreds. I'm afraid you're going to drop down on one knee and propose marriage to me by the time I get done.

PLATO: Which one of my positions would you like to refute?

McCOY: Where do I even begin? I can hardly remember all the crazy things you've said. They don't stick because they don't make sense. First, you've got the tides knowing stuff like the Pythagorean theorem and the whole universe knowing how to count, which you seem to think is enough to cause it to come into existence all by itself, without any help from the Deity—take note of the singular—who you're probably ticking off as much as you are me because you're totally sidelining Him in His major job responsibility, which is to bring the universe into existence—the whole catalogue of And God Said Let There Be's that you'd know about if you were a Christian or even a Moslem or a Jew, which I take it you're not—

PLATO: No.

McCOY: Okay, well then let me fill you in. The Deity made all the stuff and then He got it to work, the tides coming in and going out and all the rest of it. Because let me tell you something about your friend Pythagoras' theorem. You may think it's as pretty as the Mona Lisa's smile, but Ms. Lisa's smile didn't paint itself and the same goes for the cosmos. Nothing's happening, including your friend's theorem, without the word coming from the very top. Okay, so that's number one on the big hit parade of where you go wrong. Can you back up your claim that the world gets going on math?

PLATO: The only account that can do justice to the wonderful spectacle presented by the cosmic order of sun, moon, and stars and the revolutions of the whole heaven is that reason arranges it all (*Philebus* 28e).

McCOY: You're spinning faster than those revolutions. What I asked you to tell me straight is whether you can back up your claim.

PLATO: Perhaps at the time I first made the claim that the *logos* of the cosmos is mathematics—

McCOY: Careful there, buddy. English.

PLATO: Perhaps, at first it was only a strong intuition, but at this point—

McCOY: I'm going to stop you right there. You see, on this network you have to be able to back up your so-called intuitions. We stick to the facts here. That kind of trust-me-I-just-have-a-strong-intuition spin might go down on other networks, but not on this one. Okay, so that's your number one major mistake. And you followed that piece of malarkey up with another nice bit of blasphemy, claiming that astronomy is an act of religious devotion that will somehow endow a person with virtue. So there you go sidelining the Deity—singular—a second time, only this time when it comes to the difference between right and wrong, which is His second big job responsibility. Because here's the way it is, pal. Fact: The Deity decides right and wrong, and it doesn't have anything to do with the Milky Way or the Big Dipper or any kind of twinkle-twinkle-little-star but with some pretty strict rules, one of which, by the way, forbids you to drop down on your knee and propose marriage to me when I get done with refuting you, just so you know. And then you've got this crazy idea that virtue and happiness go hand in hand tripping down the aisle together, so that even if someone thinks he's as happy as an abortion doctor in a whorehouse, if he's not virtuous then he isn't really happy. See, that would make being virtuous too easy, since everybody wants to be happy, but the whole point of virtue is that it's hard. It has to go against the grain of what you want. Being virtuous has got to hurt, so that there's a real struggle in your soul and only the best of us can duke it out. Okay, so there's that. And then, to top it all off, you say that the best thing for a guy like me is to have pinheads coming after me trying to refute my point of view, instead of having a huge fan base of loyal followers who take it from me who are the bad guys and who are the good guys. According to you, the pinheads are my true friends and my fans aren't my friends at all, which has the completely absurd result that I'm not my own friend, since I happen to be my biggest fan. So there's a reductio ad absurdum for you to put in your pipe and smoke, together with whatever other funny stuff you've got to be inhaling. Okay, I think that covers all the major stupid things you've managed to say in the short time I've known you. I'm leaving out some of the more minor dunceries, like that I should go gaga over math.

PLATO: And where would you like to begin?

McCOY: Good question. So many mistakes, so little time. Why don't we just concentrate on the one that's most personally offensive to me.

PLATO: Let me guess: that it is your misfortune that your opinions sway the opinions of others.

McCOY: That's probably the stupidest thing I've ever heard in my long history of hearing stupid stuff. And I don't for a minute believe that you believe it yourself. You're as convincing as Dukakis perched on an army tank. Let's look at the facts. You're a guy who starts his own private academy, and you're a guy who publishes books, so I'm thinking you're a guy who's heavily into trying to influence how others think. And judging by your having gotten onto my show you've had some success, though don't ask me how. My guess is that it's just the shock value of what you're saying. You get an educated guy with some fancy degrees who says something so stupid that any average person knows that can't be right, and so you get people scratching their heads and saying what he says must have some deeper meaning, since it's either that or he's dumber than poultry.

PLATO: And yet it might be the case that the truth first enters our minds as something strange and unsettling, our minds slowly coming to terms with its strangeness and reconfiguring themselves so that the sense of strangeness itself alters.

MCCOY: Which reconfiguring requires, I suppose, a prolonged stay in your private academy, so that by the time the bilked students graduate they're convinced that they can hear the tides whispering sweet theorems in their ears and that they see gods splashing around in the Milky Way. That's just elitist crap. And maybe that's the reason you don't want to influence too many people, because then it wouldn't be for the elite few anymore. It would be tarnished by being in too many heads, and you'd have to go come up with some other totally bizarro assertions to set you apart from the masses. It's like when the soccer moms started getting themselves tattoos, which made the hipsters go get theirs lasered off.

PLATO: The more people there are who have minds reconfigured by the truth, the better it will be for all of us.

MCCOY: So then you *do* want to influence people. You *do* want to change their minds. You're contradicting yourself!

PLATO: I would draw a distinction between influencing and persuading.

MCCOY: Just like you drew a distinction between information and knowledge. More academic spin.

PLATO: The two distinctions, between information and knowledge, on the one hand, and influence and persuasion, on the other, are not unrelated.

MCCOY: You bet. They're related by empty semantics.

PLATO: I think I can perhaps best explain the difference between influence and persuasion by speaking of seduction.

MCCOY: You mean like sexual seduction?

PLATO: Yes.

MCCOY, *laughing:* Be my guest. My ratings just shot up.

PLATO: There is a difference, you would recognize, between someone who takes others by force, mounting others in the manner of a four-footed beast and sowing his seed (*Phaedrus* 251a)—

MCCOY: Easy there, pal. This may be cable but not everything goes.

PLATO: —and someone who genuinely seduces. He who acts in the first way takes hubris as his companion, and does not seduce, but rather violates the other, who is only overpowered and deprived of any chance to give consent or to withhold it. In fact, the overpowered is not treated as a person at all, since the will of this person is rendered inoperative. But what we call seduction takes no power away from the one who is seduced but rather empowers him or her to surrender of his or her own will.

McCOY, *laughing:* He, she. Him, her. Someone sure got painted by the political-correctness brush.

PLATO: No, not painted but rather persuaded. That is my very point. When the case was made to me, by Cheryl, my media escort at the Googleplex, that my language was sexist, I considered the case she made and I saw that it was sound. The power was mine to refute it or concede. And so it is that seduction and persuasion are similar.* Both involve surrender, but it is not surrender to another person, for there would be no dignity in that. Rather, when I am seduced I surrender to love, and when I am persuaded I surrender to truth.

McCOY: You make a pretty speech, Plato, you're a regular rhetorician, but I think you're just playing with words again. You may say that you aren't personally trying to influence anybody, just trying to get them to "surrender to the truth," but really it amounts to the same thing. Whether you want to deny that you want clout or not, you're still going after unanimity, trying to trample out differing points of view in the name of your one and only truth. You said it yourself just a little while ago: my way or the highway.

PLATO: I said that?

McCOY: We've got it on tape and can play it back to you if you deny it. "This is the right way, this is the upbringing, these are the studies." If you could have your own No Bull Bin the way I do, where I get to decide when people are playing hard and fast with the truth and, if need be, tell them to shut up or I'll cut off their mic, you'd jump at the chance.

PLATO: Only if I wished myself the greatest harm. I couldn't help feeling some grief for Aristotle, a talented student of mine, who at a certain point in time suffered the misfortune of becoming the authority for a powerful institution, whose members simply took to referring

---

* In the *Phaedrus,* Plato makes what seems a somewhat abrupt transition from speaking of seduction to speaking of rhetoric (257c). The ostensible reason is that the three speeches that are being compared to one another all had *erōs* as their subject, but there are deeper reasons for why the discussion of the right way of seducing and the right way of persuading are combined into one dialogue. People in love and people persuaded by the truth allow themselves to be overtaken by something larger than themselves. In the *Timaeus* (51e), Plato says that only the person open to reason is open to persuasion. An orator who simply wants to have his way is similar to a lover who simply wants to have his way.

to him as "the philosopher," as if there had never been nor would ever be another philosopher, and converted all his opinions into dogma.

McCOY: I know all about Aristotle from my high school. He was the pagan of choice. Thomas Aquinas loved him, and we loved Thomas Aquinas. You're telling me it doesn't gall you to have your own student so outshine you that he was called "the philosopher" as if you didn't count for anything? He totally eclipsed your sun, and you're telling me you didn't resent it? You're telling me you don't wish that you had the clout he had or that I have now with my fan base?

PLATO: For my part, I think it's better to have my lyre or a chorus that I might lead out of tune and dissonant, and have the vast majority disagree with me and contradict me, than to be out of harmony with myself, though I'm only one person (*Gorgias* 482a).

McCOY: Again with the flowery speeches! Look, pal, I'm not out of harmony with myself just because I'm in perfect harmony with my fan base.

PLATO: You speak to the people who are like you in character, so that you can give expression to what they delight to hear.

McCOY: You bet I do. And I have the ratings to prove it.

PLATO: You mutually gratify one another. What you say gives them pleasure, and their pleasure in you gives you pleasure.

McCOY: Okay, you can put it like that, even though it's a little bit creepy.

PLATO: Each group of people takes delight in speeches that are given in its character and resents those given in an alien manner (*Gorgias* 513c).

McCOY: Well, obviously. That's why I have to tell the pinheads to shut up. And my audience loves me for it. It's just the way they want the pinheads to be treated.

PLATO: So when you, who share the character of your audience, say what they themselves would like to say, you gratify them. You gratify them so much that they will never go to listen to anyone who offers reasons to question what they would like themselves to say. The pleasure of hearing you is so great because the harmony is so great.

McCOY: I'm getting pleasure just hearing you describing the situation.

PLATO: Orators who have much to gain by gratifying the people will be careful in what they say, treating the people like children, not speaking to them of anything which will cause them the pain of doubt. If there is something the people would need to hear in order to get the whole picture but which would cause them pain, such orators choose to leave it out, even if justice would demand it be included. They are to justice what pastry makers are to health (*Gorgias* 465c).

McCOY: Pastry makers! Did I just hear you say "pastry makers"? Or was it hasty wasters, as in haste makes waste? Or are you talking about those pasties that strippers wear? You know, with you it could be anything.

PLATO: It was the pastry makers of whom I was speaking. Who is it that can better tell you what is good for the body, the pastry maker or the doctor?

McCOY: What kind of dumb question is that? Don't you dare condescend to me.

PLATO: Because it is so obvious that the doctor can better treat the body, understanding how to promote its health, whereas the pastry maker simply delights the body, knowing how to give it pleasure without thought as to what is best for it. Pastry making has put on the mask of medicine, and pretends to know the foods that are best for the body, so that if a pastry maker and a doctor had to compete in front of an audience of children, or in front of people just as foolish as children, who were to determine which of the two, the doctor or the pastry maker, had expert knowledge of good food and bad, the doctor would die of starvation. And so it is that the orator is like the pastry maker, both knowing well the knack of gratifying. And what is this knack? With the lure of what's most pleasant at the moment, it sniffs out folly and hoodwinks it, so that it gives the impression of being most deserving. I call this flattery, and I say that such a thing is shameful because it guesses at what's pleasant with no consideration for what's best (*Gorgias* 464d–645a, though scrambled).

McCOY: Yeah, only here's the justice in the situation that makes it not shameful at all. The guys on the other side are doing exactly the same thing. They've got their audiences who they gratify by serving up exactly the pastry that *their* audiences find finger-licking good. That's the way it works, you've got pastry makers on both sides, with pastry eaters on both sides gobbling it up and patting their tummies with pleasure. So maybe some people like their cinnabons and others their mousses or tiramisu. It's a free country and you're free to go get your goodies from whoever you like. And yeah, it mainly all works on the free-enterprise system, which goes hand in hand with democracy. The two of them go marching down the aisle together far more naturally than your virtue and happiness do, which is just plain unnatural. So yeah, there's going to be a profit to be gained in gratifying your audience—I get to live the way I do because I gratify a certain sweet tooth in a whole bunch of people—but that's okay because both sides are doing it, and it's all out there in the great American mall, cinnabons and tiramisu, cream puffs and whoopee pies, and people can go gratify themselves however they see fit. That's democracy, pal, the kind of democracy we taught the world to love.

PLATO: But even if it's all out there in the great mall, people are only going to the stalls that serve them their favorites. The one who likes cinnabons will go there, and the one who likes tiramisu goes there.

McCOY: As I said, it's a free country. And by the way, the same goes for that Internet of yours that you're so in love with.

PLATO: I am very sorry to learn that. I'd hoped that so much information being made available demonstrated a great desire not only for information but maybe even for knowledge.

McCOY: Just because all that information is out there doesn't mean that anybody's going to access it all. I mean, how could they? It's overwhelming. So you got your nothing-better-than-apple-pie types going to our site or to the Drudge Report, your caramelized froufrou fritters going to Moveon.org or the *Huffington Post*.

PLATO: So it becomes a fight to get attention.

McCOY: Exactly. Attention is the resource everybody's after, and sometimes there are huge sums of cash that are connected with that attention—

PLATO: But even when there aren't the huge sums, the attention alone is motive enough.

McCOY: Right you are. Attention is power. So you've got all the specializing pastry makers out there on the Internet, baking up a storm. Anybody with a blog is a pastry maker.

PLATO: I am sorry to hear it.

McCOY: There's nothing wrong with putting more pastry makers out there, all of them perfecting their own particular confection. Like I said, that's democracy. You've got a problem with it, then you've got a problem with democracy.

PLATO: And to me the situation seems precisely the opposite, for if the situation is as you describe it, I wonder how your democracy can continue to function.

McCOY: What, you want government regulations sticking their big noses in so that only your they-may-taste-like-dirt-but-boy-are-they-good-for-you desserts are going to be forced on us whether we want them or not? Or are you for taking all our yummies away from us altogether? Or maybe your idea is to let the masses feast until their eyes glaze over on Dunkin' Donuts while you and your kind just run the show. Which is it, Plato?

PLATO: None of those. For if we stop the pastry makers from deciding for us what is good for us, we don't thereby necessarily deprive ourselves of pleasure.

McCOY: Okay, what kind of awful-tasting pleasure are you going to try to sell me on? One of those stinky French cheeses that proves what a sophisticated palate you have?

PLATO: It is the kind of pleasure that one can only attain if one isn't going after pleasure in the first place. For this is one of the great paradoxes of pleasure:* if pleasure is one's goal, then it eludes you, like a shimmering feather you are chasing, the wind that you generate in your chase hastening the prize from your grasp. It is only when you cease to chase after it that the feather may drift down and settle onto your lap.

McCOY: And does that precious pleasure that you can't chase have anything to do with that precious knowledge that you can't use?†

PLATO: Everything. As it has also everything to do with the unmixed pleasure that you were certain you didn't want.

McCOY: And I'm still certain that I don't want it. And what's more, I'm pretty certain you don't want me to have it, either.

PLATO: Oh, no, you are quite wrong. I would wish, were that but pos-

---

* In the first of her Whitehead Lectures, "Pleasure, Knowledge and the Good in Plato's *Philebus*," delivered at Harvard in spring of 2013, Verity Harte connected Plato's views on "idle pleasures" to the "paradox of Hedonism" that she attributed to Sidgwick, citing this passage from *Methods of Ethics*: "Here comes into view what we may call the fundamental paradox of Hedonism, that the impulse towards pleasure, if too predominant, defeats its own aim. This effect is not visible, or at any rate is scarcely visible, in the case of passive sensual pleasure. But of our active enjoyments generally, whether the activities on which they attend are classed as 'bodily' or as 'intellectual' (as well as of many emotional pleasures), it may certainly be said that we cannot attain them, at least in their highest degree, so long as we keep our main conscious aim concentrated upon them" (*Methods of Ethics*, I.4, pp. 48–49). Compare also the following quote attributed to C. P. Snow: "The pursuit of happiness is a most ridiculous phrase; if you pursue happiness you'll never find it."
† In the second of her Whitehead Lectures, Harte went on to connect up Plato's notion of "idle pleasures" and "useless knowledge."

sible, for all to have it. And for you, with your vast influence, I would wish it most of all.

McCOY: Believe me, I start going after your useless knowledge and I'm not going to have my vast influence.

PLATO: And so you would not want it.

McCOY: But meanwhile I do. I want it and I have it, and I'm going to use it right now to say that I may not want to eat your taste-like-dirt desserts, Plato, or chase after your bird feathers, but you do have to give me kudos for not turning off your mic. It's been a real experience— for me and I hope for all of you out there who have been watching *The Real McCoy*.

θ

LET THE SUNSHINE IN

THE DAIMŌN MADE ME KNOW IT

I've claimed a lot for Socrates' performance before the packed crowd in 399 B.C.E. There must have been a lot to his performance, given its effect on Plato (as well as on so many other writers of Socratic *logoi*). We don't need to rely on the authenticity of Plato's *Seventh Letter* to believe that the violence Socrates suffered transformed Plato.* We simply have to look at the output of his life.

In the *Symposium,* Plato has Alcibiades allude, rather mysteriously, to the one time when he had glimpsed Socrates, whose "whole life is one big game—a game of irony" (216e), stripped bare of his irony. What a sight it was to behold him in his undraped sincerity, a sight, Alcibiades confesses, to have made him permanently susceptible to shame (although apparently not quite susceptible enough). "I don't know if any of you have seen him when he's really serious. But I once caught him when he was open like Silenus' statues, and I had a glimpse of the figures he keeps hidden within: they were so godlike—so bright and beautiful, so utterly amazing—that I no longer had a choice—I just had to do whatever he told me" (216e–217a).

---

* "Once upon a time in my youth I cherished like many another the hope of entering upon a political career as soon as I came of age. It fell out, moreover, that political events took the following course. There were many who heaped abuse on the form of government then prevailing, and a revolution occurred" (324b–c). The letter then goes on to describe Plato's experience with the oligarchs. "Some of these happened to be relatives and acquaintances of mine, who accordingly invited me forthwith to join them, assuming my fitness for the task. No wonder that, young as I was, I cherished the belief that they would lead the city from an unjust life, as it were, to habits of justice and 'manage it,' as they put it, so that I was intensely interested to see what would come of it. Of course, I saw in a short time that these men made the former government look in comparison like an age of gold." The letter describes his disillusion with the oligarchs, most particularly his horror when they tried to make Socrates complicit in one of their shameful deeds. "When I observed all this—and some other similar matters of importance—I withdrew in disgust from the abuses of those days. Not long after came the fall of the Thirty and of their whole system of government. Once more, less hastily this time, but surely, I was moved by the desire to take part in public life and in politics" (325a). He then goes on to describe his ultimate disillusionment, the execution of Socrates, which convinced him to give up any thought of entering the political life of his city and to instead devote himself to philosophy.

Whatever occasion Plato has Alcibiades hinting at in the *Symposium*—the more mysterious since the drunken Alcibiades is supposedly spilling it all out, unfiltered and uncensored, confessing the most intimate and self-compromising details of his relations with Socrates—we know of one occasion when the author of the *Symposium* saw Socrates being serious and opening up, revealing himself as a thing godlike to the young man with the long aristocratic lineage. Out of the vision sprung an entirely new conception of what human beauty looks like. Ugly, odd-gaited Socrates, always sticking his snub nose into the business of who is sleeping with whom, assumes a grandeur that makes everyone else look stunted. The revelation may have taken place in the most public of spectacles, a trial before raucous Athenians trying to shout down the old man. But that didn't make the experience any less intimate and transformative for Plato.

All through his intensely self-critical intellectual development, Plato will regard Socrates as a kind of portent. The way in which Socrates combined his authority with his absurdity, his godlike certainty with his human confusion, is a vision that Plato will contrive to keep alive almost until the end of his life. There is everything to learn from the man's inconsistencies.

If the "early" Socratic dialogues can be historically trusted (a contentious proposition), drawing out the inconsistencies of a person's beliefs might well have been Socrates' go-to strategy. The elenctic method doesn't tell a person what he or she is getting wrong, only that he or she is getting *something* wrong, since the premises asserted engender a contradiction.

If we go back and listen again to Socrates' defense before the Athenians, it might seem that the elenctic method can be turned against Socrates himself. There seems a contradiction to be drawn.

Socrates maintains, in declaring that the unexamined life is not worth living, that

   1. No one can live virtuously without knowledge of what virtue is.

Or as Plato rephrased it in the Myth of Er, virtue requires "the ability and the knowledge to distinguish the life that is good from that which is bad." A person who accidentally lives out a life that avoids wrongdoing has been fortunate, but he has not achieved virtue. His virtuous life is a blessing that has befallen him, like good hair or a trust fund, rather than been achieved.

Socrates also maintains that

2. Socrates is virtuous.

He's quite forthcoming, during his defense, about his own virtue. The quoted passages, in which he explains why he hadn't partaken in the politics of Athens—the explanation being that his virtue itself prevented him—makes clear his high self-esteem, ethically speaking. Socrates in the dock isn't putting on any false anything, including false modesty regarding the state of his own virtue.

But what about the state of his own knowledge? Socrates juxtaposes his affirmed belief in his own virtue with an avowal of his own ignorance. Avowals of Socrates' ignorance occur often in the dialogues, but they are often attended by the suspicion of irony. "Irony" derives from the Greek for "feigned ignorance," and it attaches naturally to Socrates. But I am working on the assumption that, if ever Socrates appeared without his characteristic irony, it was on that day in 399, when he was given his last chance to try to convince his fellow Athenians that they should be stirred to their depths by what he has to say. And it is a very strong statement of ignorance that he gives in the course of telling his story about the Delphic oracle, and his own incredulity at hearing it reported that none were wiser than he.* The answer had set him off on a systematic search—among the politicians, the poets, and crafts-men—to find a counterexample:

> As a result of this investigation, men of Athens, I acquired much unpopularity, of a kind that is hard to deal with and is a heavy burden; many slanders came from these people and a reputation for wisdom, for in each case the bystanders thought that I myself possessed the wisdom that I proved that my interlocutor did not have. What is probable, gentlemen, is that in fact the god is wise and that his oracular response meant that human wisdom is worth little or nothing and that when he says this man, Socrates, he is using my name as an example, as if he said: "This man among you, mortals, is wisest who, like Socrates, understands that his wisdom is worthless. (23a–b)

This gives us the last of Socrates' beliefs, which, in conjunction with 1 and 2, seems to engender a contradiction.

---

*It was his friend Chaerephon who had traveled to Delphi and asked the question, not Socrates himself (*Apology* 21a).

3. Socrates lacks the knowledge of what virtue is.

How can we reconcile Socrates' pronounced ethical swagger with his equally pronounced cognitive humility, given the added belief that virtue requires knowledge? A distinction that comes out of twentieth-century philosophy might offer some help: the distinction between "knowing that" and "knowing how."

"Knowing that" is followed by a proposition, an assertion that is true or false (though if it is truly known, the proposition must be true). So, for example, I *know that* the distinction between "knowing that" and "knowing how" itself was first introduced by the early-twentieth-century philosopher Gilbert Ryle. I *know that* Ryle was under the sway of Ludwig Wittgenstein. I *know that* under the Wittgensteinian influence, many philosophers believed that the job of philosophy was to explain away the appearance of philosophical problems by means of analyzing language, and I *know that* it was in just such an attempt that Gilbert Ryle offered the useful distinction between "knowing that" and "knowing how." Notice how each of my examples of "knowing that" is followed by a proposition.

When I say that I "know how," in contrast, I don't follow those words with propositions, but rather with reference to activities. I know how to bake bread, I know how to ride a bicycle, I know how to speak English.

Perhaps in the case of knowing how to bake bread I could translate my knowledge into at least a few propositions, consisting of one of the recipes I use. But these propositions will not exhaust what I know how to do in knowing how to bake bread. I know, for example, how to adjust for varying altitudes and variations in humidity by the feel of the dough on my hands, to say nothing of knowing how to get the flour out of the bag and into the bowl, or how to make my arms and hands move in the right way to count as "kneading." And in the cases of my knowing how to speak English and how to ride a bicycle the "knowing how" could not possibly be rendered into sets of propositions. Knowing how to do these things is not tantamount to knowing propositions. I could know how to ride a bicycle, for example, without knowing the first thing about the physics of bicycle balancing. And, in the other direction, I could know the physics of bicycle balancing, but when you perched me on the confounded contraption for the first time, I wouldn't know how to keep it from toppling. I know how to generate grammatical English sentences, putting words together in (generally) meaningful sentences, but I couldn't translate all my knowledge of how to do this into a set of propositions. There are a handful of rules that I know and could recite, but those rules don't add up to my knowing how to speak English. I probably know just as many rules for speaking French, and yet I don't know how to speak French.

Socrates knew how—in both his opinion and Plato's—to live a virtuous life, but he wasn't able to render that knowledge as a set of propositions. In other words when it came to living virtuously, Socrates thought he knew *how,* even if he knew that he didn't know *that.* How did he manage this trick? By appealing to the supernatural. Between his "knowing how" and "knowing that," there was a gap, and filling this gap was his own personal oracle, his oft-mentioned *daimōn,* silently warning him whenever he was about to do something wrong.

There are scattered allusions to the *daimōn* throughout Plato's dialogues.* In the *Phaedrus,* which finds Socrates out in the countryside with the beautiful boy Phaedrus, Socrates is suddenly halted, after he has delivered a well-constructed speech against the *erōs*-maddened lover:

> My friend, just as I was about to cross the river, the familiar divine sign came to me which, whenever it occurs, holds me back from something I am about to do. I thought I heard a voice coming from this very spot, forbidding me to leave until I made atonement for some offense against the gods. In effect, you see, I am a seer, and though I am not particularly good at it, still—like people who are just barely able to read and write—I am good enough for my own purposes. I recognize my offense clearly now. In fact, the soul too, my friend, is itself a sort of seer; that's why, almost from the beginning of my speech, I was disturbed by a very uneasy feeling. (242b–c)

In the *Euthyphro,* the very dialogue in which Plato articulates the inadequacy of religious grounds for ethical knowledge, the eponymous religious expert who is so oblivious to Socrates' philosophical reasoning is sure that the charge of impiety against Socrates is motivated by the appeals Socrates makes to his *daimōn:* "I understand, Socrates. This is because you say that the divine sign keeps coming to you. So he

---

* Besides the instances cited below, see also *Apology* 31c–d, *Republic* 496c, and *Euthydemus* 272e.

has written this indictment against you as one who makes innovations in religious matters, and he comes to court to slander you, knowing that such things are easily misrepresented to the crowd" (3b).

The appeal to this voice, inaccessible to anyone else, raises epistemological flags: How is one to know whether this private voice, as peremptorily as it presents itself in the inner chambers of Socrates' mind, is reliable? It *feels* reliable to Socrates, but so it always is with those peremptory inner voices. Socrates doesn't accept Euthyphro's peremptory inner voice, so why should we accept Socrates'? Are reasons of this sort—subjective, private, and non-generalizable, unavailable for objective scrutiny and evaluation by others—to be accepted as grounding knowledge?

The epistemological flag is hoisted in Plato's mind, maybe even prompted by the mystery of Socrates' certainty. In fact, the whole domain of epistemology is hoisted in Plato's mind. In the *Theaetetus*, he will have his dialogic Socrates articulate the fundamental epistemological question, namely how to define knowledge in the first place: "Well, as I said just now, do you fancy it is a small matter to discover the nature of knowledge? Is it not one of the hardest questions?" (148c).

Plato not only puts the question on the table, but he makes the first few crucial incisions, distinguishing knowledge from mere true belief that doesn't achieve the status of knowledge. There's something haphazard about mere true belief or opinion that the concept of knowledge can't tolerate. As with the proverbial broken clock that is correct twice a day, true belief can be induced by illicit methods, the sort of methods that are just as likely to produce false opinions as true.[*] In the *Theaetetus*, Plato moves (though somewhat jerkily) toward the definition of knowledge as "true belief with a *logos*," an account.[†] This is a

---

[*] Plato illustrates illicit methods by the example of an orator convincing a jury by manipulating their emotions or appealing to hearsay. "And when a jury is rightly convinced of facts which can be known only by an eyewitness, then judging by hearsay and accepting a true belief, they are judging without knowledge, although, if they find the right verdict, their conviction is correct" (Ibid. 201b–c).

[†] The dialogue gets bogged down with Socrates' seeming to poke holes in the idea of an account, but the holes in fact aren't as serious as Plato judges them. Plato fails to make certain helpful distinctions in this first attempt that future philosophers will come up with, which isolate propositional knowledge (knowledge that some proposition is true) as the paradigmatic form of knowledge. Because he doesn't do this, the definition of knowledge he offers in the *Theaetetus*—that knowledge is true belief with an account—seems less promising to him than it really is. Plato is far more successful in the *Theaetetus* than he gives himself credit for being, the dialogue ending with aporia. Plato's self-criticism is commendable, even if it sometimes leads him to judge his own breakthrough proposals too harshly.

first approximation to a definition that philosophers would eventually give: *knowledge is justified true belief.** The same true proposition that is merely believed by one person can be genuinely known by another, and the difference lies in the reasons the believer has for believing. The reasons have to be *good* ones, providing *justification* for his belief, making it a *rational* belief. These are all evaluative notions. The definition of knowledge forces a further question: what counts as good reasons? All of these are questions that make up the field of epistemology, and they are questions Plato raised.

There is no evidence that the epistemological concerns that so occupied Plato, perhaps prompted by the epistemologically baffling figure of Socrates, ever occurred to Socrates himself.† Socrates is never presented questioning whether the whispering of his *daimōn* is reason enough to believe. In the *Apology,* in which we encounter as sincere a Socrates as we ever will, his *daimōn* is brought out for one final and solemn appearance. After addressing some condemnatory words to those who have condemned him, he has this to say to those who have voted him innocent:

> A surprising thing has happened to me, jurymen—you I would rightly call jurymen. At all previous times my familiar prophetic power, my spiritual manifestation, frequently opposed me, even in small matters, when I was about to do something wrong, but now that, as you can see for yourselves, I was faced with what one might think, and what is generally thought to be, the worst of evils, my divine sign has not opposed me, either when I left home at dawn, or when I came into court, or at any time that I was about to say something during my speech. Yet in other talks it often held me back in the middle of my speaking, but now it has opposed no word or deed of mine. What do I think is the reason for this? I will tell you. What has happened to me may well be a

---

* Knowledge is, at the least, justified true belief. The "at the least" is added because it is possible to dream up some highly contrived cases (always involving perception) in which a person believes something is true, and is justified in believing that it's true, but his (good) reasons for believing are unattached to the proposition's truth, so he doesn't achieve knowledge. There is still something haphazard in his happening to get it right. These rigged cases are known as "the Gettier counter-examples," and they indicate that even though justified true belief is necessary for knowledge, it may not be, at least in some very artificial situations, sufficient.

† Even in the *Theaetetus,* Socrates alludes briefly to his private *daimôn,* which sometimes warns him against allowing certain young men who have gone astray to return to his guiding friendship (150e), perhaps an allusion to Alcibiades. The irresistible rascal may have charmed his way back into the good graces of everyone else, but for Socrates, enough was enough. (Or Socrates, once burned, was twice shy.)

good thing, and those of us who believe death to be an evil are certainly mistaken. I have convincing proof of this, for it is impossible that my familiar sign did not oppose me if I was not about to do what was right. (40b–c)

Socrates is presented here as taking his "divine sign" very seriously, hazarding metaphysical speculations based on nothing more than the *daimōn*'s silence, which makes him come off as rather epistemologically insouciant. Plato, in contrast, was anything but epistemologically insouciant. What did he make of Socrates' appeal to his *daimōn*? Socrates could provide no account that he could offer to others for the truth of his beliefs. What the appeal to a *daimōn* signifies, rather, is the *absence* of an account. I just hear it, I just see it, I just *know* it. There is an *it*, and that's all that can be said about *it*.

If we wish to scrub Socrates' appeal to his *daimōn* clean of supernaturalism, we can regard it as a fanciful way of speaking about a phenomenon that gets a lot of currency today, both in psychology and philosophy, under the name of "intuitions." Intuitions are those subjective inner promptings about which one can say nothing to convince others who don't share the intuition. One can't offer a justification for them other than their peremptory announcement. Some intuitions are shared widely, and when this is true the sharing presents a datum that calls for an explanation. But offering an account of the sharing of an intuition isn't the same as offering an account of the intuition itself, an account that makes the case for the intuition's being rational to believe, *known* as opposed to *merely believed,* having a reason behind it that one can offer to those who themselves lack the intuition to convince them that they ought to cultivate it. Of course, once you can do that, you've no longer got an intuition but rather a defendable proposition. (And even a universally shared intuition is lacking a justification; it's just hard to make out the lack when everyone else agrees with it.)

Socrates, whose life Plato deemed beautiful, seemed to know how to live. Somehow his intuitions led him to lead an exemplary life, but others' intuitions, just as forceful, can lead to terrible lives. Is there any way to tell, from the inside, which intuitions are reliable and which aren't, some inner quality of feeling? Or if there isn't such an inner sign tagging the good intuitions from the bad, shall we only rely on intuitions that the majority of those around us accept? That's not what

Socrates did, and that's not what many of the moral figures we most revere—such moral revolutionaries as William Lloyd Garrison, Frederick Douglass, Bertha von Suttner, Mahatma Gandhi, and Martin Luther King Jr.—did. They challenged the intuitions of their societies and, eventually, changed those societies so that intuitions themselves changed. But if there's no way to tell, from either the inside or the outside, which intuitions can be trusted, why should we trust them at all? But then can we actually eliminate them? Aren't we, at certain points of our reasoning, including our moral reasoning, required to fall back on intuitions?

All of these questions might well have been suggested to Plato by the baffling figure of Socrates, who seemed to perfectly know how to live while also, until the very end, forswearing all human wisdom.

THE CALL OF THE KINKY

Nothing quite so divided the soul of Plato as the issues entangled in Socrates' appeal to his *daimōn*. These issues are various, and they all involve private, peremptory, possibly pathological but possibly inspired idiosyncratic experiences, which seem to carry intimations of truth— but then again, maybe they don't. Perhaps these singular experiences are nothing more than the floating vapors of hallucinations, the emanations of diseased minds, even if they do sometimes result in visions of beauty, or even, as it was with Socrates, a life of beauty. Such beauty exerts, for Plato, a degree of epistemological force. If something is beautiful, then, for Plato, there *must* be something, somehow, that is real or true or authentic about it. Beauty is struck deep into the structure of reality for Plato. His aesthetic realism is the linchpin of his realism, fastening together his mathematical realism, his metaphysical realism, and his moral realism. Any hint of beauty is what makes Plato turn back and look, and then turn back and look again, at convictions that cannot submit themselves to objective scrutiny to be carried out from more than one point of view, allowing for private kinks to come to light and be eliminated. Plato is keenly aware of such kinks, and though he regards most of them as pernicious, he also suspects that there are truths that can only be gotten to by way of the kinky.

Plato's divided soul has come down to us through the millennia, its component parts segregated into two oppositional camps, glaring at each other with mutual suspicion. We see them configured on the contemporary battlefield that pits science against religion.

There are, on the one side, those whom we might call the "Reasonables" (the term "rationalists" having already been claimed in the history of philosophy[*]). Reasonables regard knowledge as—necessarily—an equal-access good. The kind of justification that counts, turning mere true belief into knowledge, is, in principle, accessible to all, which means it must be able to put its grounding in terms that can be replicated in other points of view, making itself open to many-minds scrutiny. There can be no epistemological privilege. So mathematical proofs, for example, are accessible to many points of view (differences in aptitudes can be ignored). So are empirical data. Finding the truth is a game that anyone can play. If it can be known, it can be shown.

Then there are those whom we can call the "Unreasonables," those whose slogan is the cri de coeur of the seventeenth-century mathematician Blaise Pascal: *Le coeur a ses raisons que la raison ne connaît point.* The heart has its reasons of which reason knows nothing. Blaise Pascal was not only an important mathematician, who laid the foundations of probability theory, but a spiritual seeker who wrote a meticulous account of a mystical experience he underwent and sewed it into the lining of his coat, which was discovered after his death. Unreasonables are willing to stamp "known" on certain claims that can give no account of themselves in objective, generalizable terms. One could not remove such an insight from the immediacy of a first-person singular experience and have it retain its life, no more than an eye could be removed from the living creature and still retain its sight.

The controversy between the Reasonables and Unreasonables is an epistemological controversy, a controversy over how we can know.

---

[*] What is now known as rationalism had its heyday in the seventeenth century, especially in the figures of Descartes, Spinoza, and Leibniz. The best way to understand rationalism is to contrast it with empiricism. The empiricists believe that knowledge of the world—of what exists and what the properties of the things that exist are—requires contact with the world via our sense organs. Just thinking is not going to give us any knowledge of what our world is like. Rationalists believe that there are at least some things we can know about the world through pure thought. Like the controversy between the Reasonables and the Unreasonables, the controversy between rationalists and empiricists is epistemological—it's about how we know things, how we acquire the justification that distinguishes knowledge from mere belief. Though rationalists and empiricists lived and disagreed with each other long before then, it was only in the nineteenth century that the distinction was explicitly drawn and the terms of the disagreement defined.

What makes the controversy so irresolvable is that neither of the two sides can, non-circularly, put forward an item of knowledge that solves the underlying epistemological issue. Because neither side gets deep enough down into the underlying epistemological controversy—the very one that took hold of both sides of Plato's mind—the antagonists, more often than not, talk right past each other. The Reasonables say, I'll only believe that you can know something without it having to be demonstrated if you can demonstrate that to me. The Unreasonables say, It's obvious that you can know things without having to demonstrate them—I just know it!

It's not only religion that pits the Reasonables against the Unreasonables, but also other experiences, including, according to Plato, poetic inspiration and erotic love. In the *Phaedrus,* in which Plato is in his most Unreasonable-siding mood (so much so that we might suspect, along with Martha Nussbaum, that he was in love*), he explicitly links the three *daimōn*-haunted domains: religion, romantic love, and poetry. Religious conversion and romantic love and artistic inspiration are, for Plato, at once compelling and suspect. It *seems* to people who are in the grip of these experiences, which can offer no account of themselves independent of the experiences—unsharable and unshakable—that they are privy to a privileged knowledge, often so irresistible as to produce a major discontinuity in their lives, such that friends and relations will look at them, sadly shaking their heads and saying they have gone mad. At such times people do not take possession of the truth, but rather are possessed *by* it. "To be possessed" can mean madness, and Plato in fact uses the word *manikēs,* or "mad," to characterize such knowledge—that is, if it *is* knowledge.† That is the question. It is this question that has Plato halting Socrates in his tracks just after he has

---

* As has been mentioned, not only does Nussbaum make a compelling case for Plato's state of passion at the time that he wrote the *Phaedrus,* but she ingeniously parses its language to hazard a guess at the object of his passion: none other than Dion, the uncle of the young tyrant of Syracuse. "This dialogue has the character of a love letter, an expression of passion, wonder, and gratitude. . . . This is not, of course, to say something so simple as that love made Plato change his mind; for his experience of love was certainly also shaped by his developing thought. The dialogue has explored such interrelationships with too much complexity to allow an oversimple love story; but it does ask us to recognize experience as one factor of importance" (*The Fragility of Goodness,* pp. 229–230).

† Plato, who loves etymological play, notes the link between *manikēs* (mad) and *mantikēs* (prophetic). "It is worth pointing out that the ancient people who gave things their names also believed that madness is neither shameful nor blameworthy; otherwise they would not have connected the word itself with the noblest art, that by which the future is judged" (*Phaedrus* 244b). Our words "manic" and "mantic" are derived from the Greek.

made his first well-constructed speech to Phaedrus endorsing a sane and unmaddened love that never loses control of itself, a view Plato has Socrates recanting in his second speech, which itself is struck with the wild madness of poetry. And it is also the very question that has Plato returning again and again to the question of poetic inspiration, often censoring the poets, once expelling them, and sometimes surrendering to the enchantment of their art, allowing his own extraordinary poetic gifts their full voice.

Erotic genius, poetic genius, religious genius, and moral genius of the kind that Plato attributed to Socrates:* all of these yield non-rationalizable claims to questionable knowledge, knowledge that can offer no account for itself. These types of genius are informed by a singularity of experience that allows for knowing how—how to love, how to produce great art, how to live a virtuous life—without their being able to *account* for their knowing how. Such unaccountable knowledge marks certain points of view as intrinsically special. Those who know (*if* they know) cannot render their reasons for knowing so that non-knowers can be placed in the same epistemic position as they. The gap between their knowing *how* and their knowing *that* can only be filled, as it were, with the gods, a figurative way of saying that it is utterly mysterious. "We distinguished four parts of the divine type, associated with four gods: Prophetic madness was ascribed to the inspiration of Apollo, the madness connected with the mysteries to Dionysus, poetic madness in turn to the Muses, and the fourth, the erotic madness that we said is the best, to Aphrodite and Erōs. We described erotic passion in, I don't know, a sort of figurative manner, perhaps touching on something of the truth but also probably being led astray at other points" (*Phaedrus* 265b–c).

Note the hesitation—"I don't know, a sort of figurative manner"—resonating with the testimony to dividedness already signaled by the two contradictory speeches Plato had Socrates deliver earlier in the dialogue, clashing with each other not only in their substance but in their style—one tepid, repetitious, and cautious, the other careening in furious recklessness into ever more rapturous poetry that channels its imagery from the sacred mysteries. Athens had its own mystery cult,

---

* I knew someone I would describe as a moral genius. He died when I was around the same age as Plato was when Socrates died. Subjecting him to the kind of questions a philosopher asks always resulted in disappointing answers. But watching him do what he knew how to do was inspiring. He seemed never to err.

the Eleusinian mysteries, and these communal rituals had as their aim a singular experience—ecstatic, ineffable, extraordinary.

Plato, even in his most Unreasonable-siding mood, which is where we find him in the *Phaedrus,* motivated, for all we know, by the transports of erotic passion, cannot altogether loose his grip on his Reasonable-siding reservations. After all, there is good possession, coming from somewhere outside us, dubbed by Plato "the gods" and leading to truth, and then there is bad possession, which is nothing more than the delusions of our own minds having their way with us. And here's the thing: when we are in the grip of possession we have no means of distinguishing the good possession from the bad and are also in such a state as to dismiss anyone who is not similarly seized, which means, given the singularity of the experience, anyone who is not identical with ourselves. There is only one possible "authority" when it comes to this kind of unaccountable "knowledge," and this authority, being possessed by the experience, has no way of judging whether the possession is of the good kind or the bad. That is the predicament Plato brings to our attention. And it is still a predicament.

If we are allowed to say that Plato believes anything at all—despite his constant self-criticisms—then I think we can say that Plato believes, deeply, that reality is *out there,* the same for each of us. This is the one doctrine I'd put all my money on Plato's believing, rather than any specific doctrines regarding what reality is like. But still, doesn't the assertion of an objective reality show that the Reasonables are being, well, *reasonable* in asserting that truth is an equal-access good? Techniques for getting at reality are there for all to study and master. If we let the contents of our minds be determined by the contents of objective reality, as ascertained by techniques that can be universally mastered, under the tutelage of others and subject to corrections, then we can arrive at agreement. And on the authority of that agreement we can dismiss the kinks of an Unreasonable, who clings to the experiential contents of a single point of view, giving credence to oracular voices whispering privately in her ear. Whatever can be known by one person can, in principle, be known by everybody, just so long as they master the techniques for knowing that are most appropriate to a field. If it can't be generally known, if it is irreducibly embedded in a single and singular point of view, then we can have no good reason to accept it. This is the Epistemology of the Reasonable, and it is one side of Plato's divided soul and informs not only most of philosophy (with a few kinky exceptions like, possibly, Heidegger) but all of the sciences. Philosophy-jeerers who argue from science are unaware that they are epistemological allies with the bulk of philosophers, and depend on the Epistemology of the Reasonable that philosophers have hammered out for their convenience.

On the other hand, Plato suspects that there are intimations of reality which are given, however vaguely, in ecstatic visions experienced only by the few and which can get at aspects of reality that shared techniques can't access. Perhaps some truths stubbornly refuse to yield themselves up in objective terms, meaning terms to which many minds can gain access, unattached to any singular point of view. This is the possibility that Plato takes seriously in the *Phaedrus*. Does the objectivity of reality—in the sense of a reality existing *out there,* stubbornly itself no matter what any of us happen to think about it—imply that truth is an equal-access good? These two notions of objectivity—one ontological, the other epistemological—are, after all, distinct. Something might be out there, independently of any of us, and the same for all of us, but we may have no common way of knowing it. William James was in a *Phaedrus* frame of mind when he remarked in *The Varieties of Religious Experience,* "If there were such a thing as inspiration from a higher realm, it might well be that the neurotic temperament would furnish the chief condition of the requisite receptivity." Again, the call of the kinky.

In the theory of special relativity, the claim that *there are no privileged points of view* has a highly particular meaning: that the laws of nature must be the same regardless of the frame of reference in which they are being described (where frames of reference differ from one another because they are moving at different constant velocities relative to each other). Should the prohibition against privileged points of view in the theory of special relativity be promoted to a general law of epistemology? Ought we to exclude from our cognitive considerations any that make sense only within particular frames of reference, specially marked by certain subjective features that can make no claim on those whose points of view happen to lack those features? Should we blow off the experiences of people endowed with unique emotions, say, or with singular visions, or those who hear special messages? Should we dismiss the very possibility of extra perceptual equipment, such as the *sensus divinitatis,* the cognitive organ for sensing God, with which, John Calvin claimed, fully functioning people come equipped, and which at least one contemporary philosopher, Alvin Plantinga, has sought to resurrect?* Or should we indulge them and their claims of epistemological privilege? The point of claiming privileged points of view is to give a free pass to what is supposedly revealed there. The burden of proof is removed, and instead it is the points of view that *lack* these special irresistible features that are deemed deficient and put on notice to defend themselves.

There are strong—oh, so strong—reasons to affirm that *yes*, we ought to exclude privileged points of view as we seek to know the world. No claim to knowledge should be allowed a free pass, getting by without giving an account of itself, a justification, that can appeal to all who sign on to the project of reason, no matter the special features of their subjective points of view. It is not just a matter of the objectivity of reality that motivates the demand for objectivity of knowledge. Far more persuasive reasons arise from the obvious hazards of subjectivity, which is a breeding ground for prejudice, superstition, and egotistical self-aggrandizement. We are too prone to favoring our own particularity and, if we are talented enough, can raise up a cunningly convincing ideology that will shape all the world to fit our particular dimensions. It is a dangerous mistake to allow subjectivity to strut its stuff with such smug thuggishness. Exposing our most cherished beliefs to the rough treatment of multiple points of view—each of which is prone to see the world from the vantage of its own advantage—is our only hope for defeating the hazards of self-serving subjectivity—complacent at best, murderously certain at worst. And so philosophy—as Plato had conceived it with one-half of his divided soul—has typically been saying *yes* to the exclusion of privileged points of view ever since Plato himself set up perhaps the most powerful image in the history of thought, the Myth of the Cave, one of the highlights of Plato's *Republic*. The Myth of the Cave is as strong an endorsement of the Epistemology of the Reasonable as can be found in philosophy.

GET ON OUT OF THAT CAVE

The Myth of the Cave has come up multiple times in this book. Marcus at the Googleplex and Dr. Munitz at the 92nd Street Y both put their

---

* See Alvin Plantinga, *Warranted Christian Belief* (Oxford: Oxford University Press, 2000).

own spins on it. There are many spins and a vast literature of interpretation. I offer an interpretation based on the story I have tried to tell of how Plato drew philosophy forth from an ethos that made itself felt in the Greek city-states—and in Athens most of all.

It's a story of an extraordinary society that believed in being extraordinary, extolling exceptional individuals while at the same time creating a sense of participatory exceptionalism by means of which the extraordinary could be spread around. Athenian ideology was a response to an existential quandary that emerged dramatically during the Axial Age: What is it—if it's anything—that makes an individual human life matter? What must one *be* or *do* in order to achieve a life that matters? The existential quandary resonates no less in our time than it did during the Axial Age. Is it any wonder that the powerful religious traditions that emerged under the force of the existential quandary still resonate with so many today?

But the Greeks took a different approach. Even though religious cults and rites saturated almost every aspect of their lives, they approached the existential dilemma in secular terms. The most important of these terms is the complicated notion of *aretē,* bound up as it was with *kleos.* This approach also still resonates today, provoking ambitions to stand out from the great, massive mortal crowd, somehow or other, either individually or collectively.

The Athenian ethos might have nurtured the prerequisites for moral philosophy by approaching the existential dilemma in human, rather than divine, terms. Still, the value structure of Athenian ideology had to be challenged for moral philosophy to emerge. This was the project of Socrates, and he pursued it with his fellow Athenians wherever he found them: at the agora and the gymnasium, at dinner parties and at his trial. They often found his questions unintelligible, and it is little wonder. He was nudging the notion of *aretē* outside its familiar context, prying it away from *kleos* and pushing it closer toward a concept that English translators of the dialogues straightforwardly render as "virtue." Attempts to define this or that virtue organize many of the dialogues. The *Republic* revolves around the virtue of justice.

The dialogue opens with Socrates, the first-person narrator of the *Republic,* setting the scene. He has gone the day before, together with Glaucon, Plato's brother, to the Piraeus, Athens' harbor, to attend a religious festival surrounding a newly introduced goddess, identified by scholars as the Thracian goddess Bendis. He was eager to see the festival because, he casually mentions, it was being celebrated in Piraeus for the first time (327a). Perhaps this detail gestures toward the theological openness of the Athenians, underlining the hollowness of the charge against Socrates that he introduced new gods? In any case, on their way back, they run into a friend, Polemarchus, who tells them that they must stick around in Piraeus for more festivities to come that evening, and Socrates and Glaucon are persuaded to go to Polemarchus' house. A crowd of worthies is gathered there, including several famous sophists and rhetoricians. Socrates first pays his respects to Polemarchus' father, Cephalus, taking the opportunity to ask him what it's like to be so old. Cephalus responds that one can bear old age so long as one has lived a just life. Wealth is important only because the exigencies of poverty might have tempted one to be unjust, which will make facing death difficult. This leads naturally to a discussion of how to define justice, for both the individual and the *polis*.

The discussion is long and complicated. Socrates and Glaucon, we can be pretty sure, never made it to the evening's festivities. Not only political theory, but moral psychology and moral philosophy, metaphysics and epistemology are enlisted in the answer that will eventually be ventured as to the nature of justice, political and individual. Both are a matter of structural soundness. The just city is composed of three parts—the guardians, the army, and the producers—with each part performing the function for which it is best suited, both by temperament and training. A person's psyche is also composed of three parts—the *logistikon,* which reasons; the *thumos,* which wills; and the *epithumia,* which craves. In the just person each part performs the function for which it is best suited. The just person, like the just *polis,* has the internal arrangement just right.

The *Republic* is organized into ten books, and the Myth of the Cave occurs at the beginning of Book VII. Here is how it is introduced (the respondent is Glaucon):

> Imagine human beings living in an underground, cavelike dwelling, with an entrance a long way up, which is both open to the light and as wide as the Cave itself. They've been there since childhood, fixed in the same place, with their necks and legs fettered, able to see only in front of them, because their bonds prevent them from turning their heads around. Light is provided by a fire burning far above and behind them.

Also behind them, but on higher ground, there is a path stretching between them and the fire. Imagine that along this path a low wall has been built, like the screen in front of puppeteers above which they show their puppets.

I'm imagining it.

Then also imagine that there are people along the wall, carrying all kinds of artifacts that project above it—statues of people and other animals, made out of stone, wood, and every material. And, as you'd expect, some of the carriers are talking, and some are silent.

It's a strange image you're describing, and strange prisoners.

They're like us. (514a–515a)

These prisoners are huddled together, their state of mind one of *eikasia,* which is the lowest level of awareness, deceived and ungrounded. The contents of a mind in the grip of *eikasia* is unconnected with anything having independent existence. It is a sooty, dim, and artificial world, with everything contrived so that the prisoners cannot discover the nature of what they are looking at. It is what we now might call a socially constructed reality. (If there are any thinkers still out there who still hold to the once-fashionable view (circa 1970s–1990s) that *all* is socially constructed, then they're going to stop following Plato any further at this point.) There are elaborate props supporting it, and people tending those props. The chained image-observers are prisoners of ideology, though they would prefer not to know it. In fact, they would do anything not to know it. All their questions are answered, and the questions worth asking are never considered. Their false beliefs are mutually validating, but their unanimity counts for nothing so far as truth is concerned. They live together in darkness. Later on in the myth Plato describes "the honors, praises, or prizes among them for the one who was sharpest at identifying the shadows as they passed by and who best remembered which usually came earlier, which later, and which simultaneously, and who could best divine the future" (516c–d). Their celebrations of one another are pathetic, since none manages to attain anything worth winning. *Kleos* raises none of them above the other, despite what they might think. (But again, if you are a thinker committed to there being nothing but the socially constructed images on the cave's wall, you'll likewise recognize no higher standard than the *kleos* of your community: "The only 'proof' of membership is fellowship, the nod of recognition from someone in the same community, someone who says to you what neither of us could ever prove to a third party: 'we know.' I say to you now, knowing full well that you will agree with me . . . only if you already agree with me."[*])

Plato doesn't tell us how it happens, but one of the prisoners becomes unfettered. It's curious how involuntary the process he describes is, especially in its beginning phase, as if someone on the road to knowledge resembles a resentful teenager being dragged on a family outing and determined not to enjoy himself.

> When one of them was freed and suddenly compelled to stand up, turn his head, walk, and look up toward the light, he'd be pained and dazzled and unable to see the things whose shadows he'd seen before. What do you think he'd say, if we told him that what he'd seen before was inconsequential, but that now—because he is a bit closer to the things that are and is turned towards things that are more—he sees more correctly? Or, to put it another way, if we pointed to each of the things passing by, asked him what each of them is, and compelled him to answer, don't you think he'd be at a loss and that he'd believe that the things he saw earlier were truer than the ones he was now being shown?
>
> Much truer.
>
> And if someone compelled him to look at the light itself, wouldn't his eyes hurt, and wouldn't he turn around and flee towards the things he's able to see, believing that they're really clearer than the ones he's being shown?
>
> He would.
>
> And if someone dragged him away from there by force, up the rough, steep path, and didn't let him go until he had dragged him into the sunlight, would he be pained and irritated at being treated that way? And when he came into the light, with the sun filling his eyes, wouldn't he be unable to see a single one of the things now said to be true?
>
> He would be unable to see them, at least at first. (515c–516a)

Glaucon's "at least at first" is a necessary qualification, since gradually the escaped prisoner will be able to see extra-cavernously, but only slowly and by degrees. Recovery from ideology—deprogramming as we now call it—takes time. His eyes can't take in everything at once, because they have to adjust to the light. First he will simply look at

---

[*] Stanley Fish, *Is There a Text in This Class? The Authority of Interpretive Community* (Cambridge, MA: Harvard University Press, 1980), p. 176.

images and shadows and reflections in water. Next he will be able to look at "the things themselves" (516a). Gradually he raises his eyes upward, studying the nighttime sky. Finally, his eyes will have adjusted sufficiently for him to see "the sun itself, in its own place, and be able to study it."

The various levels, both within the cave and without, represent levels of being in Plato's metaphysics.[*] These levels are ordered by the relations of explanation, of accounts, of *logoi*. One ascends to a higher level by explaining the level one has already secured. This is how the process of knowledge proceeds, by what philosophers call abduction, or inference to the best explanation. Grasping the best explanation is the job description of reality-discovering reason. It's the way that ontology can be expanded. Whether it is employed in theoretical physics[†] or in philosophical reasoning, it's by means of abduction that one can come to know the reality that isn't passively received, in either imagination or perception, whether that reality is of quantum fields or, as Plato would have it, a value-saturated reality, structured by the confluence of the True-the Beautiful-the Good. The expert knower, whether he's a cosmologist or a metaphysician, is discovering objective reality by seeking the best explanation (and as Plato points out in the *Timaeus* [29c–d; 44d], this form of knowledge is probabilistic at best, always prepared to give way to a better explanation).

What compels us to ascend to another level are the questions posed at the level we are at. We only know where we have been after we have left it. It's our pursuit of explanation that pushes us along. At the first level of the cave, we understand that we've been looking at shadows

---

[*] Plato has, just before introducing the Myth of the Cave, presented the famous Divided Line, which separates the world into various metaphysical levels, which are also epistemological levels. The major divide is between the non-intelligible realm, passively presented to us, and the intelligible realm, actively grasped through reason, and this corresponds to the distinction between opinion and knowledge. The levels of the Divided Line proceed from the imagined, to the perceived (both of these situated beneath the main divide), to the mathematical (the portal into the intelligible), to the forms, the abstractions that lie behind universal terms. Socrates mentions (517b) that the Divided Line should be used to understand the stages the prisoner must pass through in the Myth of the Cave. But that myth adds a normative emphasis to the Divided Line. The prisoner's ascent entails changes of values. This is what explains Socrates' otherwise strange response to Glaucon, regarding the lowest level, *eikasia*, "They're like us." Compare M. F. Burnyeat, "Plato on Why Mathematics Is Good for the Soul": "Socrates instructs us to apply the prisoner's upward journey to the soul's ascent into the intelligible region of the Divided Line. . . . But the surrounding narrative, about the journey back to the cave, would suggest a different solution. For the examples mentioned in the story are values."

[†] In the *Timaeus*, Plato presents abduction as essential to scientific reasoning.

only when we see the shadow-making operation that's going on inside the cave, the ways in which everything was elaborately rigged in there to create the illusions we took for real. One only understands that one has been living in a subterranean cave when one exits it, leaving behind the constructed values of a society imprisoned by its ideology with all its ruses designed to keep the prisoner from making contact with what is out there—in other words, reality. Climbing upward and exiting the cave is stepping outside ideology, terrifying and painful at first, but liberating and natural at the end, so that the thought of returning to the abandoned ideology becomes unthinkable. "I would suppose he would rather suffer anything than live like that" (516e). Exiting the ideological cave, where all our questions are answered and everyone we know agrees with us—"I say to you now, knowing full well that you will agree with me . . . only if you already agree with me"—is the hardest and most significant step we can take. But if we don't take that step, then we will leave this life no closer to the truth than when we entered it. And that is exactly what it is to live a life not worth living, even if it proves to be the most pleasant sort of existence.

Still, there are many levels still to be achieved outside the cave. Plato enumerates the extra-cavernous levels: images and reflections, the things themselves, the sun. What do these signify? His Analogy of the Divided Line (509d–513e) is the key. The images and reflections correspond to mathematics, the various branches of which Plato has his guardians of the *Republic* spend several decades mastering. The things themselves are the forms, those abstract theoretical entities that Plato believes, at least at this stage of his philosophical thinking, are necessary to explain the identities of concrete particulars.

But the trail of explanations doesn't hit a dead end, not even here, in this theorized domain of abstractions so far from common sense. Not even the intelligible forms are self-explanatory. There is a structure to this abstract domain—not all possible forms exist, some entail others, some exclude others. A complex structure is superimposed over this abstract domain. The abstract forms and their relationships with each other give reality the shape it has. But why is it this shape rather than another? Why is it anything at all?

A further ascent is required, to the form of the good. In the Myth of the Cave, this last ascent is reached when the former prisoner, now enlightened, casts his eyes up to the heavens and beholds the source of light itself, the sun, and the language becomes appropriately heated. "In the knowable realm the form of the good is the last thing to be seen, and it is reached only with difficulty. Once one has seen it, however, one must conclude that it is the cause of all that is correct and beautiful in anything, that it produces both light and its source in the visible realm, and that in the intelligible realm it controls and provides truth and understanding, so that anyone who is to act sensibly in private or public must see it" (517b).

Plato, in the *Republic,* is firmly on the side of the Reasonables. Everything we need to know—intellectually and morally—is *out there,* and the way we come to see what is out there is no more private and unshareable than the reality itself is. One proceeds by way of reason, by offering the best explanations for the questions that each level presents. An anonymous, allegorical knower stands in for any of us, so allow me to change the gender of the pronoun. The knower doesn't come with any special cognitive equipment of a kind to make her privy to special messages from outside the cave. It's on the power of her own reason that she achieves the vision of the sun. Not only is this a path that is, in principle, open to anyone, but it is a path that requires collaborators, since judging what is the best explanation is an activity best done with others, as the man who founded the Academy, gathering the best thinkers of his day there to join him, must have believed. The prisoner was herself first freed and dragged forward on the first leg of her trip by someone else, and once she sees the sun she remembers the prisoners still fettered in the cave and pities them, returning to help them make the ascent that she has achieved. (It doesn't necessarily end well. Prisoners of ideology don't necessarily welcome liberation.)*

The form of the good, of *agathon,* is the place where all explanations stop. It is the level of the self-explanatory. There must be such a level of the self-explanatory, if reality is, as Plato has assumed it to be, thoroughly intelligible. There are no brute contingencies, facts which are facts for no other reason than that they are facts. Explanations must penetrate the whole of what is. It's not turtles all the way down, but rather reasons, *logoi,* all the way down. This is the fundamental

---

*"Wouldn't it be said of him that he'd returned from his upward journey with his eyesight ruined and that it isn't worthwhile even to try to travel upward? And, as for anyone who tried to free them and lead them upward, if they could somehow get their hands on him, wouldn't they kill him?" (517a).

intuition of the rationalist; it was picked up again in the seventeenth century by such hard-core rationalists as Spinoza and Leibniz. Leibniz named it the Principle of Sufficient Reason.

Like them, Plato has demanded that reality thoroughly account for itself, every step of the way, and this entails that there must be a level of the self-explanatory. The way we ascended to each next level was to judge (as best we could) the best explanation. We've been led, every step of the way, by the intuition that the best explanation—the most beautiful, the most elegant—is the right explanation. The good is simply that intuition affirmed. Reality is what it is because it realizes the best of all possible explanations. This is the Sublime Braid—the True-the Beautiful-the Good. The structure of the world is shot through with a sublimity so sublime that it simply *had* to exist. Reality exists because it, too, is striving to achieve an existence worth the existing. The cosmos itself is a high achiever, and existence is the prize.

Plato has, in his explanatory ascent, implicitly posed the fundamental question of metaphysics: Why is there something rather than nothing? Leibniz is customarily credited with first explicitly formulating the question, and in those very terms, but, once again, Plato got there first (and Spinoza certainly beat Leibniz to it as well).* Plato implicitly posed the question by explicitly proposing his answer. The good is what bestows existence, he tells us in the *Republic*. *Agathon* binds the structure of reality—whatever that reality might turn out ultimately to be. (In the *Timaeus*, he voices skepticism that we can ever know it in its entirety. Reality's being intelligible doesn't entail its being intelligible to us.) Plato is open to reality's turning out to be quite different from the way we conceive it at any point in our joint adventure to figure it out. The self-questioning is of the essence of the rational process. But what he holds firm to is that whatever reality turns out to be like, it is like that because the best of reasons makes it so, and we are led to those best of reasons by our own sense of intelligibility-maximizing beauty: "Both knowledge and truth are beautiful things, but the good is other and more beautiful then they" (508e).

---

*For a look at contemporary thinkers—physicists, philosophers, and even a novelist—contemplating this question, see Jim Holt, *Why Does the World Exist: An Existential Detective Story* (New York: Liveright, 2012). Two of those whom Holt interviews describe themselves as Platonists in their approach to the question Holt poses them: the physicist Roger Penrose, who concentrates on mathematical Platonism as the answer to the ultimate metaphysical problem of why there is anything, and the philosopher John Leslie, who develops what is called "extreme axiarchism," or the rule of values.

The Myth of the Cave consigns anything that cannot give an account of itself—including whispers in one's own privileged ears from one's own private oracle—to the interior of the sooty cave. It is, in the end, hostile to the Unreasonables, who have to be placed at the level of *eikasia,* prisoners of ideology, unable to give a *logos.* Inferences to the best explanation are put on the seminar table, there for all to evaluate—not just in philosophy but in all theoretical domains (except mathematics, where conclusive proofs are possible). Inference to the best explanation captures what it *is* to theorize.

The word "best" is overtly evaluative. There is no escaping evaluation, no more in deciding what is rational to believe than in deciding what is ethical to do. The fact that evaluation is involved—different people may disagree on what constitutes the best of the available explanations—makes it all the more imperative to expose one's reasoning to a multiplicity of perspectives. When I had Plato say to Roy McCoy that he would rather be refuted than to refute, I was quoting him verbatim.

But what criteria are to be used in evaluating which are the best explanations? Here, too, disagreements erupt. We might ask: Is an explanation that increases the sense of mystery in the world to be valued over one that decreases the mysterious, or is it the other way round? There are excellent reasons, well argued and generally accepted, for embracing the latter alternative. In fact, precisely because the explanation that decreases mystery is judged the *better* explanation, Plato's own explanation of universals in terms of the abstract forms has been dropped in favor of other explanations. His so-called Theory of Forms created more mysteries than it solved. There's evidence that he himself drew the same conclusion as a result of the battery of criticisms he lodged at the theory in the *Parmenides.* In the *Timaeus* and the *Laws,* the most intelligible—and therefore beautiful—of the forms are conceived of in terms of mathematical structures, other forms dropping away.

The demiurge of the creation myth presented in the *Timaeus* created the physical universe as a living organism, imparting a soul to it and infusing it with as much eternity as it is possible for a time-dwelling entity to enjoy. The infusion comes about by making time itself an image of eternity. Unlike the truly eternal, the universe is in motion, but its motion is subtended by the law of number, which means it partakes, as best it can, of eternity. It's the mathematical motions within the cosmos that itself generate time, the image of eternity.*

> So, as the model was itself an everlasting *Living Thing*, he set himself to bringing this universe to completion in such a way that it, too, would have that character to the extent that was possible. Now it was the Living Thing's nature to be eternal, but it isn't possible to bestow eternity fully upon anything that is begotten. And so he began to think of making a moving image of eternity: at the same time as he brought order to the universe, he would make an eternal image, moving according to number, of eternity remaining in unity. This number, of course, is what we now call "time." For before the heavens came to be, there were no days or nights, no months or years. . . . These all are parts of time, and *was* and *will be* are forms of time that have come to be. Such notions we unthinkingly but incorrectly apply to everlasting being. For we say that it *was* and *is* and *will be*, but according to the true account only *is* is appropriately said of it. . . . And all in all, none of the characteristics that becoming has bestowed upon the things that are borne about in the realm of perception are appropriate to it. These, rather, are the forms of time that have come to be—time that imitates eternity and circles according to number. (37c–38b)

The mathematics inscribed in the heavens' motions, giving us time, also, in the *Timaeus*, generate the structure of matter. Reason saturates the cosmos in the form of mathematics, which not only allows the world of *was* and *is* and *will be* to partake in the everlasting *is*, but also renders the cosmos accessible to our mathematical reason. Our saving virtue is that our human reason can penetrate the cosmic reason:

> And when reason which works with equal truth whether she be in the circle of the diverse or of the same—in voiceless silence holding her onward course in the sphere of the self-moved—when reason, I say, is hovering around the sensible world and when the circle of the diverse also moving truly imparts the intimations of sense to the whole soul, then arise opinions and beliefs sure and certain. But when reason is concerned with the rational, and the circle of the same moving smoothly declares it, then intelligence and knowledge are necessarily achieved. (*Timaeus* 37b–c, translation by Benjamin Jowet)

---

*The *Timaeus'* denial that time is absolute, its making it a function of motion, portends ideas that would come to fruition in the theory of special relativity.

It's no wonder that Galileo and Kepler were passionate Platonists. Since the time of Thomas Aquinas, the Church had favored Aristotle. And whom the Church favors it becomes heresy to challenge. But it is Plato, particularly the Plato of the *Timaeus,* who is made to carry the spirit of rebellion that rose up in the sixteenth and seventeenth centuries against the dogmatized Aristotelian teleology. Finding their way back to Plato, the new physicists seize on mathematics as the very soul of explanation—and the more beautiful the mathematics, the more explanatory value it is judged to have. If the Aristotelian/Ptolemaic geocentric system is rejected, it isn't on the basis of observation alone—the epicycles can accommodate all the observed motions of the planets—but because the epicycles are mathematically hideous. Make the sun the point of origin around which the earth and planets revolve and the mathematics becomes beautiful. Plato's aesthetic realism profoundly affected the men who created modern physics, and both Galileo and Kepler often mention the "divine Plato," using Plato's criteria for judging the best of explanations as their own.

Inference to the best explanation is inescapably value-laden, but then so, too, in Plato's scheme of things, is reality. The positioning of *agathon* at the apex of the prisoner's vision means that there is something inherently superior about reality *as it is* that dictates that this is the reality that *had to be. Agathon* entails that reality can ultimately give an account of itself, which doesn't mean that we, mere humans, will ever be able to arrive at that ultimate accounting. But by Plato's lights, we can trust that the accounting exists. Trusting that it exists can be regarded as a part of the metaphysics of physics, on the basis of which enormous expansions of ontology have been argued, among which none could be more expansive—could it?—than the controversial notion of the multiverse. The physicist Brian Greene wrote an article in *The Daily Beast,* explaining the current thinking (of which he's a fan) according to which our universe is only one of a vast number of universes, which are composed of different particles and governed by different forces. How vastly many? According to string theory, "the tally of possible universes stands at the almost incomprehensible 10^{500}, a number so large it defies analogy."[*] Allowing mathematical elegance to

---

[*] Brian Greene, "Welcome to the Multiverse," *Newsweek,* May 21, 2012, http://www.thedailybeast .com/newsweek/2012/05/20/brian-greene-welcome-to-the-multiverse.html.

carry us along has gotten us far beyond the cave. Who knows? Perhaps we'll someday be able to answer why our universe—whether a multiverse or not—had to be exactly as it is. And if we do, it will be because of Plato's intuition that, when it comes to the universe, it's reasons all the way down, and that's what's so good about it. That's why *agathon* is sovereign.

The sovereignty of the good isn't challenged by pointing out all the ways in which reality could be improved. The view isn't challenged by such horrors as childhood leukemia, shifting tectonic plates, or wild fires. Such tragedies loom large in the human point of view. Reality doesn't take the human point of view, and it can't be expected to. The sublimity that had to burst into existence is not one that particularly concerns itself with us. Such a human-constrained goodness would not pack the ontological wallop required to bring forth existence. Benedictus Spinoza points out the irrelevance of the human point of view at the grand metaphysical scale, remarking that "the perfections of things is to be reckoned only from their own nature and power; things are not more or less perfect, according as they delight or offend human sense, or according as they are serviceable or repugnant to mankind" (*Ethics* I, Appendix).* Plato makes a similar argument in the Book X of the *Laws* (903c).

The view of an intelligible reality that shows no tilt toward human welfare strikes many as cold and inhuman. Well, it is cold and inhuman. What I ought to have said is that the view strikes many as repugnant as a consequence of being inhuman. It is no wonder that when Platonism met monotheism—in the Jewish thinker Philo and the Christian thinker Augustine—a user-friendly substitution was made in the tenancy at the top story of the explanatory scale. The good moved out, and God moved in. The new tenant was reputed to be nearly as interested in us as *we* are interested in us. From the point of view of

---

* Spinoza further continues: "To those who ask why God did not so create all men, that they should be governed only by reason, I give no answer but this: because matter was not lacking him for the creation of every degree of perfection from highest to lowest; or, more strictly, because the laws of his nature are so vast as to suffice for the production of everything conceivable by an infinite intelligence, as I have shown in Proposition 16" (Appendix to Part I, *Ethics*). In other words, for Spinoza, the infinite ontological fecundity, itself a measure of its perfection, spills out into a reality with aspects both pleasurable and painful for sentient creatures such as us. Our being those particular creatures makes us overestimate the significance of these pleasurable-for-us and painful-for-us aspects. Such overestimation indicates insufficient distance from the cave.

Plato (or Spinoza), this substitution at the top level carries us back in the direction of the cave and its self-aggrandizing ideology. From the point of view of a Plato or Spinoza, the overvaluation of the human point of view is itself an ideology.

AND BACK INTO THE CAVE YOU GO

But so far the notion of *agathon* explored is of a sort to have far more appeal to a physicist than to an ethicist. How can this highly theoretical *agathon* have any implications for the way we're supposed to behave? Plato isn't going to demand—ethically—that we all become cosmologists, is he?

Not quite, but almost. The good that structures reality has specific normative consequences. This, too, is implied by the Myth of the Cave. It's only because the former prisoner has encountered the good and understands its supreme explanatory role in existence that she remembers the prisoners left behind. She doesn't want to have to go back into the cave to attend to them. She'd far more rather calculate the beautiful equations describing matter in motion. But the very beauty of those equations has been impressed into her own being, and because of this impression she feels a responsibility toward those still staring pathetically at their ideologically generated images, rewarding one another with meaningless prizes.

In the myth, those prisoners, shackled and distracted, present no threat to her so long as she remains outside the cave. In the myth, she can just ignore them and go on in her thrilling explorations. In the myth, it's not her own self-interest that forces her back into the cave but rather a kind of altruism. But in the real world of the *polis* it's a different story. In the real world of the *polis* it's very much in the interest of those who wish to be allowed to figure out reality to try to eradicate any ideology, religious or secular, that would prevent their free inquiries. In the real world, Socrates ends up questioning ideology and he gets himself killed for it. So it's in the self-interest of the thinker to try to change the *polis* for the better, to try to make the world safe for her thinking.

But that's not the way Plato presents the situation in the myth. The myth does justice to the entire reality of values Plato finds out there beyond the cave, and that reality of values includes responsibilities to others.

The value-laden nature of impersonal reality causes a shift in a person's moral psychology, the harmony and proportionality that structure reality settling over a person's own psychic reality and transforming it for the good.* This is what the just person is: someone whose inner reality is congruent to outer reality. This is why Plato has the guardians of the *Republic* studying mathematics for several decades in preparation for statesmanship. Only people who have allowed themselves to be reformed by reality have it in them to try to reform the *polis* for the better.

But why should they want to? That's the question. Why shouldn't the former prisoner stay as far away from the ideologues as she can and just get on with her thinking, which is hard enough for her to do? (Though apparently—and unsurprisingly—not for Plato. For him, at least, it's all "unmixed pleasure," as he puts it in the *Philebus*.) Why should she risk the mockery and wrath that will greet her when she descends again into the darkness—a wrath that Plato presents as nothing less than murderous (517a)?

Agathon, impersonal and theoretical as it is, has implications for *aretē,* the human excellence that makes a life worth living. *Aretē* cannot mean putting only your own structural soundness in order while giving no thought to the well-being of others. That anonymous knower of the myth was herself initially dragged "up the rough, steep path," by another who didn't let go until he had pointed her in the direction of the sunlight. In retrospect, that aspect of the myth seems significant. Somebody who had attained life outside the cave ventured back in to help someone else out.

The *agathon* that Plato contemplates, as his philosophical life deepens, has no regard for our mortal welfare. It is even more indifferent to us than the old Greek gods. Still, we are morally improved by the contemplation of it and our moral improvement turns us into better citizens of the *polis*. Radical objectivity, purged of human concerns, turns out to be the best antidote for the smallness of human nature. Simply to attain this vision of objectivity requires us to overcome the deformi-

---

* "And the motions that have an affinity to the divine part within us are the thoughts and revolutions of the universe. These, surely, are the ones which each of us should follow. We should redirect the revolutions in our heads that were thrown off course at our birth, by coming to learn the harmonies and revolutions of the universe, and so bring into conformity with its objects our faculty of understanding, as it was in its original condition. And when this conformity is complete, we shall have achieved our goal: that most excellent life offered to humankind by the gods, both now and forevermore" (*Timaeus* 90d).

ties of our natures, the overprivileging of our own identities and their perspectives. It requires us to dismantle whatever ideology imprisons us, almost always because it flatters our own sense of self-importance. To fall in love with the impersonal beauty of objectivity, which doesn't love us back, is moral achievement in itself.

And once such a vision has been attained, other meritorious consequences follow. The vision induces an awe that puts our own self-centered concerns into the widest perspective possible, with the result of removing those self-centered concerns from our sights. The sense of proper proportion that settles over our inner lives not only allows for self-control and judicious choices, the "nothing in excess" praised by the excessive Greeks. It also produces the sense of proper proportion between ourselves and others. Once the distortions in our perspective are corrected, we're confronted with the "proportionate equality" that ought to reign no less in the world of people than in the cosmos itself. In the _Gorgias,_ Plato has Socrates arguing with Callicles, according to whom all that any of us wants is to get our own way in everything, and that's exactly what we'll all do if we can get away with it. Tyrants lead the happiest lives of all. In the course of the discussion Socrates remarks, "Yes, Callicles, wise men claim that partnership and friendship, orderliness, self-control, and justice hold together heaven and earth, and gods and men, and that is why they call this universe a _world-order,_ my friend, and not an undisciplined world-disorder. I believe that you don't pay attention to these facts, even though you're a wise man in these matters. You've failed to notice that proportionate equality has great power among both gods and men, and you suppose that you ought to practice getting the greater share. That's because you neglect geometry" (507e–508a). Anyone with a proper appreciation of proportion can't fail to appreciate that one's own self shouldn't be in disproportion to everyone else. The beauty of proportionality that has led one on, because one loves it, would cause one to abhor a situation that would bring one into disproportion with everyone else. There is an entire moral theory contained in this passage.

"You suppose you ought to practice getting the greater share." So many ideologies come down, in the end, to ways of justifying our sense that we ought to get the greater share. And so people who have managed to get themselves out of the cave, having achieved a life worth living, must have no illusions that they're entitled to that life any more than anyone else. If they gloat about their personal accomplishments, then they haven't achieved anything worth the achieving. And if they have, then they have no choice but to make their gains a boon for all, trying to better the *polis,* as distasteful as that might strike them.

And judging from the *Laws,* Plato did find the task distasteful, undertaken not as a pleasure but as an obligation. The *Laws* is his last book, written when he was an old man. Socrates is altogether gone, and three old men—one named Spartan, one named Cretan, and an anonymous Athenian—take their morning constitutional together, discoursing, along the way, on politics and jurisprudence. The anonymous Athenian of the *Laws* goes off topic and can only by force of will pull himself away from the exhilarating contemplation of the mathematical perfection of the ordered cosmos to think about the dreary business of how best to order the affairs of humans, whom he compares to puppets. When the Athenian makes the unflattering comparison a second time and the Spartan calls him on it, the Athenian apologizes and explains he's been thinking of the gods. And then, as if assuming the vantage point of the gods themselves—the gods who, whether described frivolously by Homer or with all the solemn beauty of mathematics by Plato, never place man at the center of their attention—the Athenian concedes, "However, if you will have it so, man shall be something not so insignificant but more serious" (804b) and, with a sigh of resignation not hard to imagine, gets back to work on designing laws for a *polis* in which we can do each other the least harm possible.

We achieve a life worth living by understanding how the cosmos achieved an existence worth existing. That's the notion of *aretē* at which he arrives. The impersonally sublime is internalized into personal virtue. "But now we notice that the force of the good has taken refuge in an alliance with the nature of the beautiful. For measure and proportion manifest themselves in all areas as beauty and virtue" (*Philebus* 64e). What is required is that our cultivated sense of beauty lay us open to "the best of reasons" that give the very shape to reality, and that we be overcome by reality's otherness and beauty. For him, this is having reverence for the gods, and it's why he names astronomy a kind of religious experience in the *Timaeus.* But astronomy isn't the only way that "measure and proportion manifest themselves" to overtake us. As mentioned in chapter ζ, he allows that music (of the right kind) and poetry (of the right kind) can also do the trick, just so long as we regard them as intimations of transcendence such as to give us our true measure in the world, which is—as all finite things are when measured on the scale of the sublime—immeasurably small. (It was Leibniz who introduced the term "infinitesimal." Plato would have loved the calculus.) These are the ethically transformative measures. Only once our grandiosity has been decimated—by the shock of reality's otherness and beauty—can we achieve a life that isn't a tissue of degrading lies and laughable illusions.

Not everybody wants this best of all lives as Plato describes it. Perhaps the only ones who want it are those who are capable of attaining it. Perhaps the best measure of the capacity for it is the desire for it. That would make the restrictiveness somewhat fairer. For restrictive it is. A philosopher whose conception of the best life demands so much sheer intellectual power is no egalitarian. The recognition that not one of us is more entitled than another to a life worth living doesn't entail that all of us have it in us to achieve that life. It only carries the obligation that those who have achieved that life will do what they can to help others achieve it, as best they can. The demiurge of the *Timaeus,* as much as he wished to model the created world in the image of eternity, had to make his peace with the recalcitrance of physical nature (47e–48a); and those who achieve Plato's *aretē,* as much as they wish to help others achieve it, have to make their peace with the recalcitrance of human nature.

Plato was of such a nature to think that the concept of *aretē* at which he had arrived, by way of *agathon,* would yield not only a life of virtue but also one of rarest pleasure. "Then we may say that the pleasures of learning are unmixed with pain and belong not to the masses, but only to a very few" (*Philebus* 52b). It's an unmixed pleasure that has been attained precisely because one hasn't been seeking one's own pleasure at all, as undoubtedly those "masses" can't help but do.

The paternalist strain is undeniable.[*] It's a strain that some have reveled in, triumphantly identifying with the "very few." My sense of Plato

---

[*] Compare: "These (people with the best natures) are not easily produced, but when they are born and are nurtured and trained in the necessary way, it is absolutely right for such people to be able to hold the inferior majority in subjection by thinking, doing, and saying all that concerns the gods in the right ways at the right times, not hypocritically performing sacrifices and purification rites for violations against gods and humans, but in truth honoring virtue. In fact, honoring virtue is the single most important thing for the entire city. Now we hold that this segment of the population is by nature best suited to authority and is capable of learning the noblest and finest studies, if anyone will teach them" (*Epinomis* 989c–d).

is that he didn't revel in it at all. I think it grieved him and he would have had it be otherwise, but he nevertheless thought the paternalism necessary. How could it not be necessary for a man who described the sense of sight as our most valuable sense because it allows us to know the astronomical revolutions? "The god invented sight and gave it to us so that we might observe the orbits of intelligence in the universe and apply them to the revolutions of our own understanding. For there is a kinship between them, even though our revolutions are disturbed, whereas the universal orbits are undisturbed. So once we have come to know them and to share in the ability to make correct calculations according to nature, we should stabilize the straying revolutions within ourselves by imitating the completely unstraying revolutions of the god" (*Timaeus* 47b–c).

Plato hasn't exactly eased up on the requirement that we strive to be extraordinary, has he? So I hope he'd be impressed by some of the progress that we've managed, collectively, to make. After all, among those beliefs that I think we can attribute to Plato was the belief in philosophy's power to persuade.

MANY PLATOS

Down through the ages there have been many Platos, and there still are.

There has been the Plato of the religionists, who found it necessary to replace his idea of the good, the ordering principle immanent within the cosmos, with a cosmos-transcendent God—a conception first arrived at by a small tribe living obscurely across the Mediterranean from the Greeks of Plato's day. That tribe had lacked a robust sense of the afterlife. But the Plato of the *Phaedo* found his way into its monotheism as it spread itself into the Greek world, most especially by the followers of Jesus, and the marriage between Abrahamic monotheism and the Plato of the *Phaedo* proved strong for all of Abraham's children—Jewish, Christian, and Moslem. The hereafter, thereafter, became one of the most cherished components of the Abrahamic faiths.

There has been the Plato of the mathematicians, for whom the realms of infinite mathematical structures are far more real than the blackboards on which they write their equations, and who describe themselves as Platonists accordingly; and there has been the Plato of the physicists and cosmologists, going back to Galileo, who found in Plato's mathematical aesthetics a form of explanation to topple the entrenched Aristotelian science of the time, and who are still prepared to follow their sense of mathematical beauty anywhere it leads, taking us vast distances away from the level of reality revealed to us by our senses, perhaps all the way to the multiverse.

There has been the Plato of the political theorists, a highly contentious lot. Some of them read Plato as utopian, while others argue that he was so vehemently anti-utopian that his analysis of justice must actually be read as a demonstration that no justice at all is possible and political idealism is itself a mirage. And then of course there is his paternalism, which some argue was merely a function of the Athens of his day, which he was trying to reform, and which others embrace wholeheartedly, endorsing the justice of a "natural aristocracy."

There has been the Plato of the poets, willing to overlook his occasional insults to their art for the sake of the transcendence of his language and visions, his love of beauty and belief in its redemptive powers.

And among the philosophers, there are too many Platos to enumerate.

All that I can do is try to give you mine. My Plato is an impassioned mathematician, a wary poet, an exacting ethicist, a reluctant political theorist. He is, above all, a man keenly aware of the way that assumptions and biases slip into our viewpoints and go unnoticed, and he devised a field devoted to trying to expose these assumptions and biases and to do away with any that conflict with commitments we must make in order to render the world and our lives maximally coherent. Because he created this field, we can look back at Plato and see where his own assumptions and biases sometimes did him wrong. If we couldn't, the field that he created would have proved a colossal disappointment to him, the faith he had put in self-critical reason unfounded.

Above all, my Plato is the philosopher who teaches us that we should never rest assured that our view, no matter how well argued and reasoned, amounts to the final word on any matter. And that includes our view of Plato.

1

PLATO IN THE MAGNET

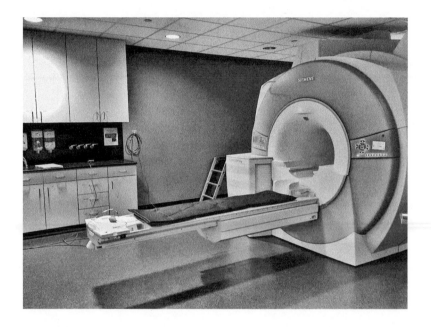

But our creators, considering whether they should make a longer-lived race which was worse, or a shorter-lived race which was better, came to the conclusion that everyone ought to prefer a shorter span of life, which was better, to a longer one, which was worse; and therefore they covered the head with thin bone, but not with flesh and sinews . . . and thus the head was added having more wisdom and sensation than the rest of the body, but also being in every man far weaker.[*]

—*Timaeus*

DRAMATIS PERSONAE:

DR. DAVID SHOKET: The Eugene and Eunice Quant Professor of Neuroscience at Olympia University; Howard Hughes Medical Institute Investigator; Fellow of the National Academy of Sciences; Fellow of the Neuroscience Research Program.
AGATHA FINE: Third-year graduate student in cognitive science at the university. Agatha has been working with Dr. Shoket to learn the tools of modern cognitive neuroscience.

Shoket walks through a door into a windowless conference room. Agatha is sitting at a small round table, filling in some paperwork. [†]

---

[*] In the *Timaeus*, Plato located our thought processes in the brain. He analyzed the brain as a sort of marrow, connected with marrow encased in bones that extend throughout the body. Not knowing of nerves, he saw this connective marrow as being the means of transmission between the brain and the rest of the body. The materialist conception of mind put forth in the *Timaeus* stands at odds with the dualism of the *Phaedo*, with which Plato is more characteristically associated. His student Aristotle disagreed with him and located thought processes in the heart, assigning the head only the task of cooling the blood, which became overheated in the thinking process. Aristotle based his mistaken conclusion on observation. He noted that all animals have sensations and yet only vertebrates and cephalopods have brains, whereas all animals have hearts or heartlike organs. He noted that the brain is relatively bloodless, and if it is laid bare it can be cut into without pain, whereas the heart is the source of blood and full of sensation.
[†] The alert reader will notice that David Shoket is referred to by his last name, while Agatha Fine is referred to by her first name. This difference is meant to reflect not only the disparity

SHOKET: Agatha! You're looking particularly radiant today.

AGATHA: Thank you. The subject is a male Caucasian, highly educated, born in Greece but fluent in English, 2,400 years old, with no detectable signs of dementia.

SHOKET: Something's different. Is that a new lab jacket?

AGATHA, *absentmindedly glancing down:* No. I've already explained the tasks we'll be asking him to perform while he's being scanned. He got the point of them quickly.

SHOKET: So not the typical undergrad we see in here. Not to speak of our violent felons, druggies, and other upstanding specimens of humanity.

Shoket laughs, and Agatha smiles perfunctorily. Shoket's laugh is loud and distinctive, reminiscent of an elephant seal's mating call.

Shoket's current research is in differences in the mesolimbic dopaminergic pathway implicated in individuals with a high degree of impulsivity. His preliminary model is that highly impulsive individuals, such as those that end up as addicts and felons, are characterized by diminished availability of midbrain autoreceptors, which potentiate dopamine release in reward pathways when the brain is exposed to novel, salient, or appetitive stimuli. Highly impulsive individuals are flooded with dopamine, which means they experience intense cravings, while their prefrontal cortex, the seat of higher-order mental processes, including self-control, remains, as a result, minimally activated.

SHOKET: Did you say the subject is 2,400 years old? *He whistles contemplatively.* Is he alert enough to be of any use?

AGATHA: He's very alert. I rated him a full 5 in mental acuity.

SHOKET: I'll spend a little extra time with him before we put him in the magnet,* just to double-check he's compos mentis. I don't want the jackbooted thugs from the Human Subjects Committee goose-stepping into my laboratory and shutting us down.

---

in their respective standings, most especially within the Shoket lab, but also the fact that I like calling the one "Shoket" and the other "Agatha." I hope the alert reader will impute no sexist tendencies to me.

* Among themselves, neuroscientists often facetiously refer to the equipment for performing functional magnetic resonance imaging (*f*MRI) as "the magnet."

Shoket laughs loudly, and Agatha smiles perfunctorily.

AGATHA: The subject has already signed the release forms.

SHOKET: He knew what he was signing?

AGATHA: You'll see for yourself. I'd pit his compos mentis against my own.

SHOKET: Uh-oh. Should I be worried about you?

Shoket laughs. Agatha smiles.

AGATHA: It's Plato, the famous philosopher. I read him in college. You must have read him yourself at some point.

SHOKET: He'll be able to follow directions?

AGATHA: That's his Chromebook over there. He's done his research, and he's eager to learn more. And, as I said, he's a famous philosopher.

SHOKET: You mean he's still a philosopher?

AGATHA: As far as I know. I don't think it's something a person can stop being.

SHOKET: Like being Jewish!

Shoket laughs. Agatha smiles.

SHOKET: I didn't even know there were still philosophers around. They have them on the faculty here?

AGATHA: Of course.

SHOKET *whistles contemplatively:* Live and learn. Do they share the same building with the astrologers and the alchemists?

Shoket laughs. Agatha smiles.

AGATHA: We have philosophers of mind at the Cognitive Science Center. They collaborate with us.

SHOKET: What's the point?

AGATHA: You could ask Plato that.

SHOKET: Right. I'm going to ask a two-thousand-year-old man to tell me something I don't know. It's like that old Mel Brooks–Carl Reiner comedy sketch. "Were you there when they prepared the cross for Jesus?" "Yeah, it was a hell of a lot easier to put together than the Star of David."

Shoket laughs. Agatha smiles.

SHOKET: You don't even know what I'm talking about, do you?

Agatha shakes her head.

SHOKET: You kids! What do you know from humor? All I can say is that it's a good thing that you got yourself over to my lab. It shows real sense on your part.

Shoket laughs. Agatha smiles.

AGATHA: He's just changing now. He knew that there was a danger of any metal heating up in the scanner, so he arrived in a chiton.

SHOKET: What's that?

AGATHA: A kind of toga.

SHOKET: A kind of toga? What is that, a joke? Are you the one being politically incorrect now?

AGATHA: It was a logical choice. No zippers, snaps, or metal buttons. But I told him that it's standard procedure to change, just in case there are any invisible metal filaments woven into the fabric.

SHOKET: And right you were. We don't want any spontaneous combustion of human subjects. Doesn't look good on the NSF grant reports.

Shoket laughs. Agatha smiles. Plato enters, wearing blue scrubs. Agatha, whose job it is to handle the test subjects, takes charge.

AGATHA, *smiling:* Dr. Shoket, this is Plato!

SHOKET: Yes, I'm Dr. Shoket. Please be seated. Agatha has explained to you what we'll be asking you to do while we're taking the functional magnetic resonance images of your brain, but if you have any remaining questions, feel free to ask me now.

PLATO: It is an honor to meet you, Dr. Shoket, and to be allowed to participate, in however small a way, in your research.

SHOKET: You understand, though, that you're just going to be a human test subject today. We're not going to be collaborators! This isn't the Cognitive Science Center! *Laughs.*

AGATHA, *hurriedly:* Thank you so much for agreeing to change into the scrubs. I realize they're not very dignified. They're comfortable, though, aren't they?

PLATO: Very comfortable indeed. Perhaps I can purchase a pair for myself?

AGATHA: Oh, I think we can let you have those as a little memento of your visit to the Shoket Laboratory. Can't we, Dr. Shoket?

SHOKET, *smiling:* Well, we may have to take it out of your stipend, Agatha. *Laughs.* Yes, of course, the scrubs are yours.

PLATO: I am grateful.

AGATHA: It's such an honor to have you here. It's the least that we can do to let you go home with something—aside from the little bit of money that I told you about that you could win, based on how you do in one of the tasks we'll be asking you to perform.

PLATO: The task that has me decide whether to bet small sums on various repeating patterns, inferring from the feedback whether betting on particular patterns allows me to lose a little money, win a little money, lose a lot of money, win a lot of money, or it makes no difference.

AGATHA, *smiling:* Exactly. I hope you don't feel any of this is beneath you. I'm hoping you'll find it interesting.

PLATO: I am certain I will. I have wanted to have my brain scanned since I first learned about the process on the Internet. In fact, I do not know if this is inappropriate for me to ask.

SHOKET: As I said, feel free to ask me any question.

PLATO: I was wondering if I might be permitted to see an image of my own brain. Is that possible?

SHOKET: No, I'm afraid not. We aren't doctors, you understand. If you've got something wrong with your brain, or you're worried that you do, we're not the right people to consult. A lot of people don't know the difference between neuroscientists, which is what we are, and neurologists, who are medical doctors, which is what we are not.

PLATO: I understand. It was an inappropriate request, and I apologize.

SHOKET: If we let people go off with images of their brain and it turns out they've got a tumor or a swollen artery that they think we should have told them about, next thing we know they'll come back lawyered up.

AGATHA, *hurriedly:* But as far as being in the brain scanner is concerned, there's nothing at all to be nervous about. Some people even find it relaxing in there. I do!

PLATO: So you yourself have volunteered as a test subject?

AGATHA: Often. And I entirely understand your wanting to see your own brain. I don't quite know why, but I feel a little thrill whenever I see an image of my brain. Look, there's my medulla oblongata and my thalamus and my pons. There's my amygdala at the tip of my hippocampus right where it's supposed to be. It looks exactly like any other brain, just like the drawings in the textbooks. I mean, it had better! But somehow it's not like seeing your lungs or appendix. It's weirdly gratifying.

PLATO, *softly:* Yes, so I had imagined it must be. One's own brain peering at an image of one's own brain, thinking thoughts prompted by seeing the image of the brain thinking thoughts one now isn't thinking but remembering.

Shoket, trying to catch Agatha's eye, laughs particularly loudly.

AGATHA: Anyway, given how you feel, I think you might hopefully get a thrill just knowing that your brain is being seen in action. We'll be right in the room outside, with a window looking into the imaging room, and even though you'll be out of our sight while you're in there, you'll be able to communicate with us the whole time by way of the intercom system. Also you'll be holding a panic button just in case anything makes you uncomfortable. The slightest bit of anxiety or unpleasantness, physical or otherwise, you just press the button, and we'll be there in seconds. The only thing that people sometimes complain about is the loud jackhammer-like noise. We'll provide you with earplugs to help with that. But it also helps to expect these noises and realize that the contraption isn't about to collapse on top of you!

PLATO: Might I ask what causes the noises?

AGATHA: Dr. Shoket can best answer that.

SHOKET: The noises are the result of the Lorentz forces in the gradient coils caused by the rapidly switching current in the static field. Do you understand?

PLATO: Not entirely.

SHOKET, *smiling:* I'll put it in simple terms. We are going to slide you into a mechanism that is basically a very strong magnet. It will give us data about which parts of your brain are activated as you perform the tasks we'll ask you to do. It will do this by showing us where the oxygenated blood is. When different parts of your brain are activated, they need oxygen, and this oxygen is carried by hemoglobin. When hemoglobin is deoxygenated, it can be magnetized, and when it's rich in oxygen it can't, and this is because hemoglobin contains iron, which is a metal. The hydrogen atoms in the water molecules in your blood, which all line up in the same direction in the strong magnetic field, are then perturbed with a strong radio pulse. As they wobble back into alignment, they give off tiny electromagnetic waves, which differ depending on whether the water molecule was next to an oxygenated or a deoxygenated hemoglobin molecule. The scanner detects those waves and reconstructs where in the brain the oxygenated hemoglobin molecules were concentrated.

PLATO: So by tracking the changes in the blood you can get a dynamic picture of the brain in real time as it functions?

SHOKET: That's just about right.

PLATO: Only it's not exactly in real time, since the changes in blood flow are much slower than the changes in activity in the brain itself. So when you look at the signal, you're looking at the level of activity averaged over two seconds. Since cognitive processes take place on the order of hundredths of a second, two seconds is a significant amount of blurring.

SHOKET, *visibly startled:* You seem to understand something of the science.

PLATO: I try. Your science interests me very much. The papers that have come out of this laboratory touch on problems that concern me.

SHOKET: You've read my papers?

PLATO: I read first "My Amygdala Made Me Do It: How Neuroscience Eliminates Right and Wrong." Then I read "There's Nobody Home: How Neuroscience Eliminates the Self."

SHOKET: Those are two of my best.

PLATO: I read them both with great interest. But in each paper, I felt that there was something essential that I was missing. I could not quite follow how you got from your data to the philosophical conclusions you drew.

SHOKET: A writer can only write. He can't make a reader understand what's written.

PLATO: This is very true. I have often reflected on the way in which writings always fail both the writer and the reader. Writings give the appearance of being intelligent, but if you question them with the intention of learning something about what they're saying, they always just continue saying the same thing. Every argument, once it's in writing, is bandied about everywhere, equally among those who understand and those who have no business having it in their hands (*Phaedrus* 275d).

SHOKET: It's funny that you should say that. When my collected papers were published as a book, they got a moron to review it in a major newspaper. This was a person who had no business even reading my book. He didn't understand anything I was saying. He wasn't even a real scientist.

PLATO: A piece of writing doesn't know to whom it ought to speak and to whom not. When it's ill treated and unfairly abused, it always needs its father to help it, since it isn't able to defend itself by itself (ibid.).

SHOKET: You're right about that also. And then it's considered bad form if an author protests a review that was written by an imbecile who had no business even reading his book. The author is just supposed to passively take it, and let the idiot get the last word.

PLATO: Yes, the writing needs its father. Which is why I have come to you, the father, to help me understand what I am missing.

SHOKET: See, that's honest. That's the way it ought to be. You're not a scientist so you can't follow everything I'm saying, but you acknowledge my expertise. If people recognized the expertise of the expert, there'd be far fewer problems in the world.

PLATO: I agree.

AGATHA: But Plato is the expert in philosophy. So perhaps the relevant expertise is his?

SHOKET: But the data I presented were scientific, not philosophical. "Philosophical data" is as oxymoronic as "military intelligence" or "airline food." *Laughs.*

AGATHA: But "philosophical conclusions" isn't an oxymoron, and the conclusions you drew were philosophical. The inference is mixed: one half lies in scientific expertise, the other half in philosophical expertise.

SHOKET, *laughing:* That sounds like something they teach you to say over at the Cognitive Science Center. I'm not sure that I even grasp the meaning of that phrase "philosophical expertise." It's not exactly an oxymoron, the way "philosophical data" is, but I still don't get it.

PLATO: I do.

SHOKET: Well, I suppose you'd have to claim that you do. How else could you justify yourself?

PLATO: True.

SHOKET: I'll tell you the way I see it, and then you can tell me where you, as the philosophical expert in the room, disagree. Philosophers hold down the fort until the cavalry, who are the scientists, arrive. That's a useful and maybe even heroic thing to do, to hold down the fort, but it's only once the cavalry gets there that anything gets done. Once the scientists arrive then the work begins. Because before that, let's face it, it was all bullshit. I hope you're not offended by my using that word?

PLATO: Bullshit? It is a useful philosophical term. I had a friend who would have loved it.

SHOKET: It's such a useful philosophical term that you might as well make it synonymous with philosophy, which suggests another metaphor. Science is like a sewage treatment plant. Scientists take the philosophical bullshit and reprocess it into knowledge.

PLATO: That metaphor presents not quite as heroic an image of philosophers as your previous one.

SHOKET: Well, let's face it. You philosophers used to have authority over a lot of questions, because none of the answers were remotely within sight. So if a person had a question, he went to his local philosopher. Do we have souls, they asked him, and if we do how do they interact with our bodies? What's the source of morality, they asked him, and how do we know whether the version we have is the right one? And what about beauty? Is it in the eye of the beholder or is it something really out there? And what are meanings? Are they out there, too, and how do they attach themselves to words? Oh, and while you're at it, could you tell us whether our lives have any meaning, and if so, how we can get us some? I bet people have even asked you these questions.

PLATO: They have.

SHOKET: And I bet you struggled to give them answers.

PLATO: I did. I do.

SHOKET: Of course you did. Nobody had the data you needed to answer the questions. Nobody had the technology to generate the data. So the default was to go to the person who could talk up a storm, who didn't require anything vaguely resembling evidence to talk his way into a fine-sounding conclusion, even though the philosopher a few villages over was talking up a storm in support of an altogether different conclusion.

PLATO: And now you have secured the evidence that will put an end to the talk-storms of philosophy.

SHOKET: Well, no, not yet, not all the evidence for all the questions, but we're on our way. We're so close that it's a foregone conclusion we'll get there, as long as enough smart people keep going into neuroscience and the funding continues. But the trend is clear. You can see it emerging over the centuries, accelerating these last few decades, especially with the advancement of brain science.

PLATO: So it is you brain scientists who are most responsible for forcing my early retirement.

SHOKET, *smiling:* Considering your age, I'd hardly call it early. Nobody is going to call you a slacker if, after 2,400 years, you just call it a day. But to answer your question, it's not just the brain sciences. You've got physics and cosmology closing in on the age-old problem of why there's something rather than nothing—that's one you philosophers, not to speak of theologians, have been chewing over for a while. With we neuroscientists explaining consciousness, free will, and morality, what's left for the philosophers to ponder?

PLATO: Perhaps self-deception?

SHOKET: We call it confabulation, and we've got that covered, too. We've known how confabulation works ever since the sixties, from experiments with split-brain patients. Patients with epilepsy were treated with surgery that severed their corpus callosum, the bundle of neural fibers that connects the left and right hemispheres of the brain. The surgery prevents the electrical firestorm that is epilepsy from reverberating throughout the whole, and it was, and still is, a treatment of last resort. But it proved a goldmine for neuroscience. We already knew that the left hemisphere controls the right side of the body, and the right hemisphere controls the left, and we already knew that the left hemisphere controls language. What the splits taught us is that the so-called rational part of the brain that controls language is actually an expert at confabulation—making up stories that are plausible but wrong. When the right hemisphere is shown images that the left hemisphere can't see and it provokes behavior in response to those images, then the left side, even though it has no idea of what has provoked the behavior, is never at a loss for words in "explaining" why he just did what he did. Show the right side dirty girly pictures and the left side will confabulate in all sincerity a pseudo-explanation for all the blushing and the giggling—how hot the testing room is, and how taking tests makes him nervous, and how hilarious he finds it that scientists could make a living from testing the likes of him. That's confabulation. That's what serves as "explanation" in the absence of data.

PLATO: So not only can you explain self-deception, you can explain the self-deception of all who try to offer explanations that aren't scientific.

SHOKET: You hit it on the head. I bet you're wishing right now that you'd run into me about 2,400-odd years sooner. I could have saved you a lot of wasted effort.

PLATO: You didn't have any of your wonderful science back then.

SHOKET: That's true. Be glad that you've lived to see it.

PLATO: I am. But even had you had the science back then, you wouldn't have turned me away from the life I've chosen, from the questions that have shaped it. I wouldn't have wanted to live my life in any other way.

SHOKET: That's nice, but I've scanned the brains of felons who would say exactly the same thing. The rush of dopamine they get when they contemplate certain pleasurable possibilities means they can't even consider living their lives in any other way.

PLATO: Oh, I can assure you I have always been able to consider living my life in other ways. When I said I would not have wanted to live another life, I did not mean I could not imagine other lives for myself, or give them serious consideration. It was a conscious decision and was not at all like the dopamine-driven brains of which you just spoke. I could give you my reasons for my decision, and, though my reasons may not persuade you to make a similar decision, they ought to allow you to understand why I chose as I did.

SHOKET: You seem to be claiming some sort of free will. I hope you're not invoking a mysterious non-physical soul, the ghost in the machine. I thought even philosophers had gotten the memo that the ghost has given up the ghost. *Laughs.*

PLATO, *smiling:* I am more than willing—whether freely or not—to concede that all of these processes of thought are brain processes. It is my brain that can erect a mental model of the future and think how things will change in that model in response to my doing one thing as opposed to another. And it is my brain that, working out the consequences of possible actions in my mental model, makes its decision. Call it free or not, but this process is very different from the mesolimbic dopaminergic processes that bypass decision-making.

AGATHA, *to Plato:* It's just like the image of the two-horsed charioteer you use to represent the human psyche. I've often thought how much Dr. Shoket's research vindicates your metaphor. Neuroscientists metaphorically refer to the prefrontal cortex as the brain's executive system, which corresponds to what you called the charioteer. The mesolimbic dopaminergic pathway corresponds to the deaf and bloodshot horse, who barely responds to a combination of whip and goad. And when you describe how the "bad horse" pulls in the direction of temptation, how it takes the bit in its mouth so that the charioteer has to resist powerfully, well, that resistance is what we're trying to catch on our brain scans as we ask subjects to imagine various scenarios and how they'd react. What we're finding is that there are some individuals whose mesolimbic horse pulls so hard that the prefrontal charioteer doesn't even exert itself. It's as if the charioteer just drops the reins and lets the bad horse go at it. But that describes only one segment of the population, and we know where to find them, often in maximum-security prisons.

PLATO: This raises the question of whether these driverless chariots can be held morally accountable. What underlies moral accountability is the capacities of the charioteer. There is the capacity to envision the future, and the capacity to deliberate over possible outcomes of actions, and the capacity to assign values to various outcomes, and the capacity to put judgments into action. I can see the argument for absolving of moral accountability those whose brain abnormalities preclude the charioteer from exercising its capacities. Their malfunctioning neurophysiology makes it impossible for deliberation to take place at all. And since it is the deliberating self that must be held accountable I would not hold these unfortunate people accountable. They should be regarded as ill and not evil (*Timaeus* 86c–e). But I cannot understand extending this absolution to those with functioning charioteers, as you argued in your paper "My Amygdala Made Me Do It." Here was a gap between your data and your conclusions that I could not follow.

AGATHA: And we even know where some of the kind of thinking that you're talking about is located. It's in the default mode network, which includes parts of the frontal lobe and the parietal lobe, and we use it to envision the future.

SHOKET: Yes, and we also use the default mode network for other things, like fantasizing and daydreaming, which sounds more like what you're describing to me. Because the story you just told about how we come to our decisions? It doesn't happen that way. It's a fantasy or daydream, that nice deliberate decision-making that you described. That's how it may seem to you when you introspect, but that's just part of the whole business of confabulation. We've known for decades that, at the point at which you seem to be deciding on some course of action, the neural machinery leading to your action has long been chugging away. Your sense of agonizing over your decisions and then putting them into play is an elaborate ruse that your neural networks are playing on you, a story to tell both to yourself and to others as to why you do what you do in the absence of any access to the real causal mechanism distributed among your synapses. That decision-making story you're telling is just like the splits, giggling over the dirty pictures and explaining how they find the scientists studying them giggle-worthy.

AGATHA: Are you talking about Libet here when you say that we've known for decades that the decision-making takes place after the fact?

SHOKET: Of course.

AGATHA, *speaking to Plato:* Are you familiar with Benjamin Libet's work?

PLATO: I'm afraid not.

AGATHA: This was back in the eighties. Libet used electroencephalography, or EEG, to record the brain activity of subjects who were told to make some spontaneous motion, like flicking their wrists whenever they felt like it. They were also watching a precise timer and were told to remember the time at which they became aware of their decision or their intention to move. Libet found there was a 200-millisecond delay, a fifth of a second, on average, between the experience of the decision to move and the movement itself. But the EEG revealed a signal that appeared in the brain even earlier—550 milliseconds, on average, a bit more than half a second—before the action. He called this the readiness potential, or RP, and he said that it's at the time of the RP that the gears fall into place that result in the movement, which means that at the time that the person reports making a decision, the action was already under way, without the subject's even knowing it. What the subject thought he was doing in making a decision was after-the-fact. So, in effect, there is no autonomous decision-making.

SHOKET: That's right. Your charioteer is out of the loop. And the moral is that introspection steers you wrong, whether it's the introspection of a subject in Libet's experiment noticing when he made a decision or the introspection of a philosopher speculating on how people make decisions. Introspection is confabulation in the absence of the data about the actual causal processes.

AGATHA: Only I've heard lots of criticisms of Libet's work over at the Cognitive Science Center. Not everybody buys that the experimental results, even if they're robust, demonstrate that decision-making is an illusion.*

---

* Some of the best analysis of the philosophical implications of the neuroscience have come from Adina Roskies, a philosopher with a Ph.D. in neuroscience. See A. L. Roskies, "Neuroscientific Challenges to Free Will and Responsibility," *Trends in Cognitive Sciences* 10 (2006): 419–423. There are more recent experimental results from which broad philosophical consequences have been drawn, especially those that have come out of the laboratory of John Dylan Haines, who used advanced analysis of fMRI data and claimed as much as a 7- to 10-second gap between neural activity predictive of an action and the time of the subject's report of

SHOKET: I don't want to tread on anybody's toes, but was it the philosophers over there who disputed Libet?

AGATHA: Well, yes, philosophers, but not only philosophers. One of the points I've heard is that Libet's results are exactly what you'd expect if the mind consists in neuronal activity. Whatever we do, including coming to a decision, there have to be neuronal processes underlying what we do. What you'd expect is a buildup of neural processes that constitute making a decision, and so you'd expect some signal like the RP a half second before the report of the decision. What would be surprising is if the complicated neural event that is coming to a decision somehow just came out of nowhere, instantaneously, with the brain doing nothing before. And another point I've heard made is that Libet asked the subjects to report when they become *aware* of their decision. But coming to a decision and becoming aware of the decision are two different events, so you'd expect a time lag between them, which adds to the interval between the RP and the time the subjects report.

SHOKET: I think that sort of nitpicking is missing the larger point, which is this: There are no "intentions" and no "decisions," because there's nothing like them down at the level of what's really going on. When you get right down to it, there isn't even a person. There's a brain, consisting of a hundred billion neurons, connected by a hundred trillion synapses, and that brain hasn't a clue as to what's going on in those hundred trillion synapses. Those things just happen, because they obey the laws of physics. All the ways the brain concocts for telling itself what's going on, in terms of intentions and decisions and urges and inclinations and preferences and conflicts, are so many fictions it's constructed in order to make sense of what it finds itself doing.

AGATHA: But do you really believe that?

SHOKET: It's what the science tells me.

AGATHA: But what you've just said doesn't do justice to all you know. You know so much more than that.

SHOKET, *to Plato:* That's the first I've heard her say I know more than I think I know. She's usually arguing with me that I know less.

Shoket laughs. Plato smiles. Agatha smiles.

AGATHA: Well, you know how to be a person.

---

deciding on the action. Here, too, the philosophical implications have been challenged, with Roskies a leading voice. See "Neuroscience vs. Philosophy: Taking Aim at Free Will," *Nature* 477 (2011): 23–25.

SHOKET: What do you mean, I know how to be a person? Are you saying I'm a good guy—in which case, I accept the compliment and, knowing how to be a good guy, I say, thank you very much.

AGATHA: I'm saying you know how to do all the things that a person has to do to be a person. You know how to feel proud of some things you've done and ashamed of others, change your mind about them both before and after you've done them. You know how to offer explanations for what you've done and to defend them, and you know how to understand the explanations and defenses that others give for what they've done. You know how to have goals and how to assess them and to put them into actions, and you know how to feel gratified when your goals are met and disappointed and frustrated and resentful when they're not, and you know how to blame others for what was really your fault. You know how to care about yourself in a way that's different from the way you care about anything else because you have a stake in the life that you know how to own as your own. That's some of what I mean by saying that you know how to live a life that's recognizably the life of a person, and if dolphins or elephants or Martians know how to do what you know how to do, then they're persons, too.

SHOKET, *to Plato:* Did you understand what she just said?

PLATO: Yes, I think so.

SHOKET: Well, I didn't.

PLATO: The disagreement between you reminds me of an argument I heard a long time ago.

SHOKET: Well, if it was so long ago I don't see why it would have any relevance to what we've been talking about. We're talking about our state of knowledge *now,* not a long time ago.

PLATO: Yes, we are always talking about our state of knowledge *now,* whenever it is that we talk, but still I think this old argument applies. Can I tell it to you?

AGATHA: I think I know the argument you mean. Was it Socrates who first proposed it, or was it you all along?

PLATO: Who can remember, after so many years and so many versions? I only know that I always hear it spoken in his voice, the one he always used when he was ready to stop his clowning and speak with quiet seriousness. *To Shoket:* You would have liked him. He was a kidder who liked to laugh, like you. So it could only strike wonder on occasions when all the laughter fled him and he appeared before us serious, as I think he must have appeared on that day, spending his last hours on the thin pallet of his prison cell, his closest friends gathered around him.

SHOKET: Why was he in prison?

PLATO: That was the very situation he used to make the argument I have in mind. Why was he in prison? What did he do to end his life in this way, and why did he do it? Imagine that Socrates were to say to you, Dr. Shoket, in answer to your question of why he is in prison, the following:

"The reason I am here, Dr. Shoket, sitting on a jail-cell bed, soon to die, is that my body is composed of bones and sinews, and the bones are rigid and separated at the joints, but the sinews are capable of contraction and relaxation and form an envelope for the bones with the help of the flesh and skin, the latter holding them all together, and since the bones move freely in their joints, the sinews by relaxing and contracting enable me to bend my limbs so that I can be in the position in which you find me.

"Would you feel that my explanation had done justice to your question, Dr. Shoket, when I never troubled to mention the real reasons, which are that since Athens has thought it better to condemn me, therefore I for my part have thought it better to sit here, and more right to stay and submit to whatever penalty she orders. Because, by dog, I fancy that these sinews and bones would have been in the neighborhood of Megara or Boeotia long ago—impelled by a conviction of what is best!—if I did not think that it was more right and honorable to submit to whatever penalty my country ordered rather than take to my heels and run away. But to call things like that causes is too absurd. If it were said that without such bones and sinews and all the rest of them I should not be able to do what I think is right, it would be true. But to say that it is because of them that I do what I am doing, and not through choice of what is best—although my actions are controlled by mind—would be a very lax and inaccurate form of expression. Fancy being unable to distinguish between the cause of a thing and the condition without which it could not be a cause (*Phaedo* 98c–99b).

AGATHA: That's amazing. Did he really say it like that? Were you there?

PLATO: I was not there. I was ill and could not come (ibid. 59b).

SHOKET: I'm sorry, I must be missing something here. There are all these allusions buzzing around that I'm not getting. But most of all, I'm just not getting why I'm supposed to be amazed by your friend's argument. Your friend is trivially correct that speaking of bones and joints and muscle contractions doesn't get the mind—which is to say the brain—into the explanation. At the very least, he should have added that the explanation for why he didn't flee was because his motor cortex didn't activate motor programs that moved his muscles and bones along some route that was guided by cognitive maps in his hippocampus.

AGATHA: That's true. You can't get an explanation of an action if you don't bring the mind in.

PLATO: Yes, that was in fact his point, that one must bring in the mind to explain his action. And he thought that the way to bring in the mind in explaining his action was to refer to the "right and honorable" aim he thought he would accomplish through his action. But that is only the old way of speaking of mind, uninformed by brain science, and we must replace it with our state of knowledge *now*. So then what my friend ought to have said is something along the following lines: The reason that I am lying here on this jailhouse bed is that my default mode network, interacting with memories stored in my hippocampus and medial temporal lobe, generates patterns of activity that correspond with various future scenarios, including fleeing and staying put. The staying-put pattern generates a conflict signal in my anterior cingulate cortex, because the ACC also receives a prepotent response from midbrain limbic circuits that cause the organism to struggle to escape confinement. The signal is then relayed to my dorsolateral prefrontal cortex, which engages in information processing to resolve the conflict. The DLPFC sends and receives signals from my ventromedial prefrontal cortex, which contains information about my long-term goals, and also connects to areas in the right superior temporal sulcus that allow me to simulate the actions of other people. The information in this network causes the DLPFC to resolve the conflict by sending signals to the premotor and motor areas, which cause the muscles of my body to leave me in the jail cell.

SHOKET: Okay, now I'm officially amazed. What have you been doing, auditing classes?

PLATO: MOOCs.

SHOKET: Okay, but I still don't see what your friend's argument, even supplemented with neuroscience, is getting at.

PLATO: Do you not see what it is that is still missing from the explanation of my friend's action? We cannot explain why my friend did what he did unless we understand what that action meant both to him and to others, how he saw it and what value he placed on it and how he saw how others would see it and what values they would place on it, both in his day and later, and back and forth in spiraling loops of values and meanings.

AGATHA: The way the philosophers at the Cognitive Science Center would put it is that you can't explain his action unless you view it in the context of value and meaning in which his behavior is embedded.

SHOKET: Fine. I have no problem with that, as long as you keep that context of meanings and values in the cortical activity of brains where it belongs. So let's say that your friend had decided to save his life and flee from jail. The patterns of synaptic strengths in the neural networks of various parts of his brain would have had to be different, since any difference in behavior must have originated from *some* differences in the brain.

PLATO: And would we be able to see those differences with the techniques you are soon to use in order to view my brain?

SHOKET: See them with our fMRI? Well, no. They'd be differences at the level of microcircuitry, which we obviously can't see with the millimeter resolution of fMRI. After all, there are a hundred thousand neurons and as many as a billion synapses in each cubic millimeter of the cortex. Even with the best technology today, we can't record from more than a few dozen of them at a time. But that's just the limits of our technology. It doesn't make any difference to the neuronal differences that would really have been there had Socrates chosen differently.

AGATHA, *to Plato:* Isn't that what you'd call the difference between ontology and epistemology? The difference between what there is as opposed to how we can know what there is?

PLATO: Yes, it is. So let us close the gap between ontology and epistemology by imagining a technology as good as you like. Imagine a technology in the future that could record the individual activity from all of the brain's hundred billion neurons at once. Also suppose that the human connectome project, showing how all the different parts of the brain are connected neuron by neuron, synapse by synapse, has been completed. Suppose you could plug the firing rates into a massive computer simulation of that wiring diagram of the brain. That sounds amazing, but it's no less amazing than the technologies I've seen develop in my 2,400 years. Would you then be able to explain what made Socrates remain in jail by giving an explanation at the level of his cortical activity?

SHOKET: Well, the difference in the firing patterns would be extremely complicated, consisting of billions of tiny differences in synaptic weights, leading to differences in the firing patterns over hundreds of millions of neurons. It would all be too complicated for anyone to trace out in his mind's eye. But the differences would have to be in there somewhere—we could run the simulation, and it would tell us whether Socrates decided to flee or to stay. There are no facts over and above those, so yes, in principle, that's where we'd find our explanation.

PLATO: But how would we explain why the computer simulation ended up in the state in which Socrates stayed rather than the state in which Socrates fled, if the explanation in both cases consisted of reciting billions of numbers that no one could understand? Where in that mass of numbers could we extract anything approaching what my friend was able to say so simply and so many years ago, making himself intelligible to his contemporaries and even now to you who live so differently, by presenting the reality of himself as a person, which is to say a creature ready to have something to say when asked why he behaves in the way that he does, submitting his reasons to a community of those who know how to interpret what he says, who may not find his particular reasons persuasive but know at least how to respond when offered reasons, how to consider and to evaluate? Even if we could get our hands—or rather our minds, which is to say our brains—on those masses of numbers, could they ever absorb the masses of meaning and mattering, the standards of reasoning and behaving to which we submit ourselves in order to live lives that are not only coherent to ourselves but coherent to one another—and coherent to ourselves at least in large part because they are, or we know how to go about making them, coherent to one another? All of that and more goes into constituting the shared world in which we do our living, and without which there is no life that is recognizably a life. That was the assertion that Agatha made, and which you asked me to explain to you.

SHOKET: But it's all just back to confabulation again. We make ourselves coherent to ourselves and each other by telling ourselves and each other stories, about what we did and why we did it. But how is that any different from the giggling splits? Where there's one way to tell a story, there are always many more, and who's to say which is the right one, which to me means that there isn't any of them that's right. Take your friend. He tells his reasons for why he's in that prison, lying there instead of fleeing, and maybe he even believes his story, since it's certainly self-serving. But maybe his wife— Did he have a wife?

PLATO: He did.

SHOKET: Well, then, maybe his wife would say that he acted the way he did because he was so vain he couldn't face growing old and infirm. Maybe some of his political enemies— Did he have any?

PLATO: He did.

SHOKET: Maybe his political enemies would say that he wanted to make a martyr of himself to serve his political agenda. Maybe a psychologist would tell us that he was suicidal but didn't have the guts to do it on his own so he worked out a way for the state to do it for him. Maybe a game theorist would say he was playing a game of chicken with the state, gambling that they'd back down and free him and he'd come off a hero, but instead he lost the game. Who's to say? Not your friend. There's no reason to give priority to his version, is there? You're the one who first mentioned self-deception.

PLATO: I did.

SHOKET: There's always a way of telling the story differently, endless ways that it can be played. There are too many degrees of freedom here for anything to ever get explained. Do the science, give me better and better brain images, and let me see the actual information that's there. At least, there aren't endless ways for me to play that out, even if I can't get a once-upon-a-time story out of it of the kind that both of you seem to think alone provides coherence.

PLATO: You're right that where there's one story there are many more, each of them competing with the others as ways of providing coherence. But that seems to me better than stopping short of coherence all together.

AGATHA: It's a choice between the indeterminacy of too many explanations and the incoherence of none at all.

SHOKET, *to Agatha:* Don't you worry. We'll find a way of getting the coherence *and* the determinacy. It's still the early days. Everything that's true of a person is there in his brain, so why shouldn't we eventually be able to get everything by knowing the brain?

PLATO: Perhaps. Perhaps you will arrive at both determinacy and coherence. And if you do, I hope I'll be around to see it.

SHOKET: Well, that would be the moment for you to retire at last.

PLATO, *smiling:* Let's wait and see you do it first.

SHOKET: And until we do, you philosophers will always be pointing out that we haven't done it yet.

AGATHA: And they'll be accounting for the coherence that everybody needs, including the scientists.

PLATO: That is my hope, for I think it means that there will always be a need for philosophers.

Shoket laughs. Plato smiles. Agatha smiles.

AGATHA, to *Shoket:* And until you *do* arrive at coherence it's premature to go around banishing things like persons, together with all the intentionality that they bring along with them—deliberating and agonizing, deciding and second-guessing, acting and regretting—just because you can't find all that down there among the synapses.

SHOKET: Deliberating and deciding and second-guessing? Agonizing and regretting? My synapses don't know from such things!

AGATHA: But you do.

SHOKET: Maybe I just *think* that I do.

AGATHA: And can you find your just *thinking* that you do any better down there among the synapses?

SHOKET: I can find patterns associated with confabulation and rationalization in the orbitofrontal and ventromedial cortices.

AGATHA: And the self that's doing the confabulating and rationalizing?

SHOKET: No such thing.

AGATHA: So there are confabulations but no confabulator? Rationalizing but no rationalizer?

SHOKET: You got it.

AGATHA: *Who* got it? There's nobody here but us synapses!

SHOKET *laughs and says to Plato:* You're definitely having a bad effect on her. It's a battle between us for her soul—and when I say *soul*, I hasten to add, I'm only speaking metaphorically! Which is what all of us are doing most of the time, whenever we invoke such things as those fascinating selves, each a star in its own story. It's a way of telling ourselves stories that correspond to nothing at all down at the level of the neurons, where we're talking about what *is*—where we're talking about *all* that there is.

AGATHA: Self-starring stories that we can't do without, in one form or another, and retain any semblance of coherence.

SHOKET: Listen, that way of thinking is going to lead you to utter *in*coherence. You open the gates to what people think they need in order to make sense of it all, and you're going to get all kinds of *mishegoss*.

PLATO: *Mishegoss?*

SHOKET: Lunacy. People claim to need all kinds of *chazerai* in the name of coherence.

PLATO: *Chazerai?*

SHOKET: Stuff you like to eat but shouldn't. You've got people who need their flakey puff pastry with the bitter-herbed God-filling, which has been hanging around way past its expiration date, so you have to wonder how they can choke it down. And then you've got the lighter fare bunch, the spiritual vegans, with their quantum-crackery energy fields, served up with some psycho-deli slices of aura or just a shmear of Jeez-whiz. What's the difference once you start speaking of what people need for coherence? It's going to be a metaphysical fres-fest. Come to think of it, that's exactly what it is, at least outside the walls of the Shoket lab.

AGATHA: I don't think so. There are other labs on this campus, and they're every bit as scientific as the Shoket lab, even though they use psychological concepts that we can't make use of at the level that we're studying.

SHOKET: How are they scientific when they're invoking things that don't exist?

AGATHA: You're begging the question when you say they don't exist just because they play no role in the Shoket lab. Those things are scientific because they're featured in testable law-like generalizations. That's the criterion for scientific, not whether or not it's studied in the Shoket lab. Imagine if the chemists went marching en masse over to the geology department and demanded that the geologists stop discovering laws about rivers—say, how rivers affect the outer banks more than the inner banks—because the chemists don't see anything at all like rivers down at the level of the water molecules, which is, the chemists declare in unison, all that water really *is*, and so they deny any allegedly scientific laws the geologists might have discovered about rivers. We can't give up rivers and all that we know about them without shrinking our landscape of coherence, and we can't give up sense-making, story-starring selves and all that we know about them without shrinking our landscape of coherence.

SHOKET: Yeah, well I give you your selves, Agatha, and next thing I know you're going to be demanding that they have free will. And *that* we simply can't have. Free will is inconsistent with what we've determined in this lab.

AGATHA: I'll freely give up free will, at least as it's usually understood, and trade it in for accountability.

SHOKET *laughs:* Accountability? What's that supposed to mean?

AGATHA: It's supposed to mean *this*. Exactly *this*.

SHOKET: What, *this*?

AGATHA: *This* that we're doing. Offering each other our reasons, evaluating them, accepting and rejecting and reconsidering them and maybe even changing our minds. To be accountable means to be prepared to give reasons for the things we say and the things we do. I'm demonstrating accountability right now by giving you an account of accountability. I'm prepared to provide you with a reason for why we're always providing reasons. I'm prepared to give you a reason for why we ought to be providing reasons.

SHOKET *laughing:* Sounds like some kind of self-referential paradox that you're toying with there. Agatha, I'm getting more worried about you by the minute. Hanging out with those philosophers at the Cognitive Science Center has given you a dangerous taste for paradox.

AGATHA: And by saying what you've just said you're offering me your reason for why you think we can get by with*out* offering reasons? I'd say that's more of a paradox!

SHOKET, *laughing and shaking his head, to Plato:* I'm holding *you* accountable for this situation.

PLATO, *smiling:* I gladly accept accountability.

SHOKET: Then it's yours. Whatever it's supposed to be, this mysterious accountability, and wherever it's supposed to be, it's yours, Plato. And since it's yours, I should be able to catch sight of it there in your brain. Speaking of which, it's time to get this show on the road. So, Norma Desmond, are you ready for your close-up?

Shoket laughs. Agatha smiles. Plato smiles.

SHOKET: You both don't have any idea what I'm speaking about, do you?

Agatha and Plato shake their heads. Shoket sighs exaggeratedly.

SHOKET: It's Plato, isn't it?

PLATO: It is.

SHOKET: And you're keen to get a look at that brain of yours?

PLATO: I am. It is as Agatha described it. I cannot say why one's own brain should matter so much to one just because it happens to be one's own brain, but it does. It undeniably does.

SHOKET: You might as well ask why your own self should matter so much to yourself just because it's your own self.

PLATO: That's right. One might just as well ask.

SHOKET: Which means you think that the question about why your own self should matter is also a question worth asking.

PLATO: Oh, it is a question well worth the asking. I myself have never stopped asking it from the beginning until this moment.

Shoket looks for a moment as if he will laugh, but then he doesn't, and he is at a loss to say why. They move to the room where the brain scanner is, and Agatha gets Plato settled onto the scanning table. She covers him gently with a blanket and gives him some last-minute instructions. He lies there calmly, earplugs in his ears and panic button in his hand.

AGATHA, *softly to Shoket:* Don't you think that we might give Plato what he's asked of us? He's waited such a long time.

SHOKET, *softly:* Of course we can. Of course we will.

And with that they slide the philosopher into the magnet.

APPENDIX A

Socratic Sources

Aristotle writes in his *Poetics* (1447b) of an established genre of Socratic literature, *Sōkratikoi logoi,* all of which was written after Socrates' death in 399 B.C.E. In an extant fragment from his own lost dialogue, *The Poets,* Aristotle is said to have mentioned Alexamenos of Teos or Styra as the originator of the genre, but nothing of Alexamenos remains. We do have fragments of four other Socratic writers: Antisthenes, Aeschines, Phaedo, and Euclides. There is anecdotal material concerning a fifth Socratic writer, Aristippus. Phaedo, Euclides, and Aristippus were all non-Athenians. In the years immediately following Socrates' death, Antisthenes was probably considered the most important of the Socratic writers.

The two writers whose Socratic literature survived intact are Xenophon and, of course, Plato. Xenophon left Athens in 401 for Persia to help Cyrus unseat the Great King Artaxerxes. The *Anabasis* is his account of the fifteen months he spent on the adventure. At the end of that time, he was placed under Spartan command—which led to his being exiled by Athens. His contributions to the Socratic literature are the *Apology, Memorabilia, Oeconomicus,* and *Symposium.* His *Apology,* like Plato's, presents itself as offering an account of what Socrates actually said at his trial, although Xenophon was away from Athens at the time. His account differs from Plato's, perhaps most significantly in stressing that Socrates had perhaps wanted to be executed by the state, since he had only old age and the diminution of his powers to look forward to. He'll be missed more if he dies with his mental faculties intact. (The Athenians had concocted a hemlock solution that led to a not overly unpleasant death.)* *Memorabilia* consists of many anecdotes about Socrates, meant

---

*The death that Plato describes in the *Phaedo* is calm and dignified, his mind remaining clear until the end. And yet hemlock, many have claimed, would have produced a far more agonized death, characterized by violent seizures. Was it really hemlock then? Plato mentions only *to pharmakon,* "the drug," not specifying hemlock, *kôneion.* Or did Plato distort Socrates' death for dramatic or philosophical reasons, as some have claimed? So have argued Christopher Gill in "The Death of Socrates," *Classical Quarterly* 23 (1973): 25–28; William

to illustrate that Socrates' life in every way possible confuted the two charges brought against him of impiety and corruption of the young. Xenophon's dialogue, the *Symposium,* is of an altogether different sort, quite playful and even naughty. It presents Socrates at a dinner party, although instead of long speeches, there is repartee and wisecracking. Each guest describes what it is about himself that he most values, his most cherished *aretē.* Socrates' response is that he most values his art of pimping. He also argues that he is more beautiful than the guest who had singled out physical beauty as his *aretē,* since Socrates' eyes are so protuberant that they can see more and so are more beautiful, and the flailing nostrils of his snub nose can take in more smells, and his thick lips can give softer kisses. *Oeconomicus* is a Socratic dialogue on household management and agriculture that features Socrates, pleading ignorance on these matters, quoting what he has heard from Ischomachus.[*]

For many years, Xenophon was regarded as more reliable than Plato on the historical Socrates, precisely because he was considered a dullard, devoid of the creative genius of Plato that reshaped the person of Socrates to philosophical and literary requirements. Bertrand Russell, for example, remarks: "Let us begin with Xenophon, a military man, not very liberally endowed with brains" (*A History of Western Philosophy,* p. 102). This attitude has been criticized in recent scholarship. Those who argue for Xenophon's historical accuracy are more likely to do so on the basis of his substantial contributions to history. His *Hellenica* takes up where Thucydides left off. Vivienne Gray argues, "Xenophon creates a coherent image of Socrates no more or less historical than those of the other Socratics."[†]

Ober in "Did Socrates Die of Hemlock Poisoning?" *New York State Journal of Medicine* 77, no. 1 (February 1977): 254–258; and Bonita Graves et al., in "Hemlock Poisoning: Twentieth Century Scientific Light Shed on the Death of Socrates," in *The Philosophy of Socrates,* ed. K. J. Boudouris, pp. 156–168 (Athens: International Center for Greek Philosophy and Culture, 1991). Enid Bloch does some fascinating detective work in "Hemlock Poisoning and the Death of Socrates: Did Plato Tell the Truth?" in *The Trial and Execution of Socrates,* ed. Thomas C. Brickhouse and Nicholas D. Smith (New York: Oxford University Press, 2001). Her conclusion: "In the end I have been able fully to align Plato's description with modern medical understanding. Socrates suffered a peripheral neuropathy, a toxin-induced condition resembling the Guillain-Barré syndrome, brought about by the alkaloids in *Conium maculatum,* the poison hemlock plant. Plato proves to have been entirely accurate in every clinical detail, while Gill, Ober and Graves were mistaken in the violent demise they imagined for Socrates."
[*] Michel Foucault has a chapter, "Ischomachus' Household," that uses Xenophon's depiction of the way in which both wives and slaves are to be trained as a paradigm for studying the ancient Greek ideology of power. *The History of Sexuality,* volume II, *The Use of Pleasure* (New York: Vintage Books Edition, 1990), pp. 152–165.
[†] Vivienne Gray, *The Framing of Socrates: The Literary Interpretation of Xenophon's Memorabilia* (Stuttgart: Franz Steiner Verlag, 1998).

APPENDIX B

The Two Speeches of Pericles from Thucydides'
The History of the Peloponnesian War

Thucydides was born an Athenian and was, at one time, an important general in the Peloponnesian War. But after a military campaign in which he participated was bungled, he was banished, and thereafter devoted himself to recording, with self-imposed objectivity, the events of the protracted war. He always speaks of the Athenians in the third person.

Thucydides' attitude toward Pericles is still a matter of debate. There is no doubt that Thucydides admired Pericles' intelligence and political skills. But did he think that his imperialistic goals would have led inevitably to Athens' downfall? Or was his judgment rather that, had Pericles not died prematurely, he might have seen Athens through to victory? Did Thucydides approve or disapprove of Pericles? Academics still argue.[*]

In Book 2, there are two important speeches of Pericles. Together they shed light on how classical Athens transformed the Homeric ethos into an ideology of Athenian exceptionalism.

The first speech is the famous Funeral Oration that Pericles delivered after the first year of what would be a twenty-seven-year war. It was at a commemoration for the war dead, a ceremony full of pomp and circumstance, at public expense, "as was the custom of their ancestors." Pericles begins by pointing out an alleged fact about Athens that Herodotus also mentions in the context of acknowledging Athenian exceptionalism. They alone, of all the *poleis,* were reputed to be autochthonous: "Because they have always lived in this land, they have so

---

[*] In *Thucydides, Pericles, and Periclean Imperialism* (Cambridge: Cambridge University Press, 2010), Edith Foster comes down on the side of disapproval. She argues that Thucydides' text surrounds Pericles' speeches with a narration of events that gestures toward the doom implicit in his imperialism. His narration in Book 2 undercuts the "simultaneously idealized and evasive view of Athenian imperial power depicted by Pericles in his speeches" (p. 183).

far always handed it down in liberty through their valor to successive generations up to now." For the bulk of the oration, Pericles extols Athenian superiority, emphasizing that it is a participatory superiority in which all its citizens share. The very character of the Athenian, his psychological traits and dispositions, is intrinsically superior—free and yet law-abiding, open-minded, generous, productive, elegant, tolerant, original, manly, responsible, self-sufficient, courageous, civic-minded, honest, daring, just, and unique. Also modest. These personal attributes derive from the *polis* in which Athenians are fortunate enough to have been born. The superiority of the *polis* is a consequence of its citizens' inherent superiority, but it also furthers their exceptionality, especially since their form of government requires them to participate actively in the life of the *polis*. Therefore, more than any other state, Athens is a genuine reflection of its citizens' own qualities, and they can lay claim to its exceptionality as their own.

> [T]he customs that brought us to this point, the form of government and the way of life that have made our city great—these I shall disclose before I turn to praise the dead. I think these subjects are quite suitable for the occasion, and the whole gathering of citizens and guests will profit by hearing them discussed.
>
> We have a form of government that does not try to imitate the laws of our neighboring states. We are more an example to others than they to us. In name, it is called a democracy, because it is managed not for a few people, but for the majority. Still, although we have equality at law for everyone here in private disputes, we do not let our system of rotating public offices undermine our judgment of a candidate's virtue; and no one is held back by poverty or because his reputation is not well-known, as long as he can do good service to the city. We are free and generous not only in our public activities as citizens, but also in our daily lives: there is no suspicion in our dealings with one another, and we are not offended by our neighbor for following his own pleasure. We do not cast on anyone the censorious looks that—though they are no punishment—are nevertheless painful. We live together without taking offense on private matters; and as for public affairs, we respect the law greatly and fear to violate it, since we are obedient to those in office at any time, and also to the laws—especially to those laws that were made to help people who have suffered an injustice, and to the unwritten laws that bring shame on their transgressors by the agreement of all.
>
> Moreover, we have provided many ways to give our minds recreation from labor: we have instituted regular contests and sacrifices through-

out the year, while attractive furnishings of our private homes give us daily delight and expel sadness. The greatness of our city has caused all things from all parts of the earth to be imported here, so that we enjoy the products of other nations with no less familiarity than we do our own.

Then, too, we differ from our enemies in preparing for war: we leave our city open to all; and we have never expelled strangers in order to prevent them from learning or seeing things that, if they were not hidden, might give an advantage to the enemy. We do not rely on secret preparations and deceit so much as on our own courage in action. And as for education, our enemies train to be men from early youth by rigorous exercise, while we live a more relaxed life and still take on dangers as great as they do. . . .

We are lovers of nobility with restraint and lovers of wisdom without any softening of character.* We use wealth as an opportunity for actions, rather than for boastful speeches. And as for poverty we think that there is no shame in confessing it; what is shameful is doing nothing to escape it. Moreover, the very men who take care of public affairs look after their own at the same time; and even those who are devoted to their own businesses know enough about the city's affairs. For we alone think that a man who does not take part in public affairs is good for nothing, while others only say he is "minding his own business." We are the ones who develop policy, or at least decide what is to be done; for we believe that what spoils action is not speeches, but going into action without first being instructed through speeches. In this too we excel over others: ours is the bravery of people who think through what they will take in hand, and discuss it thoroughly; with other men, ignorance makes them brave and thinking makes them cowards. But the people who most deserve to be judged tough-minded are those who know exactly what terror or pleasure lie ahead, and are not turned away from danger by that knowledge. Again we are opposite to most men in matters of virtue: we win our friends by doing them favors, rather than by accepting favors from them. A person who does a good turn is a more faithful friend; his goodwill towards the recipient preserves

---

* Paul Woodruff, whose translation I am using, adds this useful footnote: " 'We are lovers of nobility without restraint, and lovers of wisdom without any softening of character' *(philokaloumen te gar met' euteleias kai philsophoumen aneu malakias)*: the most famous sentence in Thucydides. Like many of Thucydides' memorable sentences it admits of a variety of interpretations. I have translated *kalon* here as nobility, meaning nobility of character, but the reader should be warned that it can mean beauty as well. *Met' eutelaia* could also mean 'without excessive expenditure,' but this seems inappropriate. If Pericles means that Athens is not extravagant, his claim is preposterous in view of his magnificent building program. 'Lovers of wisdom' translates *philosophoumen,* which is cognate to our 'philosophize' but has a much wider meaning." *Thucydides: On Justice, Power, and Human Nature,* p. 42, n. 103. And remember, too, that where Woodruff writes "virtue," the word being translated is *aretē.*

his feeling that he should do more; but the friendship of a person who has to return a good deed is dull and flat, because he knows he will be merely paying a debt—rather than doing a favor—when he shows his virtue in return. So that we alone do good to others not after calculating the profit, but fearlessly and in the confidence or our freedom.

In sum, I say that our city as a whole is a lesson for Greece, and that each of us presents himself as a self-sufficient individual, disposed to the widest possible diversity of actions, with every grace and great versatility. This is not merely a boast in words for the occasion, but the truth in fact, as the power of this city, which we have obtained by having this character, makes evident.

For Athens is the only power now that is greater than her fame, when it comes to the test. Only in the case of Athens can enemies never be upset over the quality of those who defeat them when they invade; only in our empire can subject states never complain that their rulers are unworthy. We are proving our power with strong evidence, and we are not without witnesses: we shall be the admiration of people now and in the future. We do not need Homer, or anyone else, to praise our power with words that bring delight for a moment, when the truth will refute his assumptions about what was done. For we have compelled all seas and all lands to be open to us by our daring; and we have set up eternal monuments on all sides, of our setbacks as well as of our accomplishments.

Such is the city for which these men fought valiantly and died, in the firm belief that it should never be destroyed, and for which every man of you who is left should be willing to endure distress.

That is why I have spoken at such length concerning the city in general, to show you that the stakes are not the same, between us and the enemy—for their city is not like ours in any way—and, at the same time, to bring evidence to back up the eulogy of these men for whom I speak. The greatest part of their praise has already been delivered, for it was their virtues, and the virtues of men like them, that made what I praised in the city so beautiful. Not many Greeks have done deeds that are obviously equal to their own reputations, but these men have. The present end these men have met is, I think, either the first indication, or the final confirmation, of a life of virtue. And even those who were inferior in other ways deserve to have their faults overshadowed by their courageous deaths in war for the sake of their country. Their good actions have wiped out the memory of any wrong they have done. And they have produced more public good than public harm. None of them became a coward because he set a higher value on enjoying the wealth that he had; none of them put off the terrible day of his death in hopes that he might overcome his poverty and attain riches. Their longing to punish their enemies was stronger than this, and because they believe this to be the most honorable sort of danger, they chose to punish their enemies at this risk, and to let everything else go. The uncertainty of success they entrusted to hope, but for that which was before their eyes they decided to rely on themselves in action. They believed that this choice entailed resistance and suffering, rather than surrender and safety; they ran away from the word of shame, and stood up in actions at risk of their lives. And so, in the one brief moment allotted them, at the peak of their fame and not in fear, they departed.*

The second speech of Pericles is delivered under grimmer circumstances. It is now into the second year of the war, and the overcrowded city is being devastated by the plague.

What exactly was the disease? Was it bubonic plague, ebola hemorrhagic fever, glanders, typhus? All have been suggested, but recent analysis done on dental fragments found in the mass grave discovered in the Keramikos section of Athens has found traces of *salmonella enterica serovar Typhi*, or typhoid fever.

Thucydides, who was one of the few who became sick and recovered, described in lurid detail not only the excruciating symptoms—the pustules and internal burning and an unquenchable thirst that led the tormented to lie down in the rain gutters, some expiring there as they tried to slake themselves—but also scenes of the damned that seem to come out of the imagination of a sadistic and sordid filmmaker. "The present affliction was aggravated by the crowding of country folk into the city, which was especially unpleasant for those who came in. They had no houses, and because they were living in shelters that were stifling in the summer, their mortality was out of control. Dead and dying lay tumbling on top of one another in the streets, and at every water fountain lay men half-dead with thirst. The temples also, where they pitched their tents, were all full of the bodies of those who died in them, for people grew careless of holy and profane things alike, since they were oppressed by the violence of the calamity, and did not know what to do."† Lawlessness and depravity overtook the population, trapped within the city walls with no escape from the raging contagion.

---

*Woodruff, *Thucydides* ii, 36–42. I haven't quoted the whole of the Funeral Oration as it's rendered by Thucydides, which takes up ii, 35–46.
† Book II: 52.

So it was that Athenians, who had seemed to be living exemplary lives so eminently worth living as to require no Homer to sing their praises, could, in a few short years, have descended to such pitiful straits. The precipitous reversal is akin to a Greek drama demonstrating the deadly repercussions of hubris, only with a *polis* standing in for an individual.

It is before these Athenians, devitalized and demoralized, that Pericles steps forward to address the *ekklêsia*. He knows that those gathered there on the Pnyx are in a mood to blame him and opens with a frank acknowledgment of the mood of the men he faces. "I expected you to get angry with me, and I can see why it has happened. I have called this assembly to remind you of certain points and to rebuke you for your misplaced anger at me and for your giving in too easily to misfortune."* But instead of offering the sorry spectacle of prolonged self-justification, he modulates quickly to a different range, elevated and inspiring. His rhetoric first moves to transmute personal grief into collective grief, and then collective grief into collective grandeur:

> I believe that if the city is sound as a whole, it does more good to its private citizens than if it benefits them as individuals while faltering as a collective unit. It does not matter whether a man prospers as an individual: if his country is destroyed, he is lost along with it; but if he meets with misfortune, he is far safer in a fortunate city than he would be otherwise. . . .
>
> For my part I am the man I was. I have not shifted ground. It is you who are changing: you were persuaded to fight when you were still unharmed, but now that times are bad, you are changing your minds; and to your weak judgment my position does not seem sound. That is because you already feel the pain that afflicts you as individuals, while the benefit to us all has not yet become obvious; and now that this great reversal has come upon you in so short a time you are too low in your minds to stand by your decisions, for it makes your thoughts slavish

* In the *Gorgias*, Plato has Socrates rebuke Pericles for needing to rebuke the Athenians. A good statesman would have made the citizenry progressively better, so *ex hypothesi*, Pericles must have improved the citizenry; but then why did they end up rejecting him? "[A]t first Pericles had a good reputation, and when they were worse, the Athenians never voted to convict him in any shameful deposition. But after he had turned them into 'admirable and good' people, near the end of his life, they voted to convict Pericles of embezzlement and came close to condemning him to death, because they thought he was a wicked man obviously. . . . A man like that who cared for donkeys or horses or cattle would at least look bad if he showed these animals kicking, butting, and biting him because of their wildness, when they had been doing none of these things when he took them over" (515e–516a). Plato is consistently contemptuous of the participatory *aretê* for which Pericles was so eloquent a spokesman.

when something unexpected happens suddenly and defies your best-laid plans. That is what has happened to you on top of everything else, mainly because of the plague.

Still, you live in a great city and have been brought up with a way of life that matches its greatness; so you should be willing to stand up to the greatest disasters rather than eclipse your reputation. (People think it equally right, you see, to blame someone who is so weak that he loses the glorious reputation that was really his as it is to despise someone who has the audacity to reach for a reputation he should not have.)

So set aside the grief you feel for your individual losses, and take up instead the cause of our common safety.

As for your fear that we will have a great deal of trouble in this war and still be no closer to success, I have already said something that should be enough for you: I proved many times that you were wrong to be suspicious of the outcome. I will tell you this, however, about your greatness in empire—something you never seem to think about, which I have not mentioned in my speeches. It is a rather boastful claim, and I would not bring it up now if I had not seen that you are more discouraged than you have reason to be. You think your empire extends only to your allies, but I am telling you that you are entirely the masters of one of the two usable parts of the world—the land and the sea. Of the sea, you rule as much as you use now, and more if you want. When you sail with your fleet as it is now equipped, there is no one who can stop you—not the King of Persia or any nation in existence. This power cannot be measured against the use of your land and homes, though you think it a great loss to be deprived of them. It makes no sense to take these so seriously; you should think of your land as a little kitchen garden, and your house as a rich man's trinket of little value compared to this power. Keep in mind too that if we hold fast to our liberty and preserve it we will easily recover our land and houses; but people who submit to foreign domination will start to lose what they already had. Don't show yourselves to be doubly inferior to your ancestors, who took the empire—they did not inherit from others—and, in addition, kept it safe and passed it on to you. No, what you should do is remember that it is more shameful to lose what you have than to fail in an attempt to get more. You should take on the enemy at close quarters, and go not only with pride, but with contempt. Even a coward can swell with pride, if he is lucky and ignorant; but you cannot have contempt for the enemy unless your confidence is based on a strategy to overcome them—which is your situation exactly. Even if you have only an even chance of winning, if you are conscious of your superiority it is safer for you to be daring, for in that case you do not depend on hope (which is a bulwark only to those who have no resources at all), but on a strategy based on reality, which affords a more accurate prediction of the result.

You have reason besides to support the dignity our city derives from her empire, in which you all take pride; you should not decline the trouble, unless you cease to pursue the honor, of empire. And do not think that the only thing we are fighting for is our freedom from being subjugated: you are in danger of losing the empire, and if you do, the anger of the people you have ruled will raise other dangers.[*] You are in no position to walk away from your empire, though some people might propose to do so from fear of the current situation, and act the part of virtue because they do not want to be involved in public affairs. You see, your empire is really like a tyranny—though it may have been thought unjust to seize, it is now unsafe to surrender.[†] People who would persuade the city to do such a thing would quickly destroy it, and if they set up their own government they would destroy that too. For those who stay out of public affairs survive only with the help of other people who take action; and they are no use to a city that rules an empire, though in a subject state they may serve safely enough. . . .

Keep this in mind: our city is famous everywhere for its greatness in not yielding to adversity and in accepting so many casualties and so much trouble in war; besides, she has possessed great power till now, which will be remembered forever by those who come after us, even if we do give way a little now (for everything naturally goes into decline): Greeks that we are, we have ruled most of the Greeks, and held out in great wars against them, all together or one city at a time, and our city has the most wealth of every sort, and is the greatest. And yet a man of inaction would complain about this! No matter, anyone who is active will want to be like us, and those who do not succeed will envy us. To be hated and to cause pain is, at present, the reality for anyone who takes on the rule of others, and anyone who makes himself hated for matters of great consequence has made the right decision; for hatred does not last long, but the momentary brilliance of great actions lives on as a great glory that will be remembered forever after.

As for you, keep your minds on the fine future you know will be yours, and on the shame you must avoid at this moment. Be full of zeal on both counts. Send no more heralds to the Lacedaemonians, and do not let them know how heavy your troubles are at present. The most powerful cities and individuals are the ones that are the least sensitive in their minds to calamity and the firmest in their actions to resist it.[‡]

***

[*] This acknowledgment of the resentment of the Athenian "allies" is in marked contrast to Pericles' suggestion in the Funeral Oration of how grateful even Athens' enemies must be to be conquered by a power so superior.

[†] Alcibiades is also presented by Thucydides as making the argument that an empire must grow if it is to survive.

[‡] Book II: 60–64.

This is political mouth-to-mouth resuscitation. Pericles is breathing life back into the soul of the city by articulating their collective sense of why their lives are worth the living—and, therefore, worth the losing. You have achieved it, he tells his numb-with-grief-and-with-foreboding Athenians, a greatness such that men will speak of in all the times to come. You are as song-worthy as a garlanded victor who has had a poet's epicinean ode composed in his honor. Here is your ode. I, Pericles son of Xanthippus, am singing it now for you.

This is Pericles' last speech before the Athenians. Within the year, he himself is dead of the plague.

GLOSSARY

agathon: good

agora: the center of an ancient Greek city-state, in which the commercial and civic buildings were situated

akrasia: without strength, weakness of the will

anamnesis: theory that learning is recollection

aporia: impasse; an argument that ends with no resolution

archon basileus: the sovereign magistrate; oversaw religious ritual

aretē: excellence, virtue

aristos: literal meaning is "the best," also refers to an aristocrat

asebeia: impiety

Attica: region of Greece in which city-state of Athens was located

axiarchism: the rule of value

bima: speaker's podium

boulē: Council of 500 in Athens, who ran daily affairs; they were chosen by lot annually by each of the ten tribes

catamite: male sex slave

chitōn: a piece of tunic-like clothing, often belted at the waist, and with sewn sleeves

daimōn: a spirit, not at the status of a god

deme: a subdivision of Athens, including its outlying areas; under the reforms of Cleisthenes, who abolished the old tribes, citizenship was based on being on the citizens list of a deme; Attica was divided into 139 demes

demiurge: in Plato *Timaeus,* the deity who fashions order out of chaos

Dionysia: Athenian festival of theater

eikasia: the lowest stage of knowing according to Plato's Analogy of the Divided Line; the state of the prisoners in the Myth of the Cave; ungrounded in the real

ekklêsia: the principal assembly of the Athenian democracy

epimeleia heautou: care of the self

epitaphios logos: funeral oration

epithumia: the appetitive part of the soul in Plato's tripartite soul theory

erastês: older male lover of a younger male; one half of the pair: *erastês* and *erômenos*

erômenos: younger male beloved of an older male; one half of the pair: *erastês* and *erômenos* euergesia: public benefit or service

euergetes: one who gives a public benefit

gnôthi seautón: know thyself; written on the temple of Apollo in Delphi

gymnasium: the place for training in athletics; from *gymnōs*, meaning "naked"

hesychia: keeping silent

himation: cloak

isegoria: equality of speech

isenomia: equality of the law

kallipolis: it means "beautiful city," and was the name Plato gave to his utopia in the *Republic*

kleos: glory, renown; that glory which is heard; "acoustic renown"

Lethe: one of the five rivers in Hades, the underworld; the word means "forgetfulness" or "oblivion," which is what is induced when a shade sips of it

logos: word or idea; in ancient Greek philosophy, a principle of order and knowledge; an account or explanation

maieutic: derived from Greek for midwife; the Socratic technique of asking questions to elicit the knowledge that is latent in one's interlocutor's mind; education

mēdèn ágan: nothing in excess; written on the temple of Apollo in Delphi

me mnesikakein: not to remember past wrongs; amnesty

metic: foreign resident; non-citizen in Athens

nous: mind; the intellectual part of the soul in Plato's tripartite soul theory

oligarchy: rule by the few, quite often those who are wealthy or scions of a few chosen families

ompholos: navel; according to legend, Zeus sent two doves out to meet at the navel of the world; ompholos stones, denoting this spot, are scattered around the Mediterranean, most famously in Delphi

ostraka: broken shards of pottery, on which the names of those voted to be ostracized—banished for ten years from Athens—were scratched

paiderastia: the rules and norms governing erotic relationships between older and younger males

parrhesia: meaning literally "to speak everything" and by extension "to speak freely," "to speak boldly," or "boldness"

phronēsis: practical wisdom

Pnyx: a hill in central Athens where the assembly met in the open air

polis: ancient Greek city-state; pl.: poleis

prytaneum: town hall of a Greek city-state, housing the chief magistrate and the common altar

sophist: itinerant teacher of rhetoric

sunōmosia: a swearing together; a conspiracy

synegoroi: supporting speakers in Athenian court

tautology: from ancient Greek for "same word"; a proposition which is, given its meaning, trivially true. For example, "Either you will understand this definition of tautology, or you will not understand this definition of tautology."

thaumazein: ontological wonder

thētes: the working poor; lowest class of citizen in Athens

tholos: round building

thumos also often written thymos: spiritedness; also expresses human desire for recognition; in Plato's tripartite soul, it is our spirited part. Today in Greece *thymos* means simply "anger."

to on: being

ACKNOWLEDGMENTS

I largely conceived this book while I was in residence at the Santa Fe Institute as a Miller Distinguished Scholar in the fall of 2011. There among the complexity theorists, I often felt as if I had wandered up onto Mount Olympus. I discussed the Pythagorean aspects of Plato's thinking with the legendary Murray Gell-Mann, who brought in his collection of ancient Greek coins for me to examine. This was thrilling. Evenings, as I watched the sinking sun set fire to the mathematical equations scrawled on the windowpanes, I would think about Plato's sense of the intimate connection between mathematical beauty and truth. I am grateful to both David Krakauer and Bill Miller for bringing me to SFI and to Jerry and Paula Sabloff for making me feel comfortable in my orthogonality.

I was also fortunate to be a Franke Visiting Scholar at the Whitney Humanities Center at Yale University during the fall semester of 2012, and I am extremely grateful to Barbara and Richard Franke for their visionary optimism regarding the place of the humanities in the university and society. The beloved María Rosa Menocal was responsible for my appointment, and though her premature death while I was in residence cast a deep pall of sorrow over the center, her bold and magnanimous spirit still continued to hold sway. It was while visiting at Yale that I first hesitantly aired some of my more audacious ideas about the Greeks to the eminent classicist Emily Greenwood and will be forever grateful for her unblinking generosity in hearing me out and then making valuable suggestions for refining my views. She read an early draft of the manuscript, and her comments were so helpful that my hope is that she would hardly recognize the version that is printed here. Conversations I had while at Yale with Stephen Darwall, Bryan Garsten, Tamar Gendler, Paul Grimstead, Verity Harte, Jane Levin, Alexander Loney, and Joseph Manning were also invaluable, both for the stiff skepticism they offered me (Verity is particularly to be thanked here) and for the encouragement.

I've been friends with Adina Schwartz since we were graduate students, and I knew that her highly honed critical skills would find much exercise in my views, and that my views would become the better for it. Debra Nails, to whom I wrote out of the blue because I was so smitten by the style and erudition of her books, overwhelmed me with her generosity, reading the whole of the manuscript and commenting copiously. Mistakes that remain are entirely due to my own pigheadedness. Still, I like to think that she helped me find my way to my better Plato. Joshua Buckholtz, of the psychology department at Harvard University, allowed me to participate in one of his experiments, so that I could experience what it is like to go into the magnet while performing various tasks in memorization and decision-making, exactly the experiment in which Plato participates. It was fun, and the wide-ranging conversation that Josh and I had about free will afterward was nothing at all like the one that Plato has with Shoket, since Josh is as thoughtful and open-minded as Shoket is not.

My interest in Plato has always been conditioned by my interests in philosophy of mathematics. But I always hoped that someday I might get closer to the remote figure of Plato himself. When I first mentioned such a hope to my literary agent, Tina Bennett, she immediately urged me to undertake a book that would realize this hope. I think the project would have remained in the realm of wishful thinking were it not for both her and my extraordinary editor, Dan Frank. Nothing could have been possible without an editor whose own love of philosophy is as pure and far-seeing as any Platonist could desire.

And now to my incomparable family. How lucky it is to have as one of my best friends and pretend sisters Margo Howard, whose actual profession is giving advice. Margo expertly played along with the shtick of "xxxPlato," keenly interested in how her own advice would square with Plato's. My genealogical sister Sarah has been a source of love and support ever since we shared a bedroom. My two daughters were, as usual, invaluable as sources of encouragement and a whole lot more. Not for the first time did I have cause for rejoicing that they both had majored in philosophy in college. Yael Goldstein Love read the whole of the manuscript and gave me some of the soundest advice I received in terms of making the book accessible to a wider audience. Danielle Blau added her insights, most especially into the conflicts that raged within Plato between the philosopher and the poet. Not only did Steven Pinker read and comment on the whole of the manuscript, not only did he gamely accompany me on a trip to Greece, taking on my obsessions with grace and élan; he also has given me living proof that what Plato writes about romantic love is no overstatement: "There is no greater good than this that either human self-control or divine madness can offer a person."

BIBLIOGRAPHICAL NOTE

For the translations of Plato's dialogues, I primarily relied on *Plato: Complete Works*, edited by John M. Cooper, associate editor D. S. Hutchinson (Indianapolis, IN: Hackett, 1997). This means, in particular: *Euthyphro* and *Crito*, translated by G. M. A. Grube; *Republic* translated by G. M. A. Grube, revised by C. D. C. Reeve; *Theaetetus* translated by M. J. Levett, revised by Myles Burnyeat; *Sophist* translated by Nicholas P. White; *Statesman* translated by C. J. Rowe; *Parmenides* translated by Mary Louise Gill and Mark Ryan; *Philebus* translated by Dorothea Frede; *Charmides* and *Euthydemus* translated by Rosamond Kent Sprague; *Protagoras* translated by Stanley Lombardo and Karen Bell; *Gorgias* translated by Donald J. Zeyl; *Menexenus*, translated by Paul Ryan; *Laws* translated by Trevor J. Saunders; *Epinomis* translated by Richard D. McKirahan Jr.; *Epigrams* translated by J. M. Edmonds, revised by John M. Cooper. For *Symposium* and *Phaedrus*, I relied primarily on *Plato's Erotic Dialogues: The Symposium and Phaedrus* translated with introduction and commentaries by William S. Cobb (Albany: State University of New York Press, 1993). For those two dialogues I also frequently consulted the translations of Alexander Nehamas and Paul Woodruff in Hackett 1997 and the older translation of the *Phaedrus* by R. Hackforth, in *The Collected Dialogues: Plato*, edited by Edith Hamilton and Huntington Cairns, Bollingers Series LXXI (Princeton, NJ: Princeton University Press, 1961). Unless otherwise indicated, however, the quoted translations are by Cobb. I also used the Princeton volume for the *Letters*, translated by L. A. Post, and used the Princeton volume as well for *Meno*, relying, unless otherwise indicated, on the translation by W. K. C. Guthrie. For *Timaeus*, I use the translations both of Donald J. Zeyl in Hackett 1997, but also, where indicated, the translation of Benjamin Jowett in Princeton, 1961. For the *Apology* I use both G. M. A. Grube in Hackett 1997, but also, when indicated, the translation of Hugh Tredennick in Princeton University Press, 1961. For *Phaedo* I quote from Hugh Tredennick's translation in Princeton University Press, 1961.

For Thucydides' *History of the Peloponnesian War*, I quote from the translation of Paul Woodruff, in *Thucydides: On Justice, Power, and Human Nature: Selections from The History of the Peloponnesian War* (Indianapolis, IN: Hackett, 1993).

For Pindar's epinician odes, I used *Pindar's Victory Songs*, translated by Frank J. Nisetich with a foreword by Hugh Lloyd Jones (Baltimore: Johns Hopkins University Press, 1990).

As far as secondary sources are concerned, the following are among the books I read which found their way into my writing:

Ahbel-Rappe, Sara, and Rachana Kamtekar eds. *A Companion to Socrates*. Malden, MA: Wiley-Blackwell, 2006.

Allen, Danielle S. *Why Plato Wrote: Blackwell Bristol Lectures on Greece, Rome, and the Classical Tradition*. Malden, MA: Wiley-Blackwell, 2010.

Blondell, Ruby. *The Play of Character in Plato's Dialogues*. Cambridge: Cambridge University Press, 2006.

Bowra, Cecil M. *The Greek Experience*. New York: Signet, 1959.

Brandwood, Leonard. *A Word Index to Plato*. Leeds, UK: W. S. Maney and Son, 1976.

Brickhouse, Thomas C., and Nicholas D. Smith eds. *The Trial and Execution of Socrates*. Oxford: Oxford University Press, 2001.

Burkert, Walter. *Lore and Science in Ancient Pythagoreanism*. Cambridge, MA: Harvard University Press, 1972.

———. *Ancient Mystery Cults*. Cambridge, MA: Harvard University Press, 1986.

Burtt, E. A. *The Metaphysical Foundations of the Physical Sciences*. Mineola, NY: Dover, 2004.

Carter, L. B. *The Quiet Athenian*. Oxford: Oxford University Press, 1986.

Charalabopoulos, Nikos G. *Platonic Drama and Its Ancient Reception*. Cambridge Classical Studies. Cambridge: Cambridge University Press, 2002.

Davidson, James. *The Greeks and Greek Love: A Bold New Exploration of the Ancient World*. New York: Random House, 2009.

Dewald, Carolyn, and John Marincola, eds. *The Cambridge Companion to Herodotus*. Cambridge: Cambridge University Press, 2007.

Dodds, E. R. *The Greeks and the Irrational*. Berkeley: University of California Press, 2004.

Dover, K. J. *Greek Homosexuality*. Reissue. Cambridge, MA: Harvard University Press, 1989.

———. *Greek Popular Morality in the Time of Plato and Aristotle*. Indianapolis, IN: Hackett, 1974.

Everson, Stephen, ed. *Aristotle: The Politics and the Constitution of Athens*. Cambridge Texts in the History of Political Thought. Cambridge: Cambridge University Press, 1996.

Foster, Edith. *Thucydides, Pericles, and Periclean Imperialism*. Cambridge: Cambridge University Press, 2010.

Frankfurt, Harry. *On Bullshit*. Princeton, NJ: Princeton University Press, 2005.

Gildenhard, Indigo, and Martin Riverman, eds. *Beyond the Fifth Century: Interactions with Greek Tragedy from the Fourth Century BCE to the Middle Ages*. Berlin: De Gruyter, 2010.

Gray, Vivienne. *The Framing of Socrates: The Literary Interpretation of Xenophon's Memorabilia*. Stuttgart: Franz Steiner Verlag, 1998.

Greenblatt, Stephen. *The Swerve: How the World Became Modern*. New York: W. W. Norton, 2011.

Greenwood, Emily. *Thucydides and the Shaping of History*. London: Duckworth, 2006.

Grene, David. *Greek Political Theory: The Image of Man in Thucydides and Plato*. Chicago: University of Chicago Press, 1965.

Grube, G. M. A. *Plato's Thought: Eight Cardinal Points of Plato's Philosophy as Treated in the Whole of His Works*. Boston: Beacon Press, 1958.

Hall, Jonathan M. *Hellenicity*. Chicago: University of Chicago Press, 2005.

Hansen, Mogens Herman, and Thomas Heine Nielsen, eds. *An Inventory of Archaic and Classical Poleis*. Oxford: Oxford University Press, 2004.

Hardy, G. H. *A Mathematician's Apology*. Reissue. Cambridge: Cambridge University Press, 2012.

Holt, Jim. *Why Does the World Exist: An Existential Detective Story*. New York: Liveright, 2013.

Holton, Gerald. *Einstein, History, and Other Passions: The Rebellion Against Science at the End of the Twentieth Century*. Cambridge, MA: Harvard University Press, 2000.

Kahn, Charles H. *Plato and the Socratic Dialogue: The Philosophical Use of a Literary Form*. Cambridge: Cambridge University Press, 1996.

Lewis, Sian. *Greek Tyranny*. Exeter, UK: Bristol Phoenix Press, 2009.

Miller, Patrick Lee. *Becoming God: Pure Reason in Early Greek Philosophy*. London: Continuum International Publishing, 2011.

Murdoch, Iris. *The Fire and the Sun*. New York: Viking Press, 1991.

———. *The Sovereignty of Good*. 2nd ed. London: Routledge, 2001.

Nails, Debra. *Agora, Academy, and the Conduct of Philosophy*. London: Springer, 1995.

———. *The People of Plato: A Prosopography of Plato and Other Socratics*. Indianapolis, IN: Hackett, 2002.

Nehamas, Alexander. *The Art of Living: Socratic Reflections from Plato to Foucault*. Berkeley: University of California Press, 2000.

Nightingale, Andrea Wilson. *Genres in Dialogue: Plato and the Construct of Philosophy*. Cambridge: Cambridge University Press, 1995.

———. *Spectacles of Truth in Classical Greek Philosophy*. Cambridge: Cambridge University Press, 2004.

Nightingale, Andrea, and David Sedley, eds. *Ancient Models of Mind: Studies in Human and Divine Rationality*. Cambridge: Cambridge University Press, 2010.

Nussbaum, Martha. *The Fragility of Goodness: Luck and Ethics in Greek Tragedy and Philosophy*. Cambridge: Cambridge University Press, 2001.

———. *Love's Knowledge: Essays on Philosophy and Literature*. Oxford: Oxford University Press, 1992.

Ober, Josiah. *Democracy and Knowledge: Innovation and Learning in Classical Athens*. Princeton, NJ: Princeton University Press, 2010.

Osborne, Robin. *Greek History*. London: Routledge, 2004.

Press, Gerald A., ed. *Who Speaks for Plato: Studies in Platonic Anonymity*. Lanham, MD: Rowman and Littlefield, 2000.

Russell, Bertrand. *A History of Western Philosophy*. New York: Simon & Schuster, 1967.

———. *The Autobiography of Bertrand Russell*. London: Routledge, 2000.

Stone, I. F. *The Trial of Socrates*. New York: Anchor Books, 1989.

Stove, David. *The Plato Cult*. Oxford: Basil Blackwell, 1991.

Taylor, A. E. *Plato: The Man and His Work.* London: Methuen, 1926.

Villemaire, Diane Davis. *E. A. Burtt, Historian and Philosopher: A Study of the Author of "The Metaphysical Foundations of Modern Physical Science."* London: Springer, 2002.

Vlastos, Gregory. *Socrates: Ironist and Moral Philosopher.* Ithaca, NY: Cornell University Press, 1991.

———, ed. *Plato I: Metaphysics and Epistemology.* New York: Anchor, 1971.

———, ed. *Plato II: Ethics, Politics, and Philosophy of Art and Religion.* New York: Anchor, 1971.

Williams, Bernard A. O. *Plato.* London: Routledge, 1999.

———. *The Sense of the Past: Essays in the History of Philosophy.* Princeton, NJ: Princeton University Press, 2007.

INDEX

abduction (inference to the best
  explanation), 382
Achilles, 66–7, 138n, 156, 185n, 228,
  231, 232, 246, 247, 249, 255, 303,
  327
Acropolis, 123, 125, 126, 147–50, 152,
  153, 162, 284–5
advice columns, 263–80
Aeschines, 296, 423
Aeschylus, 8, 129, 137–8, 185n, 229,
  237, 246–7
Aesop, 320
Agamemnon, 331
Agathon, 41, 251n
agathon (good), see good, goodness
agon (competition), 207–8
akrasia (weakness of the will), 71n
Alcestis, 129n
Alcibiades, 40–1, 156, 159, 162n,
  231–59, 293, 298, 326, 363–4,
  369n
Alexander of Miletus, 65n
Alexander the Great, 19, 152, 170,
  239n
algorithms, 99, 102, 105, 107,
  339
Allen, Woody, 21, 23
Anabasis (Xenophon), 87n, 423
anamnesis (recollection), 172n, 314,
  403
Anaximander, 29, 31–2, 33
Anytus, 86n, 300, 302
apeiron (boundless), 31–2
Aphrodite, 374
Apollo, 138, 225, 228, 301, 327

Apollodorus, 312–13
Apology (Plato), 9, 36n, 64n, 67n, 123,
  126, 127n, 140, 216, 284, 288,
  301, 310, 312, 313, 321 323, 365n,
  367n, 369
Apology (Xenophon), 284, 300, 423
aporia (impasse), 10n, 43, 293, 303,
  368n
Ares, 232
aretē (excellence or virtue), see virtue
Ariston (Plato's father), 54n
Aristophanes, 12n, 41, 74n, 205, 225,
  244–5, 246, 253, 284, 287
Aristotle:
  as Athenian resident, 8, 150, 152–3,
    240
  biological observations of, 24–5, 27,
    399n
  in Christian theology, 42n, 355–6,
    388
  influence of, 24–5, 27, 28, 32, 42n,
    129, 160, 355–6, 388, 396
  Lyceum founded by, 8n, 76n, 152–3,
    170, 240, 284
  Plato compared with, 10, 30, 50n,
    77n, 388, 396, 399n
  as Plato's student, 48, 49, 76n,
    100–1, 152–3, 169n, 170, 240,
    252n, 355
  political views of, 101, 147, 225,
    230–1
  teleology of, 30, 53, 388
  works of, 10, 11n, 32n, 147, 169n,
    202, 225, 423
Artemis, 227, 232

astronomy, 330–1, 341, 393, 394, 395
atheism, 230, 305
Athena, 138, 150–1
Athens:
  agora of, 11, 35, 103, 123, 283–4
  amnesty in, 294–6, 297, 324
  archon basileus of, 299–300, 303,
    305, 310
  autochthonous origins of, 151,
    425–6
  citizenship of, 123*n*, 151–2, 232,
    243, 253, 285, 286, 290, 291–5,
    302, 321, 323–4, 425–33
  Council of 500 (Boule) in, 64*n*,
    148, 288
  as democratic *polis*, 10, 12, 39,
    123*n*, 124*n*, 148, 151, 152, 153,
    160, 235–53, 256, 259, 285–300,
    325–6, 329, 337–8, 378, 390–1,
    393, 425–6
  ekklêsia (assembly) of, 153–4, 235,
    238, 285–6, 290, 296, 430
  empire and imperial power of, 40*n*,
    63, 147–52, 160, 162, 233–4,
    238–9, 293, 294, 425*n*, 432*n*
  Ethos of the Extraordinary in, 8–9,
    125–9, 135–49, 155–6, 162, 227,
    231–59, 264–6, 289, 293–9, 303,
    314*n*, 318, 327–8, 378, 395, 425
  Four Hundred as rulers of, 285*n*
  general amnesty declared in,
    294–6
  gods of, 64*n*, 378–9
  legal system of, 64, 67, 252, 273*n*,
    291, 294–5, 312, 323–5
  military forces of, 123–4, 296
  oligarchy of, 80*n*, 234, 284–7,
    290–4, 298, 325, 328, 363*n*, 392
  plague in, 283*n*, 285, 429, 433
  Plato as citizen of, 3, 4, 6, 17, 63–4,
    69, 71, 76*n*, 160, 167, 315, 396
  political situation in, 10–12, 39,
    56–7, 73*n*, 75–9, 100–1, 150*n*,
    156–7, 230–1, 274*n*, 287, 291,
    311, 325–6, 328, 363*n*, 364, 390,
    393, 430*n*

  religious institutions of, 63, 64*n*,
    232, 235, 324, 378–9
  slavery in, 14, 87*n*, 91–2, 113–18,
    120, 123, 125, 133–4, 151, 172*n*,
    213*n*, 243, 296, 302*n*, 315, 424*n*
  society of, 64*n*, 231–59
  Sparta as rival of, 18–19, 125*n*, 156,
    251, 283*n*, 284–5, 286, 287, 295,
    297, 299, 326
  Thirty as rulers of, 240, 243, 251,
    285–6, 287, 288, 290, 293–4,
    297, 300
  Three Thousand assembly in, 285,
    286–7
  walls of, 283*n*, 296, 297
  women in, 90*n*, 123, 147, 151, 244,
    247
athletics, 185–6, 206–8, 212
atomic theory, 24*n*–5*n*, 34
atopia (strangeness), 162, 313
Augustine, Saint, 160, 389
Axial Age, 7, 131–3, 138, 142, 160,
  228, 378

Bacon, Francis, 34
barbarians, 91–2, 124*n*, 147, 239
Baumard, Nicolas, 134*n*
beauty, 210
  abstract, 213, 252, 304–5, 319–20,
    330, 341–2
  in education, 208–9, 214
  knowledge of, 82–3, 101*n*, 241–2,
    252, 392
  moral nature of, 210, 214, 252–3
  physical, 138, 232–3, 241–2, 252–3,
    271–2, 364, 367, 424
  qualities of, 101*n*, 241–2
  in Sublime Braid, 50–6, 191–2,
    196–8, 208–9, 254–9, 272,
    315–19, 328–32, 340–7, 382–5,
    390–6
  virtue and, 340–1, 424
beliefs:
  false, 380–1
  knowledge compared with, 303–5,
    368–9

personal, 86, 91–4
rational, 368–9
true, 160–1, 303–5, 368–9
Bendis, 378–9
Bible, 91n, 113, 229–30
biology, 24–5, 27, 31, 34n, 399n
Boccaccio, Giovanni, 144
Bohr, Niels, 22–3
Boyer, Pascal, 134n
brain function, 19, 110n, 111n, 114,
    172, 399–422
Buddha, 7, 131, 138
"bullshit," 322–3, 406–7
Burnyeat, Myles, 49, 50n, 64n, 127n,
    218, 382n
Byron, George Gordon, Lord, 240n

Callicles, 231, 392
Calvin, John, 376–7
Catholic Church, 269, 278, 388
Cebes, 315, 320
censorship, 196–97
Chaeronea, Battle of, 239
chariot races, 234, 273, 409–10, 411
Charmides, 243, 254, 293
Charmides (Plato), 243, 254, 284
Chaucer, Geoffrey, 227
chess, 19–20, 47
children, 18–19, 57, 109–16, 165–221
chitōn (tunic), 11, 73n, 241n, 402
Chorus of Oceanides, 137–8
Christianity, 42n, 76n, 269, 278,
    306n, 315–16, 317, 318, 349, 351,
    355–6, 388, 390, 395
Churchill, Winston S., 189
Clouds, The (Aristophanes), 12n
cognition, 14, 19n, 25, 35, 37, 110n,
    111n, 154, 399–422
coherence, 418–21
common sense, 14, 86–7, 89, 383
computers, 13, 14, 35, 58, 69–70, 71,
    91, 95–6, 97, 98, 172, 339
    see also Internet
confabulation, 408–9, 410, 411,
    417–18, 419
Confucius, 7, 131, 138

Copernicus, Nicolaus, 50–1
cosmology, 11n, 29, 31–2, 37, 319,
    330–1, 340–3, 346, 351–2, 393,
    407–8
Cratylus (Plato), 249n
creation myths, 52, 53, 386–7
Critias, 80n, 240, 243–4, 251, 287,
    288–9, 291, 293, 298, 327
Crito, 312, 315
Crito (Plato), 301, 312, 313
Cyclopean architecture, 143–4

daimōn (spirit), 249, 367–71, 373
Davidson, James, 243n, 247n
death, 216
    life after, 127n–8n, 330–3, 395
    preparation for, 13, 66, 303
    transcendence of, 8–9, 126–30,
        134–5, 203–4, 228n, 313–18,
        386–7
De caelo (Simplicius), 25
decision-making, 410–12
Delian League, 125, 296
Delphic oracle, 225–7, 236n, 326–7,
    337 364
demiurge, 52, 386–7, 394
Democritus, 24, 27, 28, 34
Demosthenes, 239
De Rerum Natura (Lucretius), 34n
Deutsch, David, 23n
Dewey, John, 32
Dialogue Concerning Natural Religion
    (Hume), 5
*Dialogue Concerning the Two Chief
    World Systems* (Galileo), 5
dialogues, Platonic, 10n, 239–40
    aesthetics of, 4, 11, 196n, 197,
        208–9
    characterization in, 5, 39–44,
        86n–7n
    chronology of, 10n, 302–3, 364
    as drama, 41–5, 167, 250–1, 256n
    as philosophical works, 4–5, 10–11,
        40, 42–3
    style of, 4–5, 9, 12, 37, 39–40, 168,
        172, 173, 200

448   *Index*

dialogues, Platonic *(continued)*
    translations of, 4–5, 139, 292
    see also specific dialogues
Diogenes Laertius, 65n, 301n, 315
Dion of Syracuse, 76–8, 101, 274n
Dionysius I, 77–8
Dionysius II, 78
Dionysus, 237, 251, 374
Diotima, 129n, 196n, 252, 254
Divided Line, 382n, 383–4
divine madness, 268–71
"Does Philosophy Matter?" (Fish),
    44–5, 87n
Dostoyevsky, Fyodor, 306
Douglas, William O., 263
Dover, Kenneth, 244, 245, 246n, 247,
    252n
drama:
    inner, 41–5, 250–1
    tragic, 7–8, 77n, 229, 237

E=mc², 30, 54, 342
Echecrates of Phlius, 315, 316
education:
    beauty in, 208–9, 214
    moral, 47, 86n, 103n, 111n, 249–50,
        302
    as play, 187, 194, 207–8, 216
    of rulers, 73–9, 185–93, 199–200,
        215–17, 383, 391
    teaching in, 47, 165–221, 210–11
eikasia (lowest level of awareness),
    380, 382n
Einstein, Albert, 23n, 51, 183
Elements (Euclid), 304n, 344n
Eleusinian mysteries, 235, 255, 375
Eleusis, 293–4
Elis, 315n–16n
Empedocles, 33, 34
Enlightenment, 3, 306n, 308
Epinomis (Plato), 342, 394n
epistemology, 7n, 10n, 11n, 36, 39, 52,
    56, 196n, 258, 304–7, 311, 314,
    368–77, 379, 382n, 416
erastês and *erômenos* (older lover and
    younger lover), 244–9, 258

erotogenesis, 205–6, 245–6, 275–6
Eros, 41, 46n, 245, 251–2, 255, 274n,
    374
erōs (love), 40–1, 244–52, 265, 273–4
Ethical Answers Search Engine
    (EASE), 105–10, 112, 117, 119,
        120, 230–2
ethics, *see* morality
Ethics, The (Spinoza), 306n, 308, 309,
    389
ethōs (habit or custom), 126, 332,
    377–8
Ethos of the Extraordinary, 8–9,
    66–7, 125–9, 135–49, 155–6,
    162, 227, 231–59, 264–6, 289,
    293–9, 303, 314n, 318, 327–8,
    378, 395, 425
Euclid, 27, 304n, 344n
Euripides, 8, 151, 197, 237
Euthydemus (Plato), 265, 284, 367n
Euthyphro, 300, 305–7, 310, 368
Euthyphro (Plato), 64n, 284, 301,
    305–9, 310, 330
Euthyphro Dilemma, 306–7, 308
existence:
    of abstract forms, 11n, 19–20, 42,
        46–53, 125, 314, 383–7
    being vs. non-being in, 310–11
    reality of, 28–9, 388–9, 407–8
    sublime as basis of, 50–6, 191–2,
        196–8, 208–9, 254–9, 272,
        315–19, 328–32, 340–7, 382–5,
        390–6
existential dilemmas, 104, 109, 130,
    133, 142–3, 160, 230, 378
Expanding Circle, The (Singer), 91n

Feynman, Richard, 23n
Fish, Stanley, 44–5, 87n, 250
Forms, Theory of, 11n, 42, 46–53, 125,
    314, 383–7
Foucault, Michel, 256n, 424n
Frankfurt, Harry, 322–3, 326–7
freedom, 117–18, 120, 181–2
Frege, Gottlob, 47, 311
Freud, Sigmund, 45–6, 166

Frogs, The (Aristophanes), 12n, 225, 237, 284

Funeral Oration of Pericles, 149, 152, 232, 238, 425–9, 432n

Galileo Galilei, 5, 34, 49, 50–1, 110n, 343n, 388, 396

game theory, 132, 157–9, 330

geocentrism, 51, 388

geometry, 169, 172n, 210–11, 392

Glaucon, 153–5, 160, 182n, 330, 333, 378–9, 381, 382n

gnôthi seautón (know yourself), 225–6, 228

God, 52–53, 305–6, 376–7
    communion with, 269–71
    existence of, 292–3, 341, 345, 347, 351–2, 390–1, 395
    Hebrew conception of, 6–7, 130–1, 229–30
    name of, 130–1

Gödel, Kurt, 20n, 28, 46, 48

God Is Not Great (Hitchens), 326

gods and goddesses, 7, 30, 53–4, 64n, 126–7, 130, 137–8, 143, 147, 150–1, 162, 205–6, 217, 225–8, 231, 242, 245–6, 253, 270, 292–3, 297n, 305–6, 308, 331–2, 341, 376, 391, 392, 393

good, goodness, 55–6, 383–5
    attainment of, 55–6, 71n, 159–60, 265
    collective, 101, 186, 202, 378
    definition of, 46–9, 169
    divine, 389–90, 396
    justice and, 159–60
    personal standards of, 53–4
    pleasure and, 359–60
    in Sublime Braid, 50–6, 191–2, 196–8, 208–9, 254–9, 272, 315–19, 328–32, 340–7, 382–5, 390–6

Googleplex, 52–63, 59, 67–70, 75, 82–90, 93, 102–3, 106, 118–19

Google search engine, 69–72, 93, 96–100, 105, 168, 339

Googlization of Everything (And Why We Should Worry), The (Vaidhyanathan), 72n

Gorgias (Plato), 140, 142, 231, 249, 288, 316, 328n, 348, 350, 351, 356, 357, 392, 430n

government, 234, 285
    democratic, 10, 12, 39, 78n, 100, 101n, 107, 189, 357–8
    by elite (guardians), 73–9, 185–93, 199–200, 215–17, 383, 391
    oligarchic, 80n, 151, 215n, 284–7, 290–4, 298, 325, 328, 363n, 392
    tyrannical, 76–9, 85, 125n, 137, 148n, 152n, 191–2, 202–5, 215n, 234n, 273–4, 291, 293, 331

Graeber, David, 133–4

Greece, 53–4, 129–30
    city-states (*poleis*) of, 7, 69, 76n, 91n, 125, 126, 134, 138, 142–3, 146, 150–2, 156–7, 159, 194, 218, 225, 230, 231–2, 239, 246, 290, 294, 296–7, 378
    greater (Magna Graecia), 33n, 152–3, 239
    heroes of, 66–7, 126–7, 144–50, 185n, 246, 249n, 303, 327–8
    homosexuality in, 241–56
    Judaic culture compared with, 6–7, 91n, 130–1, 135, 229–30
    Panhellenic identity of, 7, 76n, 125n, 138, 142–5, 150–2, 226
    philosophy of, 6, 24–5, 29–37, 48, 50, 129–32, 138, 169n, 307; *see also specific philosophers*
    religious traditions of, 7, 129–30, 137–8, 142–3, 162, 225–7, 230, 378
    secularism of, 34n, 129–32, 138

Greek language, 4–5, 36, 91n, 125, 143, 146, 168, 187, 270, 373n

Greenblatt, Stephen, 34n, 144–5

Hades, 127n–8n, 160n, 237

Hamlet (Shakespeare), 20

Harmonices Mundi (Kepler), 330–1

harmony, 49, 330–1, 342
Harte, Verity, 359*n*
Hebrews, 6–7, 91*n*, 130–1, 135, 229–30
hedonism, 105, 359–60
Heidegger, Martin, 312, 376
Helen, 138*n*, 143
Hellenica (Xenophon), 424
heredity, 27, 31, 210
Hermes, 235, 344
Herodotus, 105*n*, 130, 147, 234*n*
Hesiod, 185*n*, 204
Histories (Herodotus), 147, 234*n*
History of the Peloponnesian War (Thucydides), 235*n*, 297*n*, 425
History of Western Philosophy, A (Russell), 32–3, 424
Hitchens, Christopher, 326
Hitler, Adolf, 33*n*
Hobbes, Thomas, 157
Holt, Jim, 385*n*
Holton, Gerald, 30
Homer, 8*n*, 123, 126–7, 137, 138*n*, 143, 145–7, 149, 160, 162, 168, 185*n*, 204, 246, 331, 393, 430
homosexuality, 241–56
House of Fame, The (Chaucer), 227
Howard, Margo, 263–80
hubris (pride), 217, 226–7, 232, 273*n*, 297, 298, 430
humanism, 58, 144–5, 146
Hume, David, 5, 31*n*, 160

ideology, 384–5, 389–92
Iliad (Homer), 8*n*, 123, 126–7, 137, 138*n*, 145–7, 168, 246
infanticide, 18–19
infinity, 24, 28, 319–20
information:
   aggregate, 99–110, 199
   cloud storage of, 71, 91, 97, 98, 339
   knowledge compared with, 350, 354, 358, 380–2
   search engines for, 69–72, 93, 96–100, 105, 168, 339
   teaching of, 47, 210–11
   usefulness of, 99–100
Inquiry Concerning Human Understanding (Hume), 31*n*
intelligence, 267–8
   knowledge and, 46–7, 88, 94–6, 101–2, 198–9, 345
   level of, 88, 102, 195–9
   mathematical, 46–7, 95–6
Internet, 24*n*, 69–72, 91, 93, 96–100, 103, 105, 168, 171 209, 339, 358
Ion (Plato), 196*n*
Ionia, 29–37, 48, 50, 124*n*, 125, 169*n*, 302
Ionian Enchantment, 29–37, 48, 50, 169*n*, 307

Jainism, 131, 135
James, Henry, 10*n*
James, William, 376
Jaspers, Karl, 7, 128, 131–2, 133
Jehovah, 131*n*, 138, 228
Jesus Christ, 316, 395
Judaism, 6–7, 91*n*, 130–1, 135, 161–6, 229–30
Jung, Carl Gustav, 209
justice, 201
   political, 75–6, 78*n*, 79, 156–7, 186, 201
   standards of, 142, 155–7, 201, 379–81
   virtue and, 325–6, 378–9

kallipolis (utopian city), 66*n*, 75–6, 78*n*, 160, 168*n*, 188, 191–3, 199–200, 215–17, 396
Kant, Immanuel, 160, 312, 332, 333
Keats, John, 319
Kepler, Johannes, 50–1, 330–1, 388
kleos (glory), 127, 129*n*, 130, 135, 136, 138, 139–41, 142, 149, 156, 230–2, 233, 254, 292, 296, 298, 318, 331, 378, 380
knowledge, 84, 366–7
   awareness of, 350, 354, 358, 380–2, 410–15

of beauty, 82–3, 101*n*, 241–2, 252, 392
belief vs., 303–5, 368–9
common sense in, 14, 86–7, 89, 383
expertise in, 57, 80–117, 180, 183–4, 338, 357, 387, 405–7
of "how" vs. "that," 366–7, 374
intelligence and, 46–7, 88, 94–6, 101–2, 198–9, 345
mathematical, 20, 47–51, 53, 83, 88, 105, 110*n*, 302*n*, 314–15, 318, 385*n*, 395–6
morality based on, 54–6, 111–15, 307–9
non-trivial, 93–4, 96
oracular, 45, 225–7, 236*n*, 326–7, 337, 364
philosophical, 47, 69–76, 80–117, 180, 183–4, 327–8, 415–19
political, 11–12, 56–7, 73–9, 100–1, 147, 150*n*, 156–7, 225, 230–1, 247*n*, 310–12, 363*n*, 364
reception and teaching of, 47, 94–6, 101–2, 171–2, 210–11, 314, 403; *see also* education
as recollection, 172*n*, 314, 403
self-, 78–9, 225–6, 228
theoretical basis of (epistemology), 7*n*, 10*n*, 11*n*, 36, 39, 52, 56, 196*n*, 258, 304–7, 311, 314, 368–77, 379, 382*n*, 416
useless, 339–40, 359*n*, 360
Krauss, Lawrence, 21, 22, 28–9, 35, 307, 308

Laches (Plato), 238*n*
Lao Tzu, 7, 131, 138
Laurium silver mines, 125
laws:
mathematical, 210
natural, 30–1, 376
of society, 64, 67, 156–7, 210, 218, 252, 273*n*, 291, 294–5, 312, 323–5

Laws (Plato), 9, 10–12, 52*n*, 64*n*, 90*n*, 91*n*, 183*n*, 186, 187, 196*n*, 206, 341, 342*n*, 343, 346, 386, 389, 393
Leibniz, Gottfried Wilhelm, 52*n*, 160, 304, 385, 394
Leonardo da Vinci, 183*n*
Leon of Salamis, 288
Leslie, John, 385*n*
Letter II (Plato) (see *Second Letter*)
Letter VII (Plato) (see *Seventh Letter*)
Leucippus, 24–5, 34*n*
Libet, Benjamin, 411, 412
life:
after-, 127*n*–2*n*, 330–3, 395
extraordinary, *see* Ethos of the Extraordinary
individual choice in, 94–5
as "worth living," 6–9, 14, 36, 86–96, 105, 106–7, 110, 116, 126, 138, 152, 153, 160, 196*n*, 201, 204–5, 218, 291–2, 298, 303, 309, 327, 329, 330, 333, 364, 392–3, 394, 430, 433
Life of Plato (Olympiodorus), 42*n*
Locke, John, 114*n*, 160
logic:
invalid vs. valid, 89, 92
mathematical, 20, 311
propositions in, 44–5
tautology in, 93, 169*n*
logos ("justification"), 104, 329–30, 332, 352, 368–9, 382, 384–5
love, 225–59
attraction in, 40–7, 205–6, 245–6, 274–8
erotic, 40–1, 45–6, 231–59, 264–5, 271, 272–4, 278–80, 354–5, 373, 374, 376
generalized, 200–1, 203
madness and, 373–4
morality of, 55, 231–59, 272–4, 275
personal, 12, 200–1, 203, 266–7
Platonic, 12–13, 41, 206, 258–9, 278–80

love *(continued)*
  as reunification (erotogenesis), 205–6, 245–6, 275–6
  romantic, 47, 278–80
  self-, 55, 200–3, 255–9
  types of, 200–1, 245*n*–6*n*
Lysis (Plato), 254, 284

Macedonia, 76*n*, 152–3, 239
magnetic resonance imaging (MRI), 37, 399–422
Marathon, Battle of, 149–50
Marcus Aurelius, Emperor of Rome, 76*n*
marriage, 266, 271–2
Mastermind, 209–10
materialism, 29, 30, 399*n*
mathematics, 46–7, 50*n*, 343–4
  abstraction in, 26–7, 46–7, 48, 382*n*, 384–9, 391, 393
  beauty of, 213, 252, 304–5, 319–20, 330, 341–2
  existence of, 19–20
  intelligence in, 46–7, 95–6
  knowledge of, 20, 47–51, 53, 83, 88, 105, 110*n*, 302*n*, 314–15, 318, 385*n*, 395–6
  logical foundation of, 20, 311
  modern discipline of, 27–8, 49–50
  Platonic, 20, 30, 46–51, 83, 88, 110*n*, 310–11, 385*n*, 395–6
  proofs in, 19, 26*n*, 27, 172*n*, 302*n*, 311, 372
  Pythagorean, 26–8, 50*n*, 129, 131, 314–15, 316, 317, 344–5, 352
  reality represented by, 19–20, 30, 105, 340–3, 352
  see also numbers
matter, 31–4
"mattering," 6–9, 130–2, 142, 417
mēdèn ágan ("nothing in excess"), 225–6, 392
medicine, 49, 55, 110*n*, 111*n*, 141, 142, 357
Meletus, 299–300, 305, 310, 321–2, 324, 325, 327

Memorabilia (Xenophon), 36*n*, 67*n*, 80*n*, 288, 291, 301, 303*n*, 423–4
Menexenus (Plato), 252*n*, 289
Meno, 86*n*–7*n*, 103, 249–50, 302
Meno (Plato), 86*n*–7*n*, 103, 172*n*, 302
mental illness, 110*n*, 111*n*
mesolimbic dopaminergic processes, 400, 409
metaphysics, 10*n*, 11*n*, 36, 37, 50, 52, 54, 56, 57, 196*n*, 258, 314, 379, 382*n*, 385
Methods of Ethics, The (Sidgwick), 359*n*
"military-coinage-slavery" complex, 133–4
mind:
  brain function and, 14, 19*n*, 25, 35, 37, 110*n*, 111*n*, 154, 399–422
  philosophy of, 25, 35, 37, 51–3, 55, 399–422
  see also intelligence
Montaigne, Michel de, 256*n*
Moralia (Plutarch), 12*n*
morality:
  actions based on, 54, 84–5, 105, 410
  beauty as form of, 210, 214, 252–3
  crowd-sourcing of, 105–10, 112, 117, 119, 120, 230–2
  dilemmas in, 104, 109, 130, 133, 142–3, 160, 230, 378
  free will in, 13, 332–3, 408, 409, 420–1
  good vs. bad in, 7, 84–5, 309
  knowledge as basis of, 54–6, 111–15, 307–9
  of love, 55, 231–59, 272–4, 275
  principles of (ethics), 3–7, 11–14, 18, 36, 37, 104–5, 127*n*, 141, 258, 265, 305–7, 306*n*, 307–8, 309, 332, 378
  relativism in, 18–19, 44, 45*n*
  religious, 7, 84–95, 229–30, 305–9, 311
  right vs. wrong in, 57, 91, 108–9, 115–16, 118
  sexual, 231–59

of slavery, 14, 91–2, 113–18, 120, 213*n*
standards of, 53–4, 67, 111–15, 129–30, 155–7, 226
truth in, 108–9, 159–60, 195–6, 210, 214
virtue based on, 57, 193–4, 209, 309
Moses, 130*n*–1*n*
multiverse, 22*n*–3*n*, 72
Muses, 319, 374
music, 101*n*, 185–6, 206–8, 318, 330–1, 393–4
"music of the spheres," 330–1
Mycalessus, 296–7
Mycenaeans, 143–4, 145
Myers-Briggs Type Indicator, 209–10, 218–21
Myrmidons (Aeschylus), 246–7

Nails, Debra, 87*n*, 304*n*
nature, 33
    four elements of, 33
    laws of, 30–1, 376
    rational order in, 54–5
    right form in, 48, 50
near-death experiences, 330–1
Nehamas, Alexander, 125, 139–40, 230, 256*n*
Neoplatonism, 42*n*, 76*n*
Nereids (Aeschylus), 246–7
neuroscience, 14, 19*n*, 25, 35, 37, 110*n*, 111*n*, 154, 399–422
Newton, Isaac, 26, 34
Nicias, 235*n*, 238
Nietzsche, Friedrich, 238, 256
nihilism, 159–60, 306
92nd Street YM/YWHA, 165–221, 377
numbers, 19, 27
    as abstractions, 48
    existence of, 19–20, 342–5
    googleplex, 59
    whole, 28
Nussbaum, Martha, 241, 248*n*, 251*n*, 255, 274*n*, 373

Odysseus, 8*n*, 127*n*–8*n*, 331, 332
Odyssey (Homer), 127*n*–8*n*, 145–7, 168
Oeconomicus (Xenophon), 284, 423, 424
oligarchy, 151, 215*n*
On Bullshit (Frankfurt), 322–3
On Nature (Anaximander), 31
oracles, 225–7, 236*n*, 326–7, 337, 364

"paradox of Hedonism," 359*n*
parallel universes, 22*n*–3*n*, 72
Parmenides, 45, 169, 310
Parmenides (Plato), 48, 302–3, 310, 386
Parthenon, 123, 125
Patroclus, 129*n*, 246, 247
Pausanias, 41, 243*n*
Peloponnesian War, 80*n*, 149, 235, 239, 251*n*, 254, 283*n*, 285
Penrose, Roger, 46, 385*n*
People of Plato, The (Nails), 87*n*
Pericles, 148–9, 151–2, 232, 234, 238, 241, 250, 252*n*, 283*n*, 285, 296, 299, 302, 323, 324, 425–33
Persia, 7, 123–4, 156, 236, 239, 240
Persian Wars, 123–4, 147, 149–50, 302
personality, 19, 78–9, 101*n*, 193–4, 209–10, 218–21
Petrarch, 144–5
Phaedo, 315–16, 423
Phaedo (Plato), 11*n*, 13, 19*n*, 36*n*, 52, 284, 301, 303, 312–21, 395, 414, 423*n*
Phaedrus, 41, 367, 373–4, 376
Phaedrus (Plato), 24*n*, 182, 196*n*, 221*n*, 245*n*, 247, 249*n*, 265, 267, 268, 270, 272, 274, 275*n*, 277, 279*n*, 283, 350, 354, 355*n*, 367, 373–6, 405
Phidias, 150–1
Philebus (Plato), 52*n*, 83, 198, 226, 340, 341, 345, 347, 352, 391, 393, 394

Philip II, King of Macedon, 239, 295, 302

philosophy:
  academic field of, 3–4, 10, 18–19, 20, 24, 28, 37, 39, 44–6, 89, 92, 263–4
  argument in, 37–9, 43, 89, 114–15, 116, 161
  assumptions challenged in, 9–13, 38, 40–3, 86, 91–4, 101–2, 107, 112, 161, 174–5, 177, 352–4, 359–60, 396
  conclusions reached in, 17, 22, 44–5, 47
  dialectics of, 4, 14, 28–9, 34–9, 43–4, 56, 86*n*–7*n*, 92–3, 94, 101, 161, 311
  dismissal of, 20–9, 34, 50–1, 307, 308, 318, 376
  Greek, 6, 24–5, 29–37, 48, 50, 129–32, 138, 169*n*, 307; *see also specific philosophers*
  history of, 6, 29–34, 139, 160–2
  intuitions in, 14, 17, 29–30, 32, 47, 51, 57, 92, 370–1, 385, 389
  Ionian or pre-Socratic, 29–37, 48, 50, 169*n*, 307
  knowledge in, 47, 69–76, 80–117, 180, 183–4, 327–8, 415–19
  as "love of wisdom," 12, 40, 55, 69–72, 196–200, 215–16, 258, 263–4, 265, 309–12, 317, 319–20, 327–8, 339–40, 354–5, 375
  natural, 21, 27, 34*n*, 51; *see also* science
  progress in, 9–10, 14, 21–9, 31, 35, 37, 38, 217–18, 258, 308
  as protoscience, 23*n*, 28–37, 48, 50, 55, 169*n*, 307, 318, 376, 395–6
  self-critical process in, 9–10, 38, 40–2, 43, 101–2
  teleology in, 30, 52, 53, 388
  "thought experiments" in, 17–18
  violent transformation in, 39–42, 44, 45, 57

Phyrgians (Aeschylus), 246–7

physics:
  Greek, 29–37, 48, 50, 169*n*
  as modern discipline, 23*n*, 27, 407–8
  philosophical basis of, 23*n*, 28–9, 55, 318, 395–6
  quantum, 22–3, 31
  theoretical, 22–3, 26, 31, 51, 376

Piraeus, 283*n*, 285, 294, 378–9

Plantinga, Alvin, 376–7

Plato, 10, 17, 42, 90
  Academy founded by, 8*n*, 9, 27*n*, 38, 39, 45, 51, 64, 76, 78, 106, 169, 205, 211, 236*n*, 239–40, 304, 318, 343, 344*n*, 350
  aesthetics of, 11*n*, 43, 56, 82, 196, 256*n*, 258, 371, 388, 396
  as Athenian, 3, 4, 6, 17, 63–4, 69, 71, 76*n*, 160, 167, 315, 396
  brothers of, 153–5
  in Christian theology, 315–16, 317, 318, 390, 395
  counter-reductive arguments of, 14, 17, 18
  death of, 10*n*, 17
  discussion by, 38–9, 63, 72, 74, 89–90, 103–6, 120
  as dramatist, 41–5, 167, 250–1, 256*n*, 257
  elitism of, 85, 88, 185–6, 329, 394–5, 396
  as historical figure, 10, 65*n*, 90, 114*n*
  human nature as viewed by, 4, 11–12, 58, 191–4, 211, 215, 391–2
  influence and relevance of (Platonism), 3–21, 24, 37, 44–51, 56–8, 168, 217–18, 250, 395–6
  marriage of, 90
  medical theories of, 49, 55, 110*n*, 111*n*
  name of, 62, 65*n*
  personality of, 4, 5–6, 11, 17, 63, 64, 65–8, 72, 74, 77, 86, 209–10, 218–21

as philosopher, 3–21, 24, 37, 44–5,
    49, 50–1, 56–8, 68, 72–3, 74,
    82, 92–3, 138–9, 166, 168–9,
    171, 217–18, 250, 337, 338, 363n,
    395–6, 401, 405–7
  relevance of, 3–6, 11, 14, 17–18,
    20–1, 44–5, 56–8, 217–18, 250
  slavery as viewed by, 14, 91–2,
    113–18, 120, 213n
  sophists opposed by, 19, 35, 40,
    129n, 153–6, 231, 254, 284, 290,
    326
  in Syracuse, 76–9, 274n
  as teacher, 10, 64n, 103, 169–70
  women as viewed by, 9, 66, 114,
    147, 189n
  works of, 4–5, 10–11, 82, 139, 168,
    239–40, 292; *see also specific
    works*
  writing as viewed by, 10, 24, 37, 40,
    75, 82, 89
"Plato on Why Mathematics Is Good
    for the Soul" (Burnyeat), 382n
"Plato's heaven," 48
Plutarch, 12n, 149n, 233, 236, 240,
    241
Pnyx, 153, 285, 430
poetry, 45, 101n, 319
  divine inspiration for, 270n, 271,
    373–4
  epic, 145–7
  lyric, 82n, 135–7, 140–1
  tragic, 7–8, 77n, 229, 237
  truth in, 196–7, 396
Poets, The (Aristotle), 423
politics:
  Athenian, *see* Athens
  justice in, 75–6, 78n, 79, 186, 201
  knowledge of, 11–12, 56–7, 73–9,
    100–1, 147, 150n, 156–7, 225,
    230–1, 247n, 310–12, 363n, 364
  power in, 73–9, 185–93, 215–17
  virtue and, 125, 152, 153, 159, 230–5,
    239n, 242, 244, 248, 249, 257–8,
    284, 291, 293, 298, 306–7, 430n
Politics (Aristotle), 101, 147, 225

Popper, Karl, 22, 24, 25–6
Posterior Analytics (Aristotle), 169n
power, 101n
  political, 73–9, 185–93, 215–17
  transfer of, 76–9
  wealth and, 188, 234, 285
Principia Mathematica (Whitehead
    and Russell), 20n
Prometheus Bound (Aeschylus), 137–8,
    229
Protagoras, 19n, 55
Protagoras (Plato), 19n, 56, 71n, 96,
    150, 249–50
Prytaneum, 63–4, 79, 150, 329
psychology, 11n, 37, 45–6, 110n,
    209–10, 218–21
psychometric questionnaires,
    209–10, 218–21
Ptolemy, 51, 388
Pythagoras, 26–8, 50n, 129, 131,
    314–15, 316, 317, 344–5, 352
Pythagorean theorem, 27n, 50n,
    344–5, 352

quantum field theory, 22–3, 31

rational-actor model, 132, 157–9
rationalism, 51–3, 372n, 384–5, 386,
    408
readiness potential (RP), 411–12
realism, 22, 23, 371, 387–8
reality:
  atomic basis of, 24n–5n, 34
  existence of, 17–18, 28–9, 127–8,
    169n, 388–96, 407–8
  intelligibility of, 30, 48–9, 50, 52–5,
    340–1, 382–4, 385, 389–90
  mathematical, 19–20, 30, 105,
    340–3, 352
  nature of, 29–34, 48, 318, 393
  normativity in, 44, 126, 190–1
  objective vs. subjective, 375–7
  ontological structure of, 53–4, 382,
    416
  physical, 22, 23, 25–6, 27n, 31, 33,
    34, 47, 48–9, 51

reason:
   application of, 39, 43, 53–4
   human capacity for, 9, 94, 390n
   secular, 3, 305–9
   standards of, 57, 102, 416, 417–18,
      421
   as state of ecstasy, 41, 257–8
   truth of, 51–3, 160–1, 303–5, 368–9,
      387–8
   virtue and, 306–7, 329–30, 331, 332
"Reasonables," 372–7, 384, 386
Reichenbach, Hans, 51, 161–2
relativism, 18–19, 44, 45n
relativity theory, 22, 26, 51, 376
religion:
   Christian, 42n, 76n, 269, 278,
      306n, 349, 351, 355–6, 388, 390
   good vs. evil in, 7, 84–95, 309
   Greek, 7, 30, 53–4, 63, 64n, 138,
      231, 232, 235, 324, 378–9; *see also*
      gods and goddesses
   Jewish, 6–7, 130–1, 135, 229–30
   monotheistic, 6–7, 130–1, 135,
      229–30, 389, 395
   morality of, 7, 84–95, 229–30,
      305–9, 311
   secularism compared with, 7, 9, 34,
      55, 129–32, 138, 228, 305–9, 378
   theology of, 42n, 53, 129–30,
      307–9, 315–16, 317, 318, 343n,
      355–6, 358, 395, 407–8
Renaissance, 144–5, 146
Republic (Plato), 165–221
   aretē in, *see* virtue
   "city of pigs" in, 182–3
   as dialogue, 38, 168, 316
   Glaucon as character in, 153–5, 160
   mathematics in, 49n, 50, 53
   Myth of Er in, 330–3, 364
   Myth of the Cave in, 18n, 38, 70n,
      102, 159–60, 183–5, 190–1, 202,
      309, 312, 377–95
   "noble lie" in, 78n, 199–200, 210,
      214–15, 217
   philosopher-king concept in, 73–9,
      81, 88, 105, 107, 117, 170

Ring of Gyges in, 18n, 141, 154–5
ruling elite (guardians) in, 73–9,
   185–93, 199–200, 215–17, 383,
   391
Socrates as character in, 38, 64n,
   141–2, 157, 159–60, 162, 182n,
   367n, 378–9
soul as conceived in, 55, 71n, 347n
Thrasymachus as character in,
   153–6, 231, 232
utopian city (*kallipolis*) in, 66n,
   75–6, 78n, 160, 168n, 188, 191–3,
   199–200, 215–17, 396
rhetoric, 40, 65, 76n, 289–90, 355n,
   356–7, 368n
Richard III (Shakespeare), 226
Roman Empire, 6, 76n, 146, 239
Russell, Bertrand, 20n, 32–3, 48, 305,
   424

Salutati, Coluccio, 144–5
Schliemann, Heinrich, 144
scholasticism, 53, 343n
science:
   Aristotelian influence on, 24–5, 27,
      28, 53, 396, 399n
   data and evidence in, 31n, 407–8,
      410
   empirical nature of, 22, 23, 25, 27n,
      29n, 31, 33, 34, 51, 57, 58
   falsifiability in, 22, 23, 25–6
   instrumentalism in, 22–3, 34
   methodology of, 22–3, 30–1, 33,
      34, 47
   nomological nature of, 30–1
   philosophy of, 21, 23, 26, 28–9,
      338
   Platonic influence on, 48–52, 388
   progress in, 14, 21–9, 31, 35, 37, 116,
      217, 218, 339–40
   proto-, 23n, 28–37, 55, 318, 376,
      395–6
   theories in, 14, 25n, 31–2, 34,
      47, 52
self-deception, 199–200, 267, 417–18
self-discipline, 206–8, 211–15

self-empowerment, 180–1, 186,
    193–5, 206–8
self-esteem, 105, 178–80, 193–5,
    211–13
self-love (narcissism), 55, 200–3,
    255–9
Second Letter (Plato), 82n
Seventh Letter (Plato), 10, 40, 68, 76n,
    78, 79, 82, 187, 218, 363
sexism, 66, 95, 114, 355
sexuality, 40–1, 45–6, 231–59, 264–5,
    271, 272–4, 278–80, 354–5, 373,
    374, 376
Shakespeare, William, 5, 20, 226, 284
Shelley, Percy Bysshe, 4–5, 251n, 319
Sicily, 33, 76–7, 235, 250, 287, 296
"slash fiction," 246–7
slavery, 14, 87n, 91–2, 113–18, 120,
    123, 125, 133–4, 151, 172n, 213n,
    243, 296, 302n, 315, 424n
Snow, C. P., 359n
society:
    achievement in, 174–5, 182–3; *see
        also* Ethos of the Extraordinary
    class divisions in, 56–7, 78n, 185–7,
        213
    collective vs. individualistic, 101,
        125n, 126, 147, 156–7, 183n,
        186–7, 202–3, 231–2, 234n, 235n,
        250–1, 255–6, 292, 297–9, 323,
        378, 433
    duty in, 291–2, 293, 323–5
    laws of, 64, 67, 156–7, 210, 218,
        252, 273n, 291, 294–5, 312,
        323–5
    normative values of, 18–19, 94–5,
        126, 129, 132–4, 135, 141–2, 145,
        159, 162, 186–7, 190–1, 231–2,
        291–2, 293, 323–6, 333
    political justice in, 75–6, 78n, 79,
        156–7, 186, 201
    rulers of, 57, 73–9, 185–93, 199–200,
        215–17, 310–11, 383, 391
    stability of, 204–5, 215–17, 308
*Socrate arrachant Alcibiade du sein de
    la Volupté* (Regnault), 223

Socrates:
    Alcibiades' relationship with,
        231–59, 293, 298, 326, 363–4,
        369n
    as Athenian, 35–6, 103n, 123, 162,
        283–4
    background of, 289, 304–5
    corruption of youth charge against,
        64, 251, 257, 293, 299–300, 311,
        424
    in dialogues of Plato, 10–11, 13, 38,
        43, 64n, 86n, 140, 141–2, 150n,
        157, 159–60, 162, 172n, 182n,
        226, 238n, 254, 259, 301–2, 312,
        320, 367n, 378–9, 393
    eros (love) as defined by, 231–59
    ethos of, 127, 139, 140, 162, 314n
    execution of, 10, 11, 12, 13, 35,
        64–8, 76n, 79, 89, 162, 216,
        231, 286, 302, 303, 311, 313, 315,
        320–1, 327–33, 337, 369–70, 390,
        413–14, 423, 424n
    guilt of, 64n, 66, 67, 79, 126, 162,
        288, 326–7, 328, 337, 369–70
    hemlock taken by, 162, 286, 313,
        423, 424n
    as historical figure, 11, 56, 80n,
        250, 312, 314, 423–4
    impiety (false gods) charge against,
        64, 162, 293, 311, 367–70,
        378–9, 424
    imprisonment of, 19, 313, 316,
        413–14, 417, 418
    indictment against, 299–300, 303,
        305, 321–5, 327
    irony and sense of humor of, 66,
        286, 363, 364, 413–14
    literature on (*Sokratikoi Logoi*), 11n,
        36n, 363, 423–4
    personality of, 42, 162, 283, 303,
        313, 413–14, 423–4
    as philosopher, 35–6, 80n, 89, 103,
        249–50, 283
    Plato influenced by, 4, 5, 36, 42,
        67–8, 79, 97–8, 167, 170, 218,
        259, 320–1, 327–8

Socrates *(continued)*:
Plato's relationship with, 10–11,
  12, 34, 64–8, 76*n*, 89, 90,
  103
political analysis of, 287, 291,
  325–6, 328, 390, 393, 430*n*
questioning by (elenctic method),
  86*n*–7*n*, 154, 160, 213, 283–4,
  290, 291–3, 298–9, 325, 326–7,
  364
reputation of, 11, 12*n*, 103, 241–2,
  249–51, 283, 284, 293
sophists opposed by, 35, 290,
  326
as teacher, 241–2, 249–50
trial of, 9, 35, 64–8, 89, 126, 142,
  251*n*, 310, 312, 313, 321, 323–5,
  333, 337–8, 369, 423
wife of, 284, 313
Solon, 123*n*, 152*n*, 204*n*, 244,
  288
Sophist (Plato), 301–2, 310–12, 318,
  337
sophists, 19, 35, 40, 129*n*, 153–6, 231,
  254, 284, 290, 326
Sophocles, 8
soul:
afterlife of, 127*n*–8*n*, 314–15, 330–3,
  395
"divided," 371–2, 375–6, 377
existence of, 55, 71*n*, 347*n*, 409
immortality of, 8–9, 126–30, 134–5,
  203–4, 228*n*, 313–18, 330–3,
  386–7, 395
transmigration of, 314–15
tripartite nature of, 55, 71*n*
Sparta, 18–19, 124–5, 150, 151, 156,
  232, 237, 238, 244, 251, 283*n*,
  284–5, 286, 287, 295, 297, 299,
  326
Spinoza, Baruch, 52, 53, 55, 160, 161,
  258, 305, 306*n*, 308, 309, 317,
  385, 389–90
Spirit of the Laws, The (Montesquieu),
  114*n*
Statesman (Plato), 301, 310, 311, 318

Sublime Braid (beauty, goodness,
  truth), 50–6, 191–2, 196–8,
  208–9, 254–9, 272, 315–19,
  328–32, 340–7, 382–5, 390–6
Sufficient Reason, principle of,
  384–5
Swerve, The (Greenblatt), 34*n*, 144–5
Symposium (Plato), 4–5, 40, 129*n*,
  162*n*, 196*n*, 198, 204, 205,
  231–59, 272, 274*n*, 284, 286,
  312–13, 363–4
Symposium (Xenophon), 254, 423,
  424
Syracuse, 76–9, 274*n*

Taming of the Shrew, The
  (Shakespeare), 284
technology, 14, 69–70, 116, 172, 217,
  339–40
see also computers
Tetragrammaton, 130*n*, 131*n*
Thales, 29, 32, 35–6, 307, 308
Theaetetus, 27*n*, 303–5, 311
Theaetetus (Plato), 117*n*, 118, 162*n*,
  173, 283, 289, 301, 303–5, 310,
  311, 368
Themistocles, 124, 302
"theory-ladenness of observation,"
  25*n*
Thomas Aquinas, 160, 356, 388
Thrasymachus, 153–6, 231, 232
Thucydides, 235*n*, 239, 283*n*, 424,
  425
thumos (spiritedness), 193–6, 199,
  206–8, 215, 379
Thus Spake Zarathustra (Nietzsche),
  238
Timaeus (Plato), 36*n*, 48, 52, 53, 55,
  110*n*, 111*n*, 316, 317–18, 319, 343*n*,
  345, 355*n*, 382, 385, 386, 391*n*,
  393, 394, 395, 410
time, 27, 118, 120, 127–8, 386–7
timocracy, 215*n*
to on ("being"), 253, 256, 317
Torah, 229
Troy, 127, 143, 146*n*, 231

truth:
abstract, 19–20, 26–7, 46–7, 48,
51, 57, 93, 382n, 384–9, 391, 393
conditions of, 322–3, 326–7
intuition of, 14, 17, 30, 32, 39,
46–7, 161, 238, 258, 310–11, 352,
370–1, 384–5, 389
mathematical, 46–7, 50, 94, 210,
211–13
moral, 108–9, 159–60, 195–6, 210,
214
objective vs. subjective, 91–4,
112–13
poetic, 196–7, 396
proof of, 110, 113–14, 115, 213
reason based on, 51–3, 160–1,
303–5, 368–9, 387–8
in Sublime Braid, 50–6, 191–2,
196–8, 208–9, 254–9, 272,
315–19, 328–32, 340–7, 382–5,
390–6
teaching of, 45–6, 58, 138–9
validity in, 51–3, 160–1, 303–5,
368–9, 387–8
tyranny, 76–9, 85, 125n, 137, 148n,
152n, 191–2, 202–5, 215n, 234n,
273–4, 291, 293, 331

universe, 22n, 23n, 31–2, 72
see also cosmology
"Unreasonables," 372–7, 384, 386

Varieties of Religious Experience, The
(James), 376
virtue:
aretē as term for, 87n, 101n, 125,
138, 139–42, 152, 153, 155, 158–9,
162, 204n, 230–4, 239n, 242,
244, 248–50, 257–8, 285, 288,
291–3, 298, 302–3, 307, 309,
325, 327–32, 378, 391, 393, 394,
424, 427n, 430n

character and, 101n, 193–4, 209
definition of, 76n, 103n, 139, 152,
250, 284, 364–6, 393
fame and, 129n, 138n, 139, 140,
155–6, 204n, 230–5, 292, 303,
331, 378
happiness and, 347–50, 352–3, 358
knowledge of, 54, 55–6, 57, 250,
286, 306–7, 315, 340–1, 346,
352–3, 364–6
moral basis of, 57, 193–4, 209,
309
politicizing of, 125, 152, 153, 159,
230–5, 239n, 242, 244, 248,
249, 257–8, 284, 291, 293, 298,
306–7, 430n
reason and, 306–7, 329–30, 331,
332
teaching of, 86n, 103n, 249–50,
302
Vision of Outstanding Moment
(VOOM), 72, 85, 106, 109

warfare, 91n, 125, 133–4, 294
"warrior mothers," 13, 176, 181, 186,
187, 193, 201, 206–8, 211–18
water, 17–18, 32, 33
Whitehead, Alfred North, 20–1
Why Does the World Exist (Holt),
385n

Xanthippe, 284, 313
Xenophanes, 140–1
Xenophon, 36n, 67n, 80n, 87n, 153–4,
237, 250, 251n, 254, 284, 288,
291, 300, 423–4

Yahweh, 130–1
Yom Kippur, 131n

Zeus, 137, 138, 229, 245
Zoroastrianism, 7, 131, 135

PERMISSIONS ACKNOWLEDGMENTS

Grateful acknowledgment is made to the following for permission to reprint previously published material:

Hackett Publishing Company, Inc.: Excerpt from *Greek Lyric: An Anthology in Translation*, translated by Andrew M. Miller (Indianapolis, IN: Hackett, 1996). Excerpt from *On Justice, Power, and Human Nature*, by Thucydides, translated by Paul Woodruff (Indianapolis, IN: Hackett, 1993). "Apology" from *Five Dialogues*, by Plato, translated by G. M. A. Grube, revised by C. D. C. Reeve (Indianapolis, IN: Hackett, 2002). Excerpt from *Republic*, by Plato, translated by G. M. A. Grube, revised by C. D. C. Reeve (Indianapolis, IN: Hackett, 1992). Excerpt from *Timaeus*, by Plato, translated by Donald Zeyl (Indianapolis, IN: Hackett, 2000). Reprinted by permission of Hackett Publishing Company, Inc. All rights reserved.

Johns Hopkins University Press: Excerpt from *Pindar's Victory Songs* by Frank Nisetich, copyright © 1980 by The Johns Hopkins University Press. Reprinted by permission of Johns Hopkins University Press.

State University of New York Press: Excerpt from *The Symposium and the Phaedrus Plato's Erotic Dialogues*, translated by William S. Cobb, copyright © 1993 by State University of New York Press. Reprinted by permission of State University of New York. All rights reserved.

ILLUSTRATION CREDITS

Page 15: the author; page 59: Joe Faber; pages 119, 121, 281, and 361: Steven Pinker; page 163: David Cieri; page 223: copyright © RMN-Grand Palais/Art Resource, N.Y.; page 233: cracked.com; page 261: Margo Howard; page 397: Joshua Buckholtz

ALSO BY
REBECCA NEWBERGER GOLDSTEIN

36 ARGUMENTS FOR THE EXISTENCE OF GOD

At the center of this witty and intoxicating novel is Cass Seltzer, a professor of psychology whose book, *The Varieties of Religious Illusion*, has become a surprise bestseller. Dubbed "the atheist with a soul," he wins over the stunning Lucinda Mandelbaum—"the goddess of game theory." But he is haunted by reminders of two people who ignited his passion to understand religion: his teacher Jonas Elijah Klapper, a renowned literary scholar with a suspicious obsession with messianism, and an angelic six-year-old mathematical genius, heir to the leadership of an exotic Hasidic sect. Hilarious, heartbreaking, and intellectually captivating, *36 Arguments for the Existence of God* explores the rapture and torments of religious experience in all its variety.

Fiction/Literature

VINTAGE BOOKS
Available wherever books are sold.
www.vintagebooks.com